ELEMENTARY
FLUID
MECHANICS

ELEMENTARY
FLUID
MECHANICS

Tsutomu Kambe

Institute of Dynamical Systems, Tokyo, Japan

 **World Scientific**

NEW JERSEY · LONDON · SINGAPORE · BEIJING · SHANGHAI · HONG KONG · TAIPEI · CHENNAI

Published by

World Scientific Publishing Co. Pte. Ltd.

5 Toh Tuck Link, Singapore 596224

USA office: 27 Warren Street, Suite 401-402, Hackensack, NJ 07601

UK office: 57 Shelton Street, Covent Garden, London WC2H 9HE

British Library Cataloguing-in-Publication Data
A catalogue record for this book is available from the British Library.

ELEMENTARY FLUID MECHANICS

ISBN-13 978-981-256-416-0
ISBN-10 981-256-416-0
ISBN-13 978-981-256-597-6 (pbk)
ISBN-10 981-256-597-3 (pbk)

Typeset by Stallion Press
Email: enquiries@stallionpress.com

Printed in Singapore.

Preface

This book aims to provide an elementary interpretation on *physical* aspects of fluid flows for beginners of fluid mechanics in physics, mathematics and engineering from the point of view of modern physics. Original manuscripts were prepared as lecture notes for intensive courses on Fluid Mechancis given to both undergraduate and postgraduate students of theoretical physics in 2003 and 2004 at the Nankai Institute of Mathematics (Nankai University, Tianjin) in China.

Beginning with introductory chapters of fundamental concepts of the nature of flows and properties of fluids, the text describes basic conservation equations of mass, momentum and energy in Chapter 3. The motions of viscous fluids and those of inviscid fluids are first considered in Chapters 4 and 5. Emphasizing the *dynamical* aspects of fluid motions rather than static aspects, the text describes, in subsequent chapters, various important behaviors of fluids such as waves, vortex motions, geophysical flows, instability and chaos, and turbulence. In addition to those fundamental and basic chapters, this text incorporates a new chapter on superfluid and quantized vortices because it is an exciting new area of physics, and another chapter on gauge theory of fluid flows since it includes a new fundamental formulation of fluid flows on the basis of the gauge theory of theoretical physics. The materials in this book are taken from the lecture notes of intensive courses, so that each chapter in the second half may be read separately, or handled chapter by chapter.

This book is written with the view that fluid mechanics is a branch of theoretical physics.

June 2006 *Tsutomu Kambe*

Former Professor (Physics), University of Tokyo
Visiting Professor, Nankai Institute
of Mathematics (Tianjin, China)

Contents

Chapter 1

Flows

1.1. What are flows ?

Fluid flows are commonly observed phenomena in this world. Giving
typical examples, the wind is a flow of the air and the river stream
is a flow of water. On the other hand, the motion of clouds or smoke
particles floating in the air can be regarded as visualizing the flow
that carries them. When we say **flow** of a matter, it implies usually
time development of the displacement and deformation of matter.
Namely, a number of particles compose the body of matter, and are
moving and continuously changing their relative positions, and are
evolving with time *always*. Flows are observed in diverse phenomena in addition to the wind and river above: air flows in a living
room, flows of *blood* or *respiratory air* in a body, flows of microscopic
suspension particles in a chemical test-tube, flows of *bathtub water*,
atmospheric flows and *sea currents, solar wind, gas flows in inter-
stellar space*, and so on.

From the technological aspect, vehicles such as ships, aeroplanes,
jetliners and rockets utilize flows in order to obtain thrust to move
from one place to other while carrying loads. Glider planes or soaring
birds use winds passively to get lift.

On the other hand, from the biological aspect, swimming fishes
are considered to be using water motions (*eddies*) to get thrust for
their motion. Animals such as insects or birds commonly use air
flows in order to get lift for being airborne as well as getting thrust
for their forward motion. In addition, it is understood that plant

1

seeds, or pollen, often use wind for their purposes. Every living organism has a certain internal system of physiological circulation. In general, it might not be an exaggeration to say that all the living organisms make use of flows in various ways in order to live in this world.

In general, a *material* which constantly deforms itself such as the air or water is called a **fluid**. An elastic solid is deformable as well, however its deformation stops in balance with a force acting on it. Once the body is released from the force, it recovers its original state. A plastic solid is deformed continuously during the application of a force. Once the body is free from the force, it stops deformation (nominally at least). By contrast, the fluid *keeps deforming* even when it is free from force.

A body of *fluid* is composed of innumerably many microscopic molecules. However in a macroscopic world, it is regarded as a body in which mass is continuously distributed. Motion of a fluid, i.e. *flow* of a fluid, is considered to be a mass flow involving its continuous deformation. *Fluid mechanics* studies such flows of fluids, i.e. motions of material bodies of continuous mass distribution, *under fundamental laws of mechanics*.

1.2. Fluid particle and fields

When we consider a fluid flow, it is often useful to use a discrete concept although the fluid itself is assumed to be a continuous body. A *fluid particle* is defined as a mass in a small nearly-spherical volume ΔV, whose diameter is sufficiently small from a macroscopic point of view. However, it is large enough if it is compared with the intermolecular distance, such that the total number N_Δ of molecules in the volume ΔV is sufficiently big so that the statistical description makes sense. In other words, it is a basic assumption that there exists such a volume ΔV enabling to define the concept of a fluid particle. In fact, the study of fluid mechanics is normally carried out at scales of about 10^{-3} mm or larger, whereas the intermolecular scale is 10^{-6} mm or less. At the normal temperature and pressure

(0°C and 760 mmHg), a cube of 1 mm in a gas contains about 2.7×10^{16} molecules.

We consider a monoatomic gas whose molecular mass is m, hence the mass in a small volume ΔV is $\Delta M = m N_\Delta$, and we consider the flow in the (x, y, z) cartesian coordinate frame.

Position $\mathbf{x} = (x, y, z)$ of a fluid particle is defined by the center of mass of the constituent molecules. Density of the fluid ρ is defined by dividing ΔM by ΔV, so that[1]

$$\rho(\mathbf{x}) := \frac{\Delta M}{\Delta V}. \tag{1.1}$$

A α-th molecule constituting the mass moves with its own velocity $\mathbf{u}_\alpha$, where $\mathbf{u}_\alpha$ has three components, say $(u_\alpha^x, u_\alpha^y, u_\alpha^z)$. The fluid velocity $\mathbf{v}$ at $\mathbf{x}$ is defined by the average value of the molecular velocities, in such a way

$$\mathbf{v}(\mathbf{x}) = \langle \mathbf{u}_\alpha \rangle := \frac{\sum_\alpha m_\alpha \mathbf{u}_\alpha}{\sum_\alpha m_\alpha}, \tag{1.2}$$

where $m_\alpha = m$ (by the assumption), $\sum_\alpha m_\alpha = m N_\Delta = \rho \Delta V$, and $\langle \cdot \rangle$ denotes an average with respect to the molecules concerned. The difference $\tilde{\mathbf{u}}_\alpha = \mathbf{u}_\alpha - \mathbf{v}$ is called the *peculiar velocity* or *thermal velocity*.

In the kinetic theory of molecules, the temperature T is defined by the law that the average of *peculiar* kinetic energy per degree-of-freedom is equal to $\frac{1}{2}kT$, where k is the Boltzmann constant.[2] Each molecule has three degrees of freedom for translational motion. It is assumed that

$$\left\langle \frac{1}{2}m(\tilde{u}_\alpha^x)^2 \right\rangle = \left\langle \frac{1}{2}m(\tilde{u}_\alpha^y)^2 \right\rangle = \left\langle \frac{1}{2}m(\tilde{u}_\alpha^z)^2 \right\rangle = \frac{1}{2}kT.$$

---

[1] $A := B$ denotes that A is defined by B.

[2] Boltzmann constant k is a conversion factor between *degree* (Kelvin temperature) and *erg* (energy unit), defined by $k = 1.38 \times 10^{-16}$ erg/deg.

Therefore, we have

$$\frac{3}{2}kT(\mathbf{x}) := \left\langle \frac{1}{2}m\tilde{\mathbf{u}}_\alpha^2 \right\rangle = \frac{1}{N_\Delta}\sum_\alpha \frac{1}{2}m\tilde{\mathbf{u}}_\alpha^2. \qquad (1.3)$$

On the other hand, *pressure* is a variable defined against a surface element. The pressure p exerted on a surface element ΔS is defined by the momentum flux (i.e. a force) through ΔS. Choosing the x-axis normal to the surface ΔS, the x-component of the pressure force F_x on ΔS acting from the left (smaller x) side to the right (larger x) side would be given by the flux of x-component momentum $m_\beta \tilde{u}_\beta^x$ through ΔS:

$$F_x = p(\mathbf{x})\,\Delta S = \sum_\beta (m_\beta \tilde{u}_\beta^x)\,\tilde{u}_\beta^x\,\Delta S = \Delta S \sum_{\tilde{u}^x} m(\tilde{u}^x)^2 n(\tilde{u}^x), \quad (1.4)$$

where β denotes all the molecules passing through ΔS per unit time, which are contained in the volume element $\tilde{u}_\beta^x \Delta S$, and $n(\tilde{u}^x)$ denotes the number of molecules with $\tilde{u}^x$ in a unit volume. In the kinetic theory, the factor $\sum_{\tilde{u}^x} m(\tilde{u}^x)^2 n(\tilde{u}^x)$ on the right-hand side is expressed by the following two integrals for $\tilde{u}^x > 0$ and $\tilde{u}^x < 0$ respectively:

$$\int_{\tilde{u}^x>0} m(\tilde{u}^x)^2\, Nf(\tilde{\mathbf{u}})\,\mathrm{d}^3\tilde{\mathbf{u}} + \int_{\tilde{u}^x<0} m(\tilde{u}^x)^2 Nf(\tilde{\mathbf{u}})\,\mathrm{d}^3\tilde{\mathbf{u}}, \qquad (1.5)$$

where the function $f(\tilde{\mathbf{u}})$ denotes the distribution function of the peculiar velocity $\tilde{\mathbf{u}}$, and the number of molecules between $\tilde{\mathbf{u}}$ and $\tilde{\mathbf{u}} + \mathrm{d}\tilde{\mathbf{u}}$ is given by[3]

$$Nf(\tilde{\mathbf{u}})\,\mathrm{d}^3\tilde{\mathbf{u}}, \quad \text{with} \quad \int_{\text{all } \tilde{\mathbf{u}}} f(\tilde{\mathbf{u}})\,\mathrm{d}^3\tilde{\mathbf{u}} = 1,$$

where N is the total number of molecules in a unit volume. This is interpreted as follows. The first term of (1.5) denotes that a positive momentum $m\tilde{u}^x$ ($\tilde{u}^x > 0$) is absorbed into the right side of ΔS, while the second term denotes that a negative momentum $m\tilde{u}^x$ ($\tilde{u}^x < 0$) is

---

[3]More precisely, the number of molecules of the peculiar velocity $(\tilde{u}^x, \tilde{u}^y, \tilde{u}^z)$, which takes values between $\tilde{u}^x$ and $\tilde{u}^x + \mathrm{d}\tilde{u}^x$, $\tilde{u}^y$ and $\tilde{u}^y + \mathrm{d}\tilde{u}^y$ and $\tilde{u}^z$ and $\tilde{u}^z + \mathrm{d}\tilde{u}^z$ respectively, is defined by $Nf(\tilde{u}^x, \tilde{u}^y, \tilde{u}^z)\,\mathrm{d}\tilde{u}^x\,\mathrm{d}\tilde{u}^y\,\mathrm{d}\tilde{u}^z$.

taken out from the right side of ΔS. Both means that the space on the right side has received the same amount of positive momentum. Both terms are combined into one:

$$\sum_{\beta} m\beta(\tilde{u}^x_\beta)^2 = \int_{\text{all } \tilde{\mathbf{u}}} m(\tilde{u}^x)^2 Nf(\tilde{\mathbf{u}}) \, d^3\tilde{\mathbf{u}}$$

$$= Nm\langle(\tilde{u}^x)^2\rangle = NkT, \tag{1.6}$$

since $\frac{1}{2}m\langle(\tilde{u}^x)^2\rangle = \frac{1}{2}kT$, where the average $\langle(\tilde{u}^x)^2\rangle = \langle(\tilde{u}^y)^2\rangle = \langle(\tilde{u}^z)^2\rangle$ is equal to $\frac{1}{3}\langle(\tilde{\mathbf{u}})^2\rangle$ by an isotropy assumption, and $m\frac{1}{3}\langle(\tilde{\mathbf{u}})^2\rangle$ is given by kT from (1.3). Thus, from (1.4) and (1.6), we obtain

$$p(\mathbf{x}) = NkT. \tag{1.7}$$

This is known as the **equation of state** of an ideal gas.[4]

The density $\rho(\mathbf{x})$, velocity $\mathbf{v}(\mathbf{x})$, temperature $T(\mathbf{x})$ and pressure $p(\mathbf{x})$ thus defined depend on the position $\mathbf{x} = (x, y, z)$ and the time t smoothly, since the molecular kinetic motion usually works to smooth out discontinuity (if any) by the transport phenomena considered in Chapter 2. Namely, these variables are regarded as continuous and in addition *differentiable* functions of (x, y, z, t). Such variables are called **fields**. This point of view is often called the *continuum hypothesis*.

From a mathematical aspect, *flow* of a fluid is regarded as a continuous sequence of *mappings*. Consider all the fluid particles composing a subdomain B_0 at an initial instant $t = 0$. After an infinitesimal time δt, a particle at $\mathbf{x} \in B$ moves from $\mathbf{x}$ to $\mathbf{x} + \delta\mathbf{x}$:

$$\mathbf{x} \mapsto \mathbf{x} + \delta\mathbf{x} = \mathbf{x} + \mathbf{v}\delta t + O(\delta t^2) \tag{1.8}$$

by the flow field $\mathbf{v}(\mathbf{x}, t)$. Then the domain B_0 may be mapped to $B_{\delta t}$ (say). Subsequent mapping occurs for another δt from $B_{\delta t}$ to $B_{2\delta t}$, and so on. In this way, the initial domain B_0 is mapped one after another smoothly and constantly. At a later time t, the domain

---

[4]For a gram-molecule of an ideal gas, N is replaced by $N_A = 6.023 \times 10^{23}$ (Avogadro's constant). The product $N_A K = R$ is called the *gas constant*: $R = 8.314 \times 10^7$ erg/deg. For an ideal gas of molecular weight μ_m, the equation (1.7) reduces to $p = (1/\mu_m)\rho RT$, where $\rho = mN$, $\mu_m = mN_A$ and $R = N_A k$.

B_0 would be mapped to B_t. The map might be differentiable with respect to $\mathbf{x}$, and in addition, for such a map, there is an inverse map. This kind of map is termed a *diffeomorphism* (i.e. differentiable homeomorphism).

1.3. Stream-line, particle-path and streak-line

1.3.1. *Stream-line*

Suppose that a velocity field $\mathbf{v}(\mathbf{x}, t) = (u, v, w)$ is given in a sub-domain of three-dimensional Euclidean space $\mathbb{R}^3$, and that, at a given time t, the vector field $\mathbf{v} = (u, v, w)$ is continuous and smooth at every point (x, y, z) in the domain. It is known in the theory of ordinary differential equations in mathematics that one can draw curves so that the curves are tangent to the vectors at all points. Provided that the curve is represented as $(x(s), y(s), z(s))$ in terms of a parameter s, the tangent to the curve is written as $(\mathrm{d}x/\mathrm{d}s, \mathrm{d}y/\mathrm{d}s, \mathrm{d}z/\mathrm{d}s)$, which should be parallel to the given vector field $(u(x, y, z), v(x, y, z), w(x, y, z))$ by the above definition. This is written in the following way:

$$\frac{\mathrm{d}x}{u(x, y, z)} = \frac{\mathrm{d}y}{v(x, y, z)} = \frac{\mathrm{d}z}{w(x, y, z)} = \mathrm{d}s. \tag{1.9}$$

This system of ordinary differential equations can be integrated for a given initial condition at $s = 0$, at least locally in the neighborhood of $s = 0$. Namely, a curve through the point $P = (x(0), y(0), z(0))$ is determined uniquely.[5] The curve thus obtained is called a *stream-line*. For a set of initial conditions, a family of curves is obtained. Thus, a family of stream-lines are defined at each instant t (Fig. 1.1).

---

[5]Mathematically, existence of solutions to Eq. (1.9) is assured by the continuity (and boundedness) of the three component functions of $\mathbf{v}(\mathbf{x}, t)$. For the uniqueness of the solution to the initial condition, one of the simplest conditions is the *Lipschitz condition*: $|\mathbf{v}(\mathbf{x}, t) - \mathbf{v}(\mathbf{y}, t)| \leqq K|\mathbf{x} - \mathbf{y}|$ for a positive constant K.

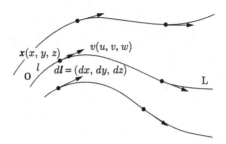

Fig. 1.1.   Stream-lines.

1.3.2.   *Particle-path (path-line)*

Next, let us take a particle-wise point of view. Choosing a fluid particle A, whose position was at $\mathbf{a} = (a, b, c)$ at the time $t = 0$, we follow its subsequent motion governed by the velocity field $\mathbf{v}(\mathbf{x}, t) = (u, v, w)$. Writing its position as $\mathbf{X}_a(t) = (X(t), Y(t), Z(t))$, equations of motion of the particle are

$$\frac{\mathrm{d}X}{\mathrm{d}t} = u(X, Y, Z, t), \quad \frac{\mathrm{d}Y}{\mathrm{d}t} = v(X, Y, Z, t),$$

$$\frac{\mathrm{d}Z}{\mathrm{d}t} = w(X, Y, Z, t). \tag{1.10}$$

This can be solved at least locally in time, and the solution would be represented as $\mathbf{X}_a(t) = \mathbf{X}(\mathbf{a}, t) = (X(t), Y(t), Z(t))$, where

$$X(t) = X(a, b, c, t), \quad Y(t) = Y(a, b, c, t),$$

$$Z(t) = Z(a, b, c, t), \tag{1.11}$$

and $\mathbf{X}_a(0) = \mathbf{a}$. For a fixed particle specified with $\mathbf{a} = (a, b, c)$, the function $\mathbf{X}_a(t)$ represents a curve parametrized with t, called the *particle path*, or a path-line. Correspondingly, the particle velocity is given by

$$\mathbf{V}_a(t) = \frac{\mathrm{d}}{\mathrm{d}t}\mathbf{X}_a(t) = \frac{\partial}{\partial t}\mathbf{X}(\mathbf{a}, t) = \mathbf{v}(\mathbf{X}_a, t). \tag{1.12}$$

This particle-wise description is often called the *Lagrangian description*, whereas the field description such as $\mathbf{v}(\mathbf{x}, t)$ for a point $\mathbf{x}$ and a time t is called the *Eulerian description*.

It is seen that the two equations (1.9) and (1.10) are identical except the fact that the right-hand sides of (1.10) include the time t. Hence if the velocity field is *steady*, i.e. **v** does not depend on t, then both equations are equivalent, implying that both stream-lines and particle-paths are identical in steady flows.

1.3.3.   *Streak-line*

In most visualizations of flows or experiments, a common practice is to introduce dye or smoke at fixed positions in a fluid flow and observe colored patterns formed in the flow field (Fig. 1.2). Smoke from a chimney is another example of analogous pattern. An instantaneous curve composed of all fluid elements that have passed the same particular fixed point P at previous times is called the *streak-line*.

Denoting the fixed point P by **A**, the fluid particle that has passed the point **A** at a previous time τ will be located at $\mathbf{X} = \mathbf{X}(\boldsymbol{a}_\tau, t)$ at a later time t where $\boldsymbol{a}_\tau$ is defined by $\mathbf{A} = \mathbf{X}(\boldsymbol{a}_\tau, \tau)$. Thus the streak-line at a time t is represented parametrically by the function $\mathbf{X}(\mathbf{A}, t - \tau)$ with the parameter τ.

If the flow field is steady (Fig. 1.3), it is obvious that the streak-line coincides with the particle-path, and therefore with the stream-line. However, if the flow field is time-dependent (Fig. 1.4), then all the three lines are different, and they appear quite differently.

1.3.4.   *Lagrange derivative*

Suppose that the temperature field is expressed by $T(\mathbf{x}, t)$ and that the velocity field is given by $\mathbf{v}(\mathbf{x}, t)$, in the way of Eulerian description. Consider a fluid particle denoted by the parameter **a** in the flow field and examine how its temperature $T_a(t)$ changes during the motion. Let the particle position be given by $\mathbf{X}_a(t) = (X, Y, Z)$ and its velocity by $\mathbf{v}_a(t) = (u_a, v_a, w_a)$. Then the particle temperature is expressed by

$$T_a(t) = T(\mathbf{X}_a, t) = T(X(t), Y(t), Z(t); t).$$

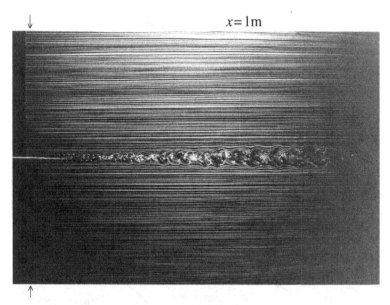

x = 1 m

Fig. 1.2. Visualization of the wake behind a thin circular cyinder (of diameter 5 mm) by a *smoke-wire* method. The wake is the central horizontal layer of
irregular smoke pattern, and the many parallel horizontal lines in the upper and
lower layers show a uniform stream of wind velocity 1 m/s from left to right. The
smoke lines originate from equally-spaced discrete points on a vertical straight
wire on the left placed at just upstream position of the cylinder (at the point of
intersection of the central horizontal white line (from the left) and the vertical
line connecting the two arrows out of the frame). Thus, all the smoke lines are
streak-lines. The illumination is from upward right, and hence the shadow line of
the cylinder is visible to downward left on the lower left side. The regular periodic pattern observed in the initial development of the wake is the Kármán vortex
street. The vertical white line at the central right shows the distance 1 m from
the cylinder. This is placed in order to show how the wake reorganizes to another
periodic structure of larger eddies. [As for the wake, see Problem 4.6 (Fig. 4.12),
and Fig. 4.7.] The photograph is provided through the courtesy of Prof. S. Taneda
(Kyushu University, Japan, 1988). $R_e = 350$ (see Table 4.2).

Hence, the time derivative of the particle temperature is given by

$$
\frac{\mathrm{d}}{\mathrm{d}t}T_a(t) = \frac{\partial T}{\partial t} + \frac{\mathrm{d}X}{\mathrm{d}t}\frac{\partial T}{\partial x} + \frac{\mathrm{d}Y}{\mathrm{d}t}\frac{\partial T}{\partial y} + \frac{\mathrm{d}Z}{\mathrm{d}t}\frac{\partial T}{\partial z}
$$

$$
= \frac{\partial T}{\partial t} + u_a\frac{\partial T}{\partial x} + v_a\frac{\partial T}{\partial y} + w_a\frac{\partial T}{\partial z}
$$

$$
= \left(\frac{\partial}{\partial t} + u\frac{\partial}{\partial x} + v\frac{\partial}{\partial y} + w\frac{\partial}{\partial z}\right)T\bigg|_{\mathbf{x}=\mathbf{X}_a}.
$$

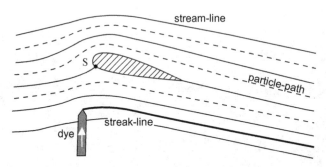

Fig. 1.3. Steady flow: stream-lines (thin solid lines), particle-path (broken lines), and streak-lines (a thick solid line).

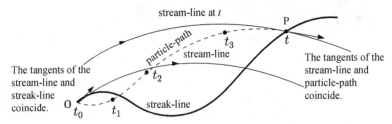

Fig. 1.4. Unsteady flow: stream-lines (thin solid lines), particle-path (a broken line), and streak-line (a thick solid line). The particle P started from the fixed point O at a time t_0 and is now located at P at t after the times t_1, t_2 and t_3.

It is convenient to define the differentiation on the right-hand side by using the operator,

$$\frac{D}{Dt} := \frac{\partial}{\partial t} + u \frac{\partial}{\partial x} + v \frac{\partial}{\partial y} + w \frac{\partial}{\partial z} = \partial_t + u \partial_x + v \partial_y + w \partial_z,$$

which is called the *convective derivative*, where $\partial_t := \partial/\partial t$, $\partial_x := \partial/\partial x$, and so on. As is evident from the above derivation, this time derivative denotes the differentiation following the particle motion. This derivative is called variously as the *material derivative, convective derivative* or *Lagrange derivative*. Thus, we have $dT_a/dt = DT/Dt \big|_{\mathbf{X}_a}$.

Let us introduce the following differential operators,

$$\nabla = (\partial_x, \partial_y, \partial_z),$$
$$\operatorname{grad} f = (\partial_x, \partial_y, \partial_z) f,$$

where $f(x, y, z)$ is a differentiable scalar function. The differential operator ∇ with three components is termed the *nabla* operator, and the vector $\operatorname{grad} f$ is called the *gradient* of the function $f(\mathbf{x})$. Using ∇ and $\mathbf{v} = (u, v, w)$, one can write

$$\frac{D}{Dt} = \partial_t + u\partial_x + v\partial_y + w\partial_z = \partial_t + \mathbf{v} \cdot \nabla, \qquad (1.13)$$

where the dot denotes the inner product [Appendix A.2 and see (A.7)].

Suppose that we have a physical (scalar) field $Q(\mathbf{x}, t)$ such as density ρ, temperature T, etc. Its time derivative following the motion of a fluid particle is given by

$$\frac{DQ}{Dt} = \partial_t Q + (\mathbf{v} \cdot \nabla)Q. \qquad (1.14)$$

If the value is invariant during the particle motion, we have

$$\frac{DQ}{Dt} = 0. \qquad (1.15)$$

1.4.   Relative motion

Given a velocity field $\mathbf{v}(\mathbf{x}, t)$, each fluid element moves subject to straining deformation and local rotation. This is shown as follows.

1.4.1.   *Decomposition*

In order to represent such a local motion mathematically, we consider a relative motion of fluid in a neighborhood of an arbitrarily chosen point $P = \mathbf{x} = (x_1, x_2, x_3)$, where the velocity is $\mathbf{v} = (v_1, v_2, v_3)$. Writing the velocity of a neighboring point $Q = \mathbf{x} + \mathbf{s}$ as $\mathbf{v} + \delta\mathbf{v}$ at the same instant t, we have

$$\delta \mathbf{v} = \mathbf{v}(\mathbf{x} + \mathbf{s}, t) - \mathbf{v}(\mathbf{x}, t) = (\mathbf{s} \cdot \nabla)\mathbf{v} + O(s^2),$$

by the Taylor expansion with respect to the separation vector $\mathbf{s} = (s_1, s_2, s_3)$. Writing this in components, we have the following matrix equation,

$$\begin{pmatrix} \delta v_1 \\ \delta v_2 \\ \delta v_3 \end{pmatrix} = \begin{pmatrix} \partial_1 v_1 & \partial_2 v_1 & \partial_3 v_1 \\ \partial_1 v_2 & \partial_2 v_2 & \partial_3 v_2 \\ \partial_1 v_3 & \partial_2 v_3 & \partial_3 v_3 \end{pmatrix} \begin{pmatrix} s_1 \\ s_2 \\ s_3 \end{pmatrix}, \qquad (1.16)$$

where $\partial_k v_i = \partial v_i / \partial x_k$. This can be also written as[6]

$$\delta v_i = \sum_{k=1}^{3} s_k \, \partial_k v_i = s_k \, \partial_k v_i. \qquad (1.17)$$

The term $\partial_k v_i$ can be decomposed into a symmetric part e_{ik} and an anti-symmetric part g_{ik} in general (Stokes (1845), [Dar05]), defined by

$$e_{ik} = \frac{1}{2}(\partial_k v_i + \partial_i v_k) = e_{ki}, \qquad (1.18)$$

$$g_{ik} = \frac{1}{2}(\partial_k v_i - \partial_i v_k) = -g_{ki}. \qquad (1.19)$$

Then one can write as $\partial_k v_i = e_{ik} + g_{ik}$. Using e_{ik} and g_{ik}, the velocity difference δv_i is decomposed as $\delta v_i = \delta v_i^{(s)} + \delta v_i^{(a)}$, where

$$\delta v_i^{(s)} = e_{ik} \, s_k, \qquad (1.20)$$

$$\delta v_i^{(a)} = g_{ik} \, s_k. \qquad (1.21)$$

These components represent two fundamental modes of relative motion, which we will consider in detail below.

---

[6]The summation convention is assumed here, which takes a sum with respect to the repeated indices such as k. Henceforth, summation is meant for such indices without the symbol $\sum_{k=1}^{3}$.

1.4.2. *Symmetric part (pure straining motion)*

The symmetric part is written as

$$
\begin{pmatrix} \delta v_1^{(s)} \\ \delta v_2^{(s)} \\ \delta v_3^{(s)} \end{pmatrix} = \begin{pmatrix} e_{11} & e_{12} & e_{13} \\ e_{12} & e_{22} & e_{23} \\ e_{13} & e_{23} & e_{33} \end{pmatrix} \begin{pmatrix} s_1 \\ s_2 \\ s_3 \end{pmatrix}. \tag{1.22}
$$

Any symmetric (real) matrix can be made diagonal by a coordinate transformation (called the orthogonal transformation, see the footnote 7) to a principal coordinate frame. Using capital letters to denote corresponding variables in the principal frame, the expression (1.22) is transformed to

$$
\begin{pmatrix} \delta V_1^{(s)} \\ \delta V_2^{(s)} \\ \delta V_3^{(s)} \end{pmatrix} = \begin{pmatrix} E_{11} & 0 & 0 \\ 0 & E_{22} & 0 \\ 0 & 0 & E_{33} \end{pmatrix} \begin{pmatrix} S_1 \\ S_2 \\ S_3 \end{pmatrix}. \tag{1.23}
$$

The diagonal elements E_{11}, E_{22}, E_{33} are the eigenvalues of e_{ik} by the orthogonal transformation $(s_1, s_2, s_3) \to (S_1, S_2, S_3)$.[7] The length is invariant, $|\mathbf{s}| = |\mathbf{S}|$, and in addition, the trace of matrix is invariant:

$$
\begin{aligned}
E_{11} + E_{22} + E_{33} &= e_{11} + e_{22} + e_{33} \\
&= \partial_1 v_1 + \partial_2 v_2 + \partial_3 v_3 := \operatorname{div} \mathbf{v}. \tag{1.24}
\end{aligned}
$$

In the principal frame, we have

$$
\delta \mathbf{V}^{(s)} = (E_{11} S_1, E_{22} S_2, E_{33} S_3), \tag{1.25}
$$

---

[7]One may write the transformation as $\mathbf{s} = A\mathbf{S}$ and $\delta \mathbf{v}^{(s)} = A\delta \mathbf{V}^{(s)}$, where A is a 3×3 transformation matrix. Then, substituting these into (1.22): $\delta \mathbf{v}^{(s)} = e\mathbf{s}$, one obtains $\delta \mathbf{V}^{(s)} = A^{-1} e A \mathbf{S} = E\mathbf{S}$ (where $E = A^{-1} e A$), which corresponds to (1.23). Every orthogonal transformation makes the length invariant by definition, i.e. $|\mathbf{s}|^2 = s_i s_i = A_{ik} S_k A_{il} S_l = A_{ki}^T A_{il} S_k S_l = S_k S_k = |\mathbf{S}|^2$, where A^T is the transpose of A. Namely, the orthogonal transformation is defined by $A^T A = A A^T = I = (\delta_{ik})$ (a unit matrix). Hence, $A^T = A^{-1}$. There exists an orthogonal transformation A which makes $A^{-1} e A$ diagonal. We have $\operatorname{Tr}\{A^{-1} e A\} = \operatorname{Tr}\{A^T e A\} = A_{ji} e_{jk} A_{ki} = \delta_{jk} e_{jk} = \operatorname{Tr}\{e\}$.

namely, the velocity component $\delta V_i^{(s)}$ of the symmetric part is proportional to the displacement S_i in the respective axis. This motion $\delta \mathbf{v}^{(s)}$ is termed the *pure straining* motion, and the symmetric tensor e_{ik} is termed the *rate-of-strain tensor*. The trace div $\mathbf{v}$ gives the relative rate of volume change (see Problem 1.2).

1.4.3. Anti-symmetric part (local rotation)

The anti-symmetric part is written as

$$(g_{ij}) = \begin{pmatrix} 0 & -g_{21} & g_{13} \\ g_{21} & 0 & -g_{32} \\ -g_{13} & g_{32} & 0 \end{pmatrix} = \frac{1}{2} \begin{pmatrix} 0 & -\omega_3 & \omega_2 \\ \omega_3 & 0 & -\omega_1 \\ -\omega_2 & \omega_1 & 0 \end{pmatrix}.$$

(1.26)

where we set $\omega_1 = 2g_{32}$, $\omega_2 = 2g_{13}$, $\omega_3 = 2g_{21}$. In this way, a vector $\boldsymbol{\omega} = (\omega_1, \omega_2, \omega_3)$ is introduced. Using the original definition, we have

$$\boldsymbol{\omega} = (\partial_2 v_3 - \partial_3 v_2, \ \partial_3 v_1 - \partial_1 v_3, \ \partial_1 v_2 - \partial_2 v_1).$$

(1.27)

This is nothing but the curl of the vector $\mathbf{v}$ defined by (A.14) in Appendix A.3, and denoted by

$$\boldsymbol{\omega} = \text{curl } \mathbf{v} = \nabla \times \mathbf{v}, \quad \omega_i = \varepsilon_{ijk} \partial_j v_k.$$

(1.28)

The above relation (1.26) between g and ω suggests the following[8]:

$$g_{ij} = -\frac{1}{2} \varepsilon_{ijk} \omega_k.$$

(1.29)

Now, from (1.21) and (1.29), the anti-symmetric part is

$$\delta v_i^{(a)} = g_{ij} s_j = -\frac{1}{2} \varepsilon_{ijk} \omega_k s_j = \varepsilon_{ikj} \left(\frac{1}{2} \omega_k \right) s_j.$$

In the vector notation, using (A.12), this is written as

$$\delta \mathbf{v}^{(a)} = \frac{1}{2} \boldsymbol{\omega} \times \mathbf{s}.$$

(1.30)

---

[8]For the definition of ε_{ijk}, see Appendix A.1. For example, we have $g_{12} = -\frac{1}{2}(\varepsilon_{121}\omega_1 + \varepsilon_{122}\omega_2 + \varepsilon_{123}\omega_3) = -\frac{1}{2}\omega_3$.

This component of relative velocity describes a rotation of the angular velocity $\frac{1}{2}\boldsymbol{\omega}$. Although $\boldsymbol{\omega}$ depends on $\mathbf{x}$, it is independent of the displacement vector $\mathbf{s}$. Namely, every point $\mathbf{s}$ in the neighborhood of $\mathbf{x}$ rotates with the same angular velocity. Thus, it is found that $\delta\mathbf{v}^{(a)}$ represents *local rigid-body rotation*.

In summary, it is found that *the local relative velocity* $\delta\mathbf{v}$ *consists of a pure straining motion* $\delta\mathbf{v}^{(s)}$ *and a local rigid-body rotation* $\delta\mathbf{v}^{(a)}$.

1.5. Problems

Problem 1.1 Pattern of ink-drift

Suppose that some amount of water is contained in a vessel, and the water is set in motion and its horizontal surface is in smooth motion. Let a liquid-drop of Chinese ink be placed quietly on the flat horizontal surface maintaining a flow with some eddies. The ink covers a certain compact area of the surface.

After a while, some ink pattern will be observed. If a sheet of plain paper (for calligraphy) is placed quietly on the free surface of the water, a pattern will be printed on the paper, which is called the *ink-drift printing* (Fig. 1.5). This pattern is a snap-shot at an instant and consists of a number of curves. What sort of lines are the curves printed on the paper? Are they stream-lines, particle-paths or streak-lines, or other kind of lines?

Problem 1.2 Divergence operator div

Consider a small volume of fluid of a rectangular parallelepiped in a flow field of fluid velocity $\mathbf{v} = (v_x, v_y, v_z)$. The fluid volume V changes under the straining motion. Show that the time-rate of change of volume V per unit volume is given by the following,

$$\frac{1}{V}\frac{dV}{dt} = \operatorname{div}\mathbf{v} = \frac{\partial v_x}{\partial x} + \frac{\partial v_y}{\partial y} + \frac{\partial v_z}{\partial z}. \tag{1.31}$$

Fig. 1.5. Ink-drift printing.

Problem 1.3 Acceleration of a fluid particle

Given the velocity field $\mathbf{v}(\mathbf{x}, t)$ with $\mathbf{x} = (x, y, z)$ and $\mathbf{v} = (u, v, w)$. Show that the velocity and acceleration of a fluid particle are given by the following expressions:

$$\frac{\mathrm{D}}{\mathrm{D}t}\mathbf{x} = \mathbf{v}, \quad (\text{Sec. } 12.6.2) \tag{1.32}$$

$$\frac{\mathrm{D}}{\mathrm{D}t}\mathbf{v} = \partial_t \mathbf{v} + (\mathbf{v} \cdot \nabla)\mathbf{v}. \tag{1.33}$$

Chapter 2

Fluids

2.1. Continuum and transport phenomena

The motion of a fluid is studied on the basis of the fundamental principle of mechanics, namely the conservation laws of mass, momentum and energy. For a state of matter to which the *continuum hypothesis* (Sec. 1.2) can be applied, macroscopic motions of the matter (a fluid) are less sensitive to whether the structure of matter is discrete or continuous. In the continuum representation of fluids, the effect of actual discrete molecular motion is taken into account as transport phenomena such as diffusion, viscosity and thermal conductivity in equations of motion. In fluid mechanics, all variables, such as mass density, momentum, energy and thermodynamic variables (pressure, temperature, entropy, enthalpy, or internal energy, etc.), are regarded as continuous and differentiable functions of position and time.

Equilibrium in a material is represented by the property that the thermodynamic state-variables take uniform values at all points of the material. In this situation, each part of the material is assumed to be in equilibrium mechanically and thermodynamically with the surrounding medium. However, in most circumstances where real fluids are exposed, the fluids are hardly in equilibrium, but state variables vary from point to point. When the state variables are not uniform, there occurs exchange of physical quantities dynamically and thermodynamically. Usually when external forcing is absent, the matter is brought to an equilibrium in most circumstances by the exchange. This is considered to be due to the molecular structure or due to random interacting motion of uncountably many molecules. The entropy law is a typical one in this regard.

For conservative quantities, the exchange of variables make perfect sense. Because, with conservative variables, it is possible to connect the decrease of some quantity at a point to the increase of the same quantity at another point, and the exchange is understood as the *transfer*. This type of exchange is called the *transfer phenomenon*, or *transport phenomenon*. Therefore, corresponding to the three conservation laws mentioned above, we have three transfer phenomena: mass diffusion, momentum diffusion and thermal diffusion.

2.2. Mass diffusion in a fluid mixture

Diffusion in a fluid mixture occurs when composition varies with position. Suppose that the concentration of one constituent β of matter is denoted by C which is the mass proportion of the component with respect to the total mass ρ in a unit volume. Hence, the mass density of the component is given by ρC, and C is assumed to be a differentiable function of point $\mathbf{x}$ and time t: $C(\mathbf{x}, t)$.

Within the mixture, we choose an arbitrary surface element δA with its unit normal $\mathbf{n}$. Diffusion of the component β through the surface δA occurs from one side to the other and *vice versa*. However, owing to the nonuniformity of the distribution $C(\mathbf{x})$, there is a net transfer as a balance of the two counter fluxes (Fig. 2.1). Let us write the net transfer through $\delta A(\mathbf{n})$ toward the direction $\mathbf{n}$ per unit time as

$$\mathbf{q}(\mathbf{x}) \cdot \mathbf{n}\, \delta A(\mathbf{n}), \qquad (2.1)$$

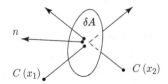

Fig. 2.1.   Two counter fluxes.

assuming that it is proportional to the area δA, and a vector $\mathbf{q}$ can be defined at each point $\mathbf{x}$, where $\mathbf{q}$ is called the *diffusion flux*.

From the above consideration, the diffusion flux $\mathbf{q}$ will be related to the concentration C. Since the flux $\mathbf{q}$ should be zero if the concentration is uniform, $\mathbf{q}$ would depend on the concentration gradient or derivatives of C. Provided that the concentration gradient is small, the flux $\mathbf{q}$ would depend on the gradient linearly with the proportional constant k_C as follows:

$$\mathbf{q}^{(C)} = -k_C \operatorname{grad} C = -k_C \left(\partial_x C, \, \partial_y C, \, \partial_z C \right), \qquad (2.2)$$

where k_C is the coefficient of mass diffusion. This is regarded as a mathematical assumption that higher-order terms are negligible when the flux is represented by a Taylor series with respect to derivatives of C. From the aspect of molecular motion, a macroscopic scale is much larger than the microscopic intermolecular distance, so that the concentration gradient in usual macroscopic problems would be very small from the view point of the molecular structure. The above expression (2.2) is valid in an isotropic material. In an anisotropic medium, the coefficient should be a tensor k_{ij}, rather than a scalar constant k_C. The coefficient k_C is positive usually, and the diffusion flux is directed from the points of larger C to those of smaller C, resulting in attenuation of the degree of C nonuniformity.

Next, in order to derive an equation governing C, we choose an arbitrary volume V bounded by a closed surface A in a fluid mixture at rest (Fig. 2.2), and observe the volume V with respect to the frame of the center of mass. So that, there is no macroscopic motion. The total mass of the component β in the volume V is $M_\beta = \int_V \rho C \, dV$ by the definition of C. Some of this component will move out of V

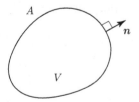

Fig. 2.2. An arbitrary volume V.

by the diffusion flux **q** through the bounding surface A, the total amount of outward diffusion is given by

$$\oint_A \mathbf{q}^{(C)} \cdot \mathbf{n} \, dA = - \oint_A k_C \, \mathbf{n} \cdot \operatorname{grad} C \, dA,$$

where **n** is unit outward normal to the surface element dA. This outward flux gives the rate of decrease of the mass M_β (per unit time). Hence we have the equation for the rate of increase of M_β:

$$\frac{d}{dt} \int_V \rho C \, dV = \int_V \frac{\partial}{\partial t} (\rho C) \, dV = \oint_A k_C \, \mathbf{n} \cdot \nabla C \, dA,$$

where the time derivative is placed within the integral sign since the volume element dV is fixed in space, and grad is replaced by ∇. Applying the Gauss's theorem (see Sec. 3.1 and Appendix A.6) transforming the surface integral into a volume integral, we obtain

$$\int_V \left[\frac{\partial}{\partial t} (\rho C) - \operatorname{div}(k_C \nabla C) \right] dV = 0.$$

Since this relation is valid for any volume V, the integrand must vanish identically. Thus we obtain

$$\frac{\partial}{\partial t} (\rho C) = \operatorname{div}(k_C \nabla C). \tag{2.3}$$

In a fluid at rest in equilibrium, no net translation of mass is possible. Therefore, the total mass ρ in unit volume is kept constant.[1] Moreover, the diffusion coefficient k_C is assumed to be constant. In this case, the above equation reduces to

$$\frac{\partial C}{\partial t} = \lambda_C \Delta C, \quad \lambda_C = \frac{k_C}{\rho}, \tag{2.4}$$

where Δ is the Laplacian operator,

$$\Delta := \nabla^2 = \frac{\partial^2}{\partial x^2} + \frac{\partial^2}{\partial y^2} + \frac{\partial^2}{\partial z^2}.$$

---

[1]When the diffusing component is only a small fraction of total mass, the density ρ may be regarded as constant even when the frame is not of the center of mass.

Equation (2.4) is the **diffusion equation**, and λ_C is the *diffusion coefficient*.

2.3. Thermal diffusion

Transport of the molecular random kinetic energy (i.e. the *heat energy*) is called *heat transfer*. A molecule in a gas carries its own kinetic energy. The average kinetic energy of molecular random velocities is the thermal energy, which defines the temperature T (Sec. 1.2).

Choosing an imaginary surface element δA in a gas, we consider such molecules moving from one side to the other, and those *vice versa*. If the temperatures on both sides are equal, then the transfer of thermal energy (from one side to the other) cancels out with the counter transfer, and there is no net heat transfer. However, if the temperature T depends on position $\mathbf{x}$, obviously there is a net heat transfer. The flow of heat through the surface element δA will be written in the form (2.1), where the vector $\mathbf{q}$ is now called the *heat flux*. In liquids or solids, heat transport is caused by collision or interaction between neighboring molecules.

Analogously with the mass diffusion, the heat flux will be represented in terms of the temperature gradient as

$$\mathbf{q} = -k \operatorname{grad} T, \tag{2.5}$$

where k is termed the *thermal conductivity*. The second law of thermodynamics (for the entropy) implies that the coefficient k should be positive (see Sec. 4.2).

The equation corresponding to (2.3) is written as

$$\rho C_p \frac{\partial T}{\partial t} = \operatorname{div}(k\nabla T),$$

since the increase of heat energy is given by $\rho C_p \Delta T$ for a temperature increase ΔT, where C_p is the specific heat per unit mass at constant pressure.

Corresponding to (2.4), the *equation of thermal conduction* is given by

$$\frac{\partial T}{\partial t} = \lambda_T \Delta T, \quad \lambda_T = \frac{k}{\rho c_p}, \tag{2.6}$$

in a fluid at rest, where λ_T is the *thermal diffusivity*. Equation (2.6) is also called the *Fourier's equation* of thermal conduction.

2.4. Momentum transfer

Transfer of molecular momentum emerges as an internal friction. A fluid with such an internal friction is said to be *viscous*. The momentum transfer is caused by molecules carrying their momenta, or by interacting force between molecules. The concentration and temperature considered above were scalars, however momentum is a vector. This requires some modification in the formulation of momentum transfer.

Macroscopic velocity **v** of a fluid at a point **x** in space is defined by the velocity of the center of mass of a fluid particle located at **x** instantaneously. Constituent molecules in the fluid particle are moving randomly with velocities $\tilde{\mathbf{u}}_\alpha$ (Sec. 1.2). Let us consider the transport of the ith component of momentum. Instead of the expression (2.1), the ith momentum transfer through a surface element $\delta A(\mathbf{n})$ from the side I (to which the normal **n** is directed) to the other II is defined (Fig. 2.3) as

$$q_{ij}\, n_j\, \delta A(\mathbf{n}), \tag{2.7}$$

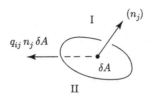

Fig. 2.3. Momentum transfer through a surface element $\delta A(\mathbf{n})$.

where $q_{ij} n_j = \sum_{j=1}^{3} q_{ij} n_j$. The tensor quantity q_{ij} represents the ith component of momentum passing per unit time through a unit area normal to the jth axis. The dimension of q_{ij} is equivalent to that of force per unit area, and such a quantity is termed a *stress tensor*. The stress associated with nonuniform velocity field $\mathbf{v}(\mathbf{x})$ is characterized by a tangential force-component to the surface element considered, and called the *viscous stress*. It can be verified that the stress tensor must be symmetric (Problem 2.3):

$$q_{ij} = q_{ji}.$$

Concerning the momentum transfer, there is another significant difference from the previous cases of the transfer of concentration or temperature. Suppose that the fluid is at rest and is in both mechanical and thermodynamical equilibrium. Hence, variables are distributed uniformly in space. Let us pay attention to a neighborhood on one side of a surface element $\delta A(\mathbf{n})$ where the normal vector $\mathbf{n}$ is directed. In the case of heat, the heat flux escaping out of δA is balanced with the flux coming in through δA, resulting in vanishing net flux in the equilibrium. How about in the case of momentum? The negative momentum (because it is anti-parallel to $\mathbf{n}$) escaping from from the side I out of δA would be expressed as "vanishing of negative momentum" Q, while the positive momentum coming into the side I through δA would be expressed as "emerging of positive momentum" P which is a contraposition of the previous statement. Hence, both are same and we have twice the positive momentum gain P (stress). However, on the other side of the surface δA, the situation is reversed and we have twice the loss of P. Thus, both stresses counter balance. This is recognized as the *pressure*.

In a fluid at rest, the momentum transfer is given by

$$q_{ij} = -p\delta_{ij}, \quad \text{hence } q_{ij} n_j = -pn_i, \tag{2.8}$$

(see Eq. (4.1)), where δ_{ij} is the Kronecker's delta and the minus sign is due to the definition of q_{ij} (see the footnote to Sec. 4.1 and Appendix A.1 for δ_{ij}). In a uniform fluid, the pressure is always normal to the surface chosen (Fig. 2.4). Total pressure force acting

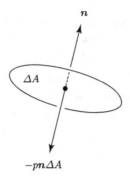

$$n$$

$$\Delta A$$

$$-pn\Delta A$$

Fig. 2.4.   Pressure stress.

on a fluid particle is given by

$$-\oint_{S_p} pn_i \mathrm{d}A, \tag{2.9}$$

where S_p denotes the surface of a small fluid particle.

In the transport phenomena considered above such as diffusion of mass, heat or momentum, the net transfers are in the direction of diminishing nonuniformity (Sec. 4.2). The coefficients of diffusivity, thermal conductivity and viscosity in the representation of fluxes are called the *transport coefficients*.

2.5.   An ideal fluid and Newtonian viscous fluid

The flow of a viscous fluid along a smooth solid wall at rest is characterized by the property that the velocity vanishes at the wall surface. If the velocity far from the wall is large enough, the profile of the tangential velocity distribution perpendicular to the surface has a characteristic form of a thin layer, termed as a *boundary layer*.

Suppose there is a parallel flow along a flat plate with the velocity far from it being U in the x direction, the y axis being taken perpendicular to the plate, and the flow velocity is represented by $(u(y), 0)$ in the (x, y) cartesian coordinate frame. The flow field represented as $(u(y), 0)$ is called a parallel *shear flow*. Owing to this shear flow, the plate is acted on by a *friction force* due to the flow. If the friction force per unit area of the plate is represented as

$$\sigma_f = \mu \left. \frac{du}{dy} \right|_{y=0}, \tag{2.10}$$

where μ is the coefficient of shear viscosity, this is called the *Newton's law of viscous friction* (see Problem 2.1).

This law can be extended to the law on an internal imaginary surface of the flow. Consider an internal surface element B perpendicular to the y-axis located at an arbitrary y position (Fig. 2.5). The unit normal to the surface B is in the positive y direction. The internal friction force on B from the upper to lower side has only the x-component. If the friction $\sigma^{(s)}$ per unit area is written as

$$\sigma^{(s)} = \mu \frac{d}{dy} u(y), \quad (= q_{xy}), \tag{2.11}$$

then the fluid is called the *Newtonian fluid*. The friction $\sigma^{(s)}$ per unit area is called the *viscous stress*, in particular, called the *shear stress* for the present shear flow, and it corresponds to q_{xy} of (2.7). The stress is also termed as a *surface force*. The pressure force given by (2.7) and (2.8) in the previous section is another surface force. The pressure stress has only the normal component to the surface $\delta A(\mathbf{n})$, whereas the viscous stress has a tangential component and a normal component (in general compressible case).

One can consider an idealized fluid in which the shear viscosity μ vanishes everywhere. Such a fluid is called an *inviscid fluid*, or an *ideal fluid*. In the flow of an inviscid fluid, the velocity adjacent to the solid wall does not vanish in general, and the fluid has nonzero tangential

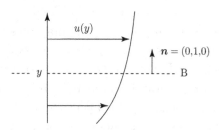

Fig. 2.5. Momentum transfer through an internal surface B.

slip-velocity at the wall. On the other hand, the flow velocity of a viscous fluid vanishes at the solid wall. This is termed as *no-slip*.

Thus, the **boundary conditions** of the velocity **v** on the surface of a body at rest are summarized as follows:

Viscous fluid: $\mathbf{v} = 0$ (no-slip),   (2.12)

Inviscid fluid: *nonzero tangential velocity*   (slip-flow).   (2.13)

The inviscid fluid is often called an *ideal fluid* (or sometimes a *perfect fluid*), in which the surface force has only normal component.

In this textbook, the ideal fluid denotes a fluid characterized by the property that *all transport coefficients of viscosity and thermal conductivity vanish*. Since all the transport coefficients vanish, macroscopic flows of an ideal fluid is separated from the microscopic irreversible dissipative effect arising from atomic thermal motion.

2.6. Viscous stress

For a Newtonian fluid, the viscous stress is given in general by

$$\sigma_{ij}^{(v)} = 2\mu D_{ij} + \zeta D \delta_{ij}, \tag{2.14}$$

where μ and ζ are coefficients of viscosity, and

$$D_{ij} := e_{ij} - \frac{1}{3} D \delta_{ij} = \frac{1}{2}(\partial_i v_j + \partial_j v_i) - \frac{1}{3}(\partial_k v_k)\,\delta_{ij} \tag{2.15}$$

$$D := \partial_k v_k = e_{kk} = \operatorname{div} \mathbf{v}. \tag{2.16}$$

The tensor D_{ij} is readily shown to be traceless. In fact, $D_{ii} = e_{ii} - \frac{1}{3} D \delta_{ii} = D - \frac{1}{3} D \cdot 3 = 0$ since $\delta_{ii} = 3$. It may be said that D_{ij} is a deformation tensor associated with a straining motion which keeps the volume unchanged. The expression (2.14) of the viscous stress can be derived from a general linear relation between the stress $\sigma_{ij}^{(v)}$ and the rate-of-strain tensor e_{ij} for an isotropic fluid, in which the number of independent scalar coefficients is only two — μ and ζ (see Problem 2.4).

Substituting the definitions of D_{ij} and e_{ij} of (1.18), the Newtonian *viscous stress* is given by

$$\sigma_{ij}^{(v)} = \mu \left(\partial_i v_j + \partial_j v_i - (2/3) D \delta_{ij} \right) + \zeta D \delta_{ij}, \qquad (2.17)$$

where the coefficient μ is termed the coefficient of *shear viscosity*, while ζ the bulk viscosity (or the second viscosity).

If the surface $\delta A(\mathbf{n})$ is inclined with its normal $\mathbf{n} = (n_x, n_y, n_z)$, the viscous force $F_i^{(v)}$ acting on $\delta A(\mathbf{n})$ at $\mathbf{x}$ from side I (to which the normal $\mathbf{n}$ is directed) to II is given by

$$F_i^{(v)} \delta A(\mathbf{n}) = \sigma_{ij}^{(v)} n_j \delta A(\mathbf{n}). \qquad (2.18)$$

For each component, we have

$$F_x^{(v)} = \sigma_{xx}^{(v)} n_x + \sigma_{xy}^{(v)} n_y + \sigma_{xz}^{(v)} n_x z,$$

$$F_y^{(v)} = \sigma_{yx}^{(v)} n_x + \sigma_{yy}^{(v)} n_y + \sigma_{yz}^{(v)} n_x z,$$

$$F_z^{(v)} = \sigma_{zx}^{(v)} n_x + \sigma_{zy}^{(v)} n_y + \sigma_{zz}^{(v)} n_x z.$$

Example 1. Parallel shear flow. Let us consider a parallel shear flow with velocity $\mathbf{v} = (u(y), 0, 0)$. We immediately obtain $D = \operatorname{div} \mathbf{v} = \partial u / \partial x = 0$. Moreover, all the components of the tensors D_{ij} vanish except $D_{xy} = D_{yx} = \frac{1}{2} u'(y)$. Therefore, the viscous stress reduces to

$$\sigma_{xy}^{(v)} = \mu \, u'(y) = \sigma_{yx}^{(v)},$$

which is equivalent to the expression (2.11). Thus, it is seen that the expression (2.14) or (2.17) gives a generalization of the viscous stress of the Newtonian fluid. Non-Newtonian fluid is one in which the stress is not expressed in this form, sometimes it is nonlinear with respect to the strain tensor e_{ij}, or sometimes it includes elasticity.

Example 2. Uniform rotation. The viscous stress $\sigma_{ij}^{(v)}$ vanishes identically in the uniform rotation where the vorticity is uniform. The uniform rotation is represented by

$$\boldsymbol{\omega} = \nabla \times \mathbf{v} = (\Omega_1, \Omega_2, \Omega_3) = [\text{a constant vector}]. \qquad (2.19)$$

Corresponding velocity $\mathbf{v} = (v_j)$ is given by

$$v_j = \frac{1}{2} \varepsilon_{jlm} \Omega_l \, x_m = \frac{1}{2} (\boldsymbol{\Omega} \times \mathbf{x})_j.$$

Differentiating this with respect to x_i, we obtain $\partial_i v_j = \frac{1}{2} \varepsilon_{jlm} \Omega_l \, \delta_{mi} = \frac{1}{2} \varepsilon_{ijl} \, \Omega_l$. Similarly, we have $\partial_j v_i = \frac{1}{2} \varepsilon_{jil} \Omega_l$. Hence, we have

$$\partial_i v_j + \partial_j v_i = \frac{1}{2} \left(\varepsilon_{ijl} + \varepsilon_{jil} \right) \Omega_l = 0 \qquad (2.20)$$

by the anti-symmetric property $\varepsilon_{jil} = -\varepsilon_{ijl}$. Furthermore, this means

$$D = \operatorname{div} \mathbf{v} = \partial_k v_k = \frac{1}{2} \varepsilon_{kkl} \, \Omega_l = 0. \qquad (2.21)$$

Using (2.20) and (2.21), it is found that the viscous stress $\sigma_{ij}^{(v)}$ of (2.17) vanishes identically in the motion of uniform rotation $\boldsymbol{\omega} = $ a constant vector.

2.7. Problems

Problem 2.1 Viscous friction

Suppose that a Poiseuille flow of a Newtonian viscous fluid (with shear viscosity μ) is maintained in a channel between two parallel walls under a pressure gradient in x direction, where the walls are located at $y = 0$ and $y = d$ in the (x, y) plane [Fig. 2.6(a)]. The fluid velocity is represented by the form $(u(y), 0)$, where

$$u(y) = \frac{4U}{d^2} \, y(d - y), \qquad (2.22)$$

with the maximum velocity U (see Sec. 4.6.1) at the center line $y = d/2$. Determine the viscous friction force F per unit distance along the channel.

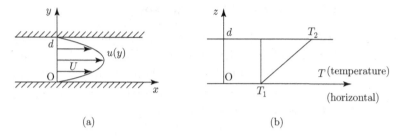

Fig. 2.6. (a) Velocity distribution $u(y)$; (b) Temperature distribution $T(z)$.

Problem 2.2 Steady thermal conduction

Suppose that a horizontal fluid layer of thickness d is at rest between two horizontal walls and that the temperatures of the lower and upper walls are maintained at T_1 and T_2, respectively, where $T_2 > T_1$, and that the z axis is taken in the vertical direction with the lower wall at $z = 0$ [Fig. 2.6(b)]. Find the equation governing the steady temperature distribution $T(z)$, and determine the temperature $T(z)$ by solving the equation, which is given as

$$T(z) = T_1 + (T_2 - T_1)\frac{z}{d}. \tag{2.23}$$

(The problem of steady thermal conduction will be considered in Sec. 9.4.2 again.)

Problem 2.3 Initial value problem of diffusion equation

An initial value problem of a function $u(x, t)$ for an infinite range of a space coordinate x $(-\infty < x < \infty)$ is posed as follows:

$$\partial_t u = \lambda \partial_x^2 u \quad (t > 0), \tag{2.24}$$

$$u|_{t=0} = u_0(x), \tag{2.25}$$

where λ is a constant diffusion coefficient. Show that the following solves the above initial value problem:

$$u(x, t) = \frac{1}{2\sqrt{\pi \lambda t}} \int_{-\infty}^{\infty} u_0(\xi) \exp\left[-\frac{(x - \xi)^2}{4\lambda t}\right] d\xi. \tag{2.26}$$

Problem 2.4 Symmetry of stress tensor

Suppose that a stress tensor σ_{ij} is acting on the surface of an infinitesimal cubic volume of side a (with its edges parallel to the axes x, y, z) in a fluid of density ρ. Considering the balance equation of angular momentum for the cube and taking the limit as $a \to 0$, verify that the stress tensor must be symmetric:

$$\sigma_{ij} = \sigma_{ji}.$$

Problem 2.5 Stress and strain

Consider a flow field $v_k(\mathbf{x})$ of an *isotropic* viscous fluid, and suppose that there is a general linear relation between the viscous stress $\sigma_{ij}^{(v)}$ and the rate-of-strain tensor e_{kl}:

$$\sigma_{ij}^{(v)} = A_{ijkl}\, e_{kl}.$$

where e_{kl} is defined by (1.18). In an isotropic fluid, the coefficients A_{ijkl} of the fourth-order tensor are represented in terms of isotropic tensors (see (2.8)), which are given as follows:

$$A_{ijkl} = a\, \delta_{ij}\delta_{kl} + b\, \delta_{ik}\delta_{jl} + c\, \delta_{il}\delta_{jk}, \tag{2.27}$$

where a, b, c are constants. Using this form and the symmetry of the stress tensor (Problem 2.3), derive the expression (2.14) for the viscous stress $\sigma_{ij}^{(v)}$:

$$\sigma_{ij}^{(v)} = 2\mu D_{ij} + \zeta D\delta_{ij}.$$

Chapter 3

Fundamental equations
of ideal fluids

Fluid flows are represented by fields such as the velocity field $\mathbf{v}(\mathbf{x}, t)$, pressure field $p(\mathbf{x}, t)$, density field $\rho(\mathbf{x}, t)$, temperature field $T(\mathbf{x}, t)$, and so on. The field variables denote their values at a point $\mathbf{x}$ and at a time t. The position vector $\mathbf{x}$ is represented by (x, y, z), or equivalently (x_1, x_2, x_3) in the cartesian frame of reference. Fluid particles move about in the space with a velocity $d\mathbf{x}/dt = \mathbf{v}(\mathbf{x}, t)$.

Flow field evolves with time according to fundamental conservation laws of physics. There are three kinds of conservation laws of mechancis, which are conservation of *mass, momentum* and *energy*.[1] In fluid mechanics, these conservation laws are represented in terms of field variables such as $\mathbf{v}$, p, ρ, etc. Since the field variables depend on (x, y, z) and t, the governing equations are of the form of partial differential equations.

---

[1]The conservation laws in mechanics result from the fundamental homogeneity and isotropy of space and time. From the homogeneity of time in the Lagrangian function of a closed system, the conservation of energy is derived. From the *homogeneity* of space, the conservation of *momentum* is derived. Conservation of *angular momentum* (which is not discussed in this chapter) results from *isotropy* of space. Conservation of mass results from the invariance of the relativistic Lagrangian in the Newtonian limit (Appendix F.1). See Chap. 12, or [LL75], [LL76].

3.1. Mass conservation

The law of mass conservation is represented by the following Euler's *equation of continuity*, which reads

$$\frac{\partial \rho}{\partial t} + \frac{\partial (\rho u)}{\partial x} + \frac{\partial (\rho v)}{\partial y} + \frac{\partial (\rho w)}{\partial z} = 0, \tag{3.1}$$

where ρ is the fluid density and $\mathbf{v}(\mathbf{x}, t) = (u, v, w)$ the velocity. Using the differential operator div of the vector analysis, this is written as

$$\partial_t \rho + \mathrm{div}(\rho \mathbf{v}) = 0, \tag{3.2}$$

and derived in the following way.

Take a certain volume V_0 fixed in space arbitrarily (Fig. 3.1), and choose a volume element dV within V_0. Fluid mass in the volume dV is given by ρdV, and the total mass is its integral over the volume V_0,

$$M_0(t) = \int_{V_0} \rho dV. \tag{3.3}$$

The fluid density ρ depends on t, and in addition, the fluid itself moves around with velocity $\mathbf{v}$. Therefore, the total mass M_0 varies with time t, and its rate of change is given by

$$\frac{\mathrm{d}}{\mathrm{d}t} M_0(t) = \frac{\partial}{\partial t} \int_{V_0} \rho dV = \int_{V_0} \frac{\partial \rho}{\partial t} \, dV, \tag{3.4}$$

where the partial differential operator $\partial/\partial t$ is used on the right-hand side. Since we are considering fixed volume elements dV in space, the time derivative $\partial_t \int \rho dV$ can be replaced by $\int (\partial_t \rho) dV$.

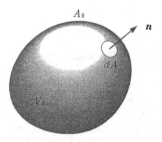

Fig. 3.1.   Volume V_0.

The change of total mass M_0 is caused by inflow or outflow of fluid through surface A_0 bounding V_0. The amount of fluid flowing through a surface element dA with the unit normal $\mathbf{n}$ is given by

$$\rho v_n dA = \rho \mathbf{v} \cdot \mathbf{n} dA,$$

where v_n is the normal component of the velocity[2] and $v_n dA$ denotes the volume of fluid passing through dA per unit time (Fig. 3.2). The normal $\mathbf{n}$ is taken to be directed outward from V_0, so that $\rho v_n dA$ denotes the mass of fluid flowing out of the volume V_0 per unit time. Total out-flow of the fluid mass per unit time is

$$\oint_{A_0} \rho\, v_n dA = \oint_{A_0} \rho \mathbf{v} \cdot \mathbf{n} dA = \oint_{A_0} \rho v_k n_k dA.$$

This integral over the closed surface A_0 is transformed into a volume integral by Gauss's divergence theorem[3]:

$$\oint_{A_0} \rho v_k n_k dA = \int_{V_0} \frac{\partial}{\partial x_k}(\rho v_k)\, dV = \int_{V_0} \text{div}(\rho \mathbf{v})\, dV, \qquad (3.5)$$

where the div operator is defined by (1.31). This gives the rate of decrease of total mass in V_0, that is equal to $-dM_0/dt$. Thus, on

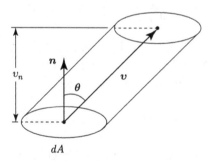

Fig. 3.2. Mass flux through dA.

---

[2] $v_n = \mathbf{v} \cdot \mathbf{n} = v_k n_k = |\mathbf{v}||\mathbf{n}| \cos\theta$, where $|\mathbf{n}| = 1$ and θ is the angle between $\mathbf{v}$ and $\mathbf{n}$.

[3] The rule for transforming the surface integral into a volume integral with the bounding surface is as follows: the term $n_k\, dA$ in the surface integral is replaced by the volume element dV and the differential operator $\partial/\partial x_k$ acting on the remaining factor in the integrand of the surface integral.

addition of the two terms, the right-hand sides of (3.4) and (3.5), must vanish:

$$\int_{V_0} \frac{\partial \rho}{\partial t} dV + \int_{V_0} \operatorname{div}(\rho \mathbf{v}) dV = \int_{V_0} \left[\frac{\partial \rho}{\partial t} + \operatorname{div}(\rho \mathbf{v}) \right] dV = 0. \quad (3.6)$$

This is an identity and must hold for any choice of volume V_0 within the fluid. Hence, the integrand inside [ ] must vanish point-wise.[4] Thus, we obtain the equation of continuity (3.2), which is also written as

$$\frac{\partial}{\partial t} \rho + \frac{\partial}{\partial x_k} (\rho v_k) = 0. \quad (3.7)$$

The second term of (3.7) is decomposed as

$$\frac{\partial}{\partial x_k} (\rho v_k) = v_k \partial_k \rho + \rho \partial_k v_k = (\mathbf{v} \cdot \nabla)\rho + \rho \operatorname{div} \mathbf{v},$$

where $\partial_k = \partial/\partial x_k$. Using the Lagrange derivative D/Dt of (1.13), the continuity equation (3.7) is rewritten as

$$\frac{D\rho}{Dt} + \rho \operatorname{div} \mathbf{v} = 0. \quad (3.8)$$

Notes: (i) If the density of each fluid particle is *invariant* during the motion, then $D\rho/Dt = 0$. In this case, the fluid is called *incompressible*. If the fluid is incompressible, we have

$$\operatorname{div} \mathbf{v} = \partial_x u + \partial_y v + \partial_z w = 0. \quad (3.9)$$

This is valid even when ρ is not uniformly constant.
(ii) *Uniform density*: The same equation (3.9) is also obtained from (3.2) by setting ρ to be a constant.
    Thus, Eq. (3.9) implies both cases (i) and (ii).

---

[4]Otherwise, the integral does not always vanish, e.g. when choosing V_0 where the integrand [ ] is not zero.

3.2. Conservation form

It would be instructive to remark a general characteristic feature of Eq. (3.7). Namely, it has the following structure:

$$\frac{\partial}{\partial t} D + \frac{\partial}{\partial x_k} (F)_k = Q, \qquad (3.10)$$

where D is a density of some physical field and $(F)_k$ is the kth component of corresponding flux F, and Q is a source generating the field D (Fig. 3.3). If D is the mass density ρ and $(F)_k$ is the mass flux ρv_k passing through a unit surface per unit time (and there is no mass source $Q = 0$), then Eq. (3.10) reduces to the continuity equation (3.7). The equation of the form (3.10) is called the *conservation form*, in general.

3.3. Momentum conservation

The conservation of momentum is the fundamental law of mechanics. This conservation law results from the homogeneity of space with respect to the Lagrangian function in Newtonian mechanics. However, we write down firstly the equation of motion for a fluid particle of mass $\rho \delta V$ in the form, $(\rho \delta V)(\text{acceleration}) = (\text{force})$.[5]

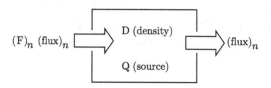

Fig. 3.3.   Conservation form.

---

[5]This relation is the Newton equation of motion itself for a point mass (a discrete object). Fundamental equations and conservation equations for continuous fields is the subject of Chapter 12 (gauge principle for flows of ideal fluids).

Thereafter, it will be shown that this is equivalent to the conservation of momentum.

3.3.1. *Equation of motion*

In an ideal fluid, the surface force is only the pressure (2.8) or (2.9) (there is no viscous stress). Choosing a volume V_0 as before and denoting its bounding surface by A_0, the total pressure force on A_0 is given by

$$- \oint_{A_0} p n_i \mathrm{d}A = - \int_{V_0} \frac{\partial}{\partial x_i} p \mathrm{d}V = - \int_{V_0} (\mathrm{grad}\, p)_i \mathrm{d}V, \qquad (3.11)$$

where the middle portion is obtained by applying the rule described in the footnote in Sec. 3.1. By this extended Gauss theorem, the surface integral is transformed to the volume integral. It is remarkable to find that the force on a small fluid particle of volume δV is given by the pressure gradient,

$$-\mathrm{grad}\, p\, \delta V = -(\partial_x p,\, \partial_y p,\, \partial_z p)\, \delta V.$$

In addition to the surface force just given, there is usually a volume force which is proportional to the mass $\rho \delta V$. Let us write this as $(\rho \delta V)\, \mathbf{f}$, which is often called an *external force*. A typical example is the gravity force, $(\rho \delta V)\, \mathbf{g}$ with $\mathbf{g}$ as the acceleration of gravity.

Once the velocity field $\mathbf{v}(\mathbf{x}, t)$ is given, the acceleration of the fluid particle is written as $D\mathbf{v}/Dt$ (see (1.33)). The Newton equation of motion for a fluid particle is written as

$$(\rho \delta V) \frac{D}{Dt} \mathbf{v} = -\mathrm{grad}\, p\, \delta V + (\rho \delta \mathbf{V})\, \mathbf{f}. \qquad (3.12)$$

Dividing this by $\rho \delta V$ and using the expression given in (1.33), we obtain

$$\partial_t \mathbf{v} + (\mathbf{v} \cdot \nabla)\mathbf{v} = -\frac{1}{\rho} \mathrm{grad}\, p + \mathbf{f}. \qquad (3.13)$$

This is called **Euler's equation of motion.**[6] The second term $(\mathbf{v} \cdot \nabla)\mathbf{v}$ on the left-hand side is also written as $(\mathbf{v} \cdot \mathrm{grad})\mathbf{v}$. This is of the second order with respect to the velocity vector $\mathbf{v}$, and often called a *nonlinear* term, or *advection* term. The nonlinearity is responsible for complex behaviors of flows. This term is also called an *inertia term* (however, the term $\partial_t \mathbf{v}$ is related to fluid inertia as well).

In the component representation, the ith component of the equation is

$$\partial_t v_i + v_k \partial_k v_i = -\frac{1}{\rho} \partial_i p + f_i. \tag{3.14}$$

If the external force is the *uniform gravity* represented by $\mathbf{f} = (0, 0, -g)$ where the acceleration of gravity g is constant and directed towards the negative z axis, each component is written down as follows:

$$\begin{aligned}
\partial_t u + u \partial_x u + v \partial_y u + w \partial_z u &= -\rho^{-1} \partial_x p, \\
\partial_t v + u \partial_x v + v \partial_y v + w \partial_z v &= -\rho^{-1} \partial_y p, \\
\partial_t w + u \partial_x w + v \partial_y w + w \partial_z w &= -\rho^{-1} \partial_z p - g.
\end{aligned} \tag{3.15}$$

Euler's equation of motion is given another form. To see it, the following vector *identity* (A.21) is useful:

$$\mathbf{v} \times (\nabla \times \mathbf{v}) = \nabla \left(\frac{1}{2} \mathbf{v}^2 \right) - (\mathbf{v} \cdot \nabla)\mathbf{v}. \tag{3.16}$$

---

[6]The paper of Leonhard Euler (1707–1783) was published in the proceedings of the *Royal Academy Prussia* (1757) in Berlin with the title, *Principes généraux du mouvement des fluides* (General principles of the motions of fluids). The form of equation Euler actually wrote down is that using components such as (3.15) (with x, y components of external force), determined from the principles of mechanics (as carried out in the main text). Before that, he showed in the same paper the equation of continuity exactly of the form (3.1). The *equation of continuity* had been derived in his earlier paper (Principles of the motions of fluids, 1752) by assuming that continuity of the fluid is never interrupted. It is rather surprising to find that not only Euler (1757) derived an integral for potential flows of the form (5.29) of Chap. 5, but also the z-component of the vorticity equation (3.32) of Sec. 3.4.1 was derived in 1752. Thus, Euler presented an essential part of modern fluid dynamics of ideal fluids. However, his contribution should be regarded as one step toward the end of a long process, carried out by Bernoulli family, d'Alembert, Lagrange, etc. in the 18th century [Dar05].

Eliminating the advection term $(\mathbf{v} \cdot \nabla)\mathbf{v}$ in (3.13) with the help of (3.16) and introducing the definition $\boldsymbol{\omega} := \nabla \times \mathbf{v}$ called the vorticity, the equation of motion (3.13) is rewritten as

$$\partial_t \mathbf{v} + \boldsymbol{\omega} \times \mathbf{v} = -\nabla \frac{\mathbf{v}^2}{2} - \frac{1}{\rho} \nabla p + \mathbf{f}. \tag{3.17}$$

Consider a particular case when the density is uniform, i.e. $\rho = \rho_0$ (constant), and that the external force has a potential χ, i.e. $\mathbf{f} = -\nabla \chi$. Then the above equation reduces to

$$\partial_t \mathbf{v} + \boldsymbol{\omega} \times \mathbf{v} = -\nabla \left(\frac{\mathbf{v}^2}{2} + \frac{1}{\rho_0} p + \chi \right). \tag{3.18}$$

The gravity potential of uniform acceleration g is given by

$$\chi = gz. \tag{3.19}$$

3.3.2. *Momentum flux*

Euler's equation of motion just derived represents the conservation of momentum. In fact, consider the momentum included in the volume element δV, which is given by $(\rho \delta V)\mathbf{v}$. Therefore, $\rho \mathbf{v}$ is the momentum in a unit volume, namely the *momentum density*. Its time derivative is

$$\partial_t(\rho v_i) = (\partial_t \rho)v_i + \rho \partial_t v_i.$$

The two time derivatives $\partial_t \rho$ and $\partial_t v_i$ can be eliminated by using (3.7) and (3.14), and we obtain

$$\begin{aligned} \partial_t(\rho v_i) &= -\partial_k(\rho v_k)v_i - \rho v_k \partial_k v_i - \partial_i p + \rho f_i \\ &= -\partial_k(\rho v_k v_i) - \partial_i p + \rho f_i. \end{aligned} \tag{3.20}$$

The pressure gradient term on the right-hand side is rewritten as

$$\partial_i p = \partial_k(p \delta_{ik}),$$

by using Kronecker's delta δ_{ij} (A.1). Substituting this and rearranging (3.20), we obtain

$$\partial_t(\rho v_i) + \partial_k P_{ik} = \rho f_i, \tag{3.21}$$

where the tensor P_{ik} is defined by

$$P_{ik} = \rho v_i v_k + p \delta_{ik}. \tag{3.22}$$

Equation (3.21) represents the *conservation of momentum*, which is seen to have the form of Eq. (3.10). The quantity ρv_i is the ith component of the momentum density (vector), while P_{ik} denotes the momentum flux (tensor).[7] The difference from the previous equation of mass conservation is that there is a source term ρf_i on the right-hand side. This is natural because the external force is a source of momentum in the true meaning.

Why is the term on the right-hand side of (3.21) a source (production) of momentum will become clearer, if we integrate the equation over a volume V_0. In fact, we have

$$\frac{\partial}{\partial t} \int_{V_0} \rho v_i \mathrm{d}V + \int_{V_0} \frac{\partial}{\partial x_k} P_{ik} \mathrm{d}V = \int_{V_0} \rho f_i \mathrm{d}V.$$

The second integral term on the left-hand side can be transformed to a surface integral by using the rule of the Gauss theorem, i.e. replace $\partial_k(X)\,\mathrm{d}V$ with $X n_k \mathrm{d}A$. Moving this term to the right-hand side, we obtain

$$\frac{\partial}{\partial t} \int_{V_0} \rho v_i \mathrm{d}V = - \int_{A_0} P_{ik} n_k \mathrm{d}A + \int_{V_0} \rho f_i \mathrm{d}V. \tag{3.23}$$

This is interpreted as follows. The left-hand side denotes the rate of change of total momentum included in volume V_0, whereas the first integral on the right-hand side represents the momentum flowing out of the bounding surface A_0 per unit time, and the second integral is the total force acting on V_0 which is nothing but the rate of production of momentum per unit time in mechanics. Thus, the Euler equation of motion represents the momentum conservation in conjunction with the continuity equation.

---

[7]The mass density was a scalar, whereas the momentum density is a vector. Correspondingly, the flux P_{ik} is a second-order tensor.

An important point brought to light is the following expression,

$$P_{ik}n_k\mathrm{d}A = \rho v_i v_k n_k\mathrm{d}A + p\delta_{ik}n_k\mathrm{d}A, \qquad (3.24)$$

where (3.22) is used. This represents the ith component of total momentum flux passing through the surface element $\mathrm{d}A$. The first term is the macroscopic momentum flux, whereas the second represents a microscopic momentum flux. In Sec. 1.2 we saw that the microscopic momentum flux gives the expression for pressure of an ideal gas.

3.4. Energy conservation

The conservation of energy is a fundamental law of physics. An ideal fluid is characterized by the absence of viscosity and thermal diffusivity. Because of this property, the motion of a fluid particle is *adiabatic*. It will be shown that this is consistent with the conservation of energy.

3.4.1. *Adiabatic motion*

In an ideal fluid, there is neither heat generation by viscosity nor heat exchange between neighboring fluid particles. Therefore the motion of an ideal fluid is *adiabatic*, and the entropy of each fluid particle is *invariant* during its motion. Denoting the entropy per unit mass by s, the adiabatic motion of a fluid particle is described as

$$\frac{\mathrm{D}}{\mathrm{D}t}s = \partial_t s + (\mathbf{v} \cdot \nabla)s = 0. \qquad (3.25)$$

It is said, the motion is *isentropic*. This equation is transformed to

$$\partial_t(\rho s) + \mathrm{div}(\rho s\mathbf{v}) = 0, \qquad (3.26)$$

by using the continuity equation (3.2). If the entropy value was uniform initially, the value is invariant thereafter. This is called a *homentropic* flow. Then the equations will be simplified.

Denoting the internal energy and enthalpy per unit mass by e and h, respectively and the specific volume by $V = 1/\rho$, we have a thermodynamic relation $h = e + pV$. For an infinitesimal change of state, two thermodynamic relations are written down as

$$de = Tds - pdV = Tds + (p/\rho^2)\, d\rho, \qquad (3.27)$$

$$dh = Tds + Vdp = Tds + (1/\rho)\, dp, \qquad (3.28)$$

[LL80, Chap. 2], where T is the temperature.

For a homentropic fluid motion, we have $ds = 0$, and therefore $dh = dp/\rho$. Thus, we obtain

$$(1/\rho)\, \text{grad}\, p = \text{grad}\, h.$$

Substituting this into (3.13) and assuming the external force is conservative, i.e. $\mathbf{f} = -\text{grad}\, \chi$, we obtain

$$\partial_t \mathbf{v} + (\mathbf{v} \cdot \nabla)\mathbf{v} = -\text{grad}\, h - \text{grad}\, \chi. \qquad (3.29)$$

Equation (3.17) reduces to

$$\partial_t \mathbf{v} + \boldsymbol{\omega} \times \mathbf{v} = -\text{grad}\left(\frac{\mathbf{v}^2}{2} + h + \chi\right). \qquad (3.30)$$

This is valid for compressible flows, while Eq. (3.18) is restricted to the case of $\rho = \text{const}$. Taking the curl of this equation, the right-hand side vanishes due to (A.25) of Appendix A.5, and we obtain

$$\partial_t \boldsymbol{\omega} + \text{curl}(\boldsymbol{\omega} \times \mathbf{v}) = 0, \quad \boldsymbol{\omega} = \text{curl}\, \mathbf{v}. \qquad (3.31)$$

Thus, we have obtained an equation for the vorticity $\boldsymbol{\omega}$, which includes only $\mathbf{v}$ since $\boldsymbol{\omega} = \text{curl}\, \mathbf{v}$ (p and other variables are eliminated). In this equation, the entropy s is assumed a constant, although ρ is variable.

Using (A.24) of Appendix A.4, this is rewritten as

$$\partial_t \boldsymbol{\omega} + (\mathbf{v} \cdot \text{grad})\boldsymbol{\omega} + (\text{div}\, \mathbf{v})\boldsymbol{\omega} = (\boldsymbol{\omega} \cdot \text{grad})\mathbf{v}. \qquad (3.32)$$

Note that the first two terms are unified to $D\boldsymbol{\omega}/Dt$ and the third term is written as $-(1/\rho)(D\rho/Dt)\boldsymbol{\omega}$ by (3.8). Dividing (3.32) by ρ,

$$\frac{1}{\rho}\frac{D\boldsymbol{\omega}}{Dt} - \frac{\boldsymbol{\omega}}{\rho^2}\frac{D\rho}{Dt} = \frac{1}{\rho}(\boldsymbol{\omega} \cdot \text{grad})\mathbf{v}.$$

This is transformed to a compact form:

$$\frac{D}{Dt}\left(\frac{\boldsymbol{\omega}}{\rho}\right) = \left(\frac{\boldsymbol{\omega}}{\rho} \cdot \text{grad}\right)\mathbf{v}. \tag{3.33}$$

3.4.2. *Energy flux*

Let us consider an energy flux just as the momentum flux in Sec. 3.3.2. Total energy in a volume element δV is given by the sum of the kinetic and internal energies as

$$\rho\delta V\left(\frac{1}{2}v^2 + e\right) = \rho E\delta V, \quad E := \frac{1}{2}v^2 + e.$$

Hence, the energy density (per unit volume) is given by ρE. We consider its time derivative,

$$\frac{\partial}{\partial t}\left(\frac{1}{2}\rho v^2 + \rho e\right) = \frac{1}{2}v^2(\partial_t\rho) + \rho v_i\partial_t v_i + \partial_t(\rho e). \tag{3.34}$$

The first two terms come from $\partial_t(\rho v^2/2)$. By using (3.7) and (3.14),

$$\partial_t\left(\frac{1}{2}\rho v^2\right) = -\frac{1}{2}v^2\partial_k(\rho v_k) - \rho v_i v_k\partial_k v_i - v_i\partial_i p + \rho v_i f_i$$

$$= -\partial_k\left[\rho v_k\left(\frac{1}{2}v^2\right)\right] - \rho v_k\partial_k h + \rho T v_k\partial_k s + \rho v_k f_k, \tag{3.35}$$

where in the last equality the thermodynamic relation (3.28) is used to eliminate dp.

Next, to the remaining term $\partial_t(\rho e)$ of (3.34), we apply the thermodynamic relation (3.27) and obtain

$$d(\rho e) = ed\rho + \rho de = hd\rho + \rho Tds,$$

where $h = e + p/\rho$. Therefore, time differentiation leads to

$$\partial_t(\rho e) = h\partial_t\rho + \rho T\partial_t s = -h\partial_k(\rho v_k) + \rho T\partial_t s. \tag{3.36}$$

Summing up Eqs. (3.35) and (3.36) yields

$$\partial_t\left(\frac{1}{2}\rho v^2 + \rho e\right) = -\partial_k\left[\rho v_k\left(\frac{1}{2}v^2\right)\right] - \partial_k(\rho v_k h) + \rho T(D/Dt)s + \rho v_k f_k.$$

Using the entropy equation (3.25), we finally obtain

$$\partial_t \left[\rho \left(\frac{1}{2} v^2 + e \right) \right] + \partial_k \left[\rho v_k \left(\frac{1}{2} v^2 + h \right) \right] = W, \tag{3.37}$$

where $W = \rho v_k f_k$ denotes the rate of work by the external force. This is the equation of *conservation of energy* in the form of (3.10). When there is no external force, the right-hand side W vanishes (Fig. 3.4).

It is remarked that the energy density is given by the sum of kinetic energy and internal energies, whereas the energy flux $(F)_k$ is

$$(F)_k = \rho v_k \left(\frac{1}{2} v^2 + h \right),$$

namely e is replaced by h for the flux F.

Integrating (3.37) over volume V_0 and transforming the volume integral of the second term into the surface integral, we have

$$\partial_t \int_{V_0} \left(\frac{1}{2} \rho v^2 + \rho e \right) dV = - \oint_{A_0} \left(\frac{1}{2} \rho v^2 + \rho e + p \right) v_k n_k dA$$

$$+ \int_{V_0} W dV. \tag{3.38}$$

This indicates that the rate of increase of the total energy in V_0 is given by the sum of the energy inflow $- \left(\frac{1}{2} \rho v^2 + \rho e \right) v_k n_k$ through the bounding surface and the rate of work by the pressure $-p v_k n_k$, in addition to the rate of work W by the external force.

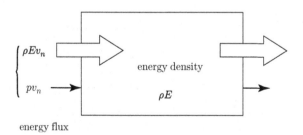

Fig. 3.4. Energy density $E = \frac{1}{2} v^2 + e$, and energy flux $Q_n = \rho E v_n + p v_n = \rho(\frac{1}{2} v^2 + h) v_n$.

3.5. Problems

Problem 3.1 One-dimensional unsteady flow

Write down the three conservation equations of mass, momentum and energy for one-dimensional unsteady flows (in the absence of external force) with x as the spatial coordinate and t the time when the density is $\rho(x,t)$, velocity is $\mathbf{v} = (u(x,t), 0, 0)$, and so on.

Comment: **Lagrange's form of equation of motion**

Lagrangian representation of position of a fluid particle $\mathbf{a} = (a, b, c)$ at time t is defined by $\mathbf{X}(\mathbf{a}, t)$ in (1.11) with its velocity $\mathbf{V}_a(t) = \partial_t \mathbf{X}(\mathbf{a}, t)$. Then, the x component of equation of motion (3.12) can be written as $\partial_t^2 \mathbf{X}(\mathbf{a}, t) = -\partial_x p + f_x$. We multiply this by $\partial x/\partial a$. Similarly, multiplying y and z components of Eq. (3.12) and summing up the three expressions, we obtain

$$\partial_t^2 \mathbf{X} \cdot \frac{\partial \mathbf{X}}{\partial a} = -\frac{\partial p}{\partial a} + \mathbf{f} \cdot \frac{\partial \mathbf{X}}{\partial a}. \tag{3.39}$$

This is the Lagrange's form of equation of motion (1788), corresponding to the Eulerian version (3.13).

Defining the Lagrangian coordinates $\mathbf{a} = (a_i)$ by the particle position at $t = 0$, the continuity equation for the particle density $\rho_a(t)$ is given by

$$\rho_a(t) \frac{\partial(\mathbf{x})}{\partial(\mathbf{a})} = \rho_a(0), \tag{3.40}$$

where $\partial(\mathbf{x})/\partial(\mathbf{a}) = \det(\partial \mathbf{X}_j/\partial a_i)$ is the Jacobian determinant.

Chapter 4

Viscous fluids

4.1. Equation of motion of a viscous fluid

In the previous chapters, we learned the existence of surface forces to describe fluid motion in addition to the volume force such as gravity. The surface force is also termed as the stress and represented by a tensor. A typical stress is the pressure $p\delta_{ij}$,[1] which is written as

$$\sigma_{ij}^{(p)} = -p\delta_{ij} = - \begin{pmatrix} p & 0 & 0 \\ 0 & p & 0 \\ 0 & 0 & p \end{pmatrix}. \tag{4.1}$$

The pressure force acts perpendicularly to a surface element $\delta A(\mathbf{n})$ and is represented as $-pn_i dA$ (see (2.8), (2.9)). Another stress was the viscous stress (internal friction) $\sigma_{ij}^{(v)}$ considered in Sec. 2.6, which has a tangential force to the surface $\delta A(\mathbf{n})$ as well. A typical one is the shear stress in the parallel shear flow in Sec. 2.5.

The conservation of momentum of flows of an ideal fluid is given by (3.21) and (3.22), which reads

$$\partial_t(\rho v_i) + \partial_k P_{ik} = \rho f_i, \tag{4.2}$$

where $P_{ik} = \rho v_i v_k + p\delta_{ik}$ is the momentum flux tensor for an ideal-fluid flow. In order to obtain the equation of motion of a viscous fluid, a viscous stress should be added to the "ideal" momentum flux P_{ik},

---

[1]The tensor of the form δ_{ij} is called an *isotropic tensor*. For any vector A_i, the transformation $\delta_{ij}A_j$ is A_i.

which is written as $-\sigma_{ij}^{(v)}$ representing irreversible *viscous* transfer of momentum.[2] Thus we write the momentum flux tensor in a viscous fluid in the form,

$$P_{ik} = \rho v_i v_k + p_{ik} - \sigma_{ik}^{(v)}, \tag{4.3}$$

$$\sigma_{ik} = -p\delta_{ik} + \sigma_{ik}^{(v)}, \tag{4.4}$$

where σ_{ij} is the *stress tensor* written as q_{ij} in Sec. 2.4, and $\sigma_{ij}^{(v)}$ the *viscous stress tensor*, defined by (2.17) for a Newtonian fluid.

Substituting this in (4.2) and eliminating some terms by using the continuity equation, we obtain the equation of motion of a Newtonian viscous fluid. The result is that a new term of the form $\partial\sigma_{ik}^{(v)}/\partial x_k$ is added to the right-hand side of the Euler equation (3.14):

$$\rho(\partial_t v_i + v_k \partial_k v_i) = -\partial_i p + \frac{\partial}{\partial x_k}\sigma_{ik}^{(v)}, \tag{4.5}$$

where the external force f_i is omitted.

If the viscosity coefficients μ and ζ are regarded as constants,[3] the viscous force is written as

$$\frac{\partial}{\partial x_k}\sigma_{ik}^{(v)} = \mu\left(\nabla^2 v_i + \frac{\partial}{\partial x_i}(\partial_k v_k) - (2/3)\frac{\partial}{\partial x_i}D\right) + \zeta\frac{\partial}{\partial x_i}D$$

$$= \mu\nabla^2 v_i + \left(\frac{1}{3}\mu + \zeta\right)\partial_i D, \tag{4.6}$$

$$\nabla^2 = \frac{\partial^2}{\partial x_k^2} = \partial_x^2 + \partial_y^2 + \partial_z^2,$$

where (2.17) is used with $D = \mathrm{div}\,\mathbf{v}$ and ∇^2 the Laplacian. Substituting (4.6) in (4.5) and writing it in vector form, we have

$$\rho[\partial_t \mathbf{v} + (\mathbf{v} \cdot \nabla)\mathbf{v}] = -\nabla p + \mu\nabla^2\mathbf{v} + \left(\frac{1}{3}\mu + \zeta\right)\nabla(\mathrm{div}\,\mathbf{v}). \tag{4.7}$$

---

[2]The minus sign is added as a matter of definition.
[3]In general, the viscosity coefficients μ and ζ depend on thermodynamic variables such as density, pressure, temperature, etc.

This is called the *Navier–Stokes equation.*[4]

The equation becomes considerably simpler if the fluid is regarded as incompressible, so that $\text{div}\,\mathbf{v} = 0$. Denoting the uniform density by ρ_0, the equation of motion is

$$\partial_t \mathbf{v} + (\mathbf{v} \cdot \nabla)\mathbf{v} = -\frac{1}{\rho_0}\nabla p + \nu\nabla^2\mathbf{v} + \mathbf{f} \tag{4.8}$$

together with $\text{div}\,\mathbf{v} = 0$, where the external force $\mathbf{f}$ is added on the right-hand side. Equation (4.8) is also called the Navier–Stokes equation. The coefficient $\nu = \mu/\rho_0$ is termed the *kinematic viscosity,* while μ itself is called the *dynamic viscosity.* The viscous stress in an incompressible fluid ($D = \text{div}\,\mathbf{v} = 0$) is characterized by a single viscosity μ,

$$\sigma_{ik}^{(v)} = \mu\left(\partial_k v_i + \partial_i v_k\right). \tag{4.9}$$

Table 4.1 shows the dynamic viscosity μ and kinematic viscosity ν of various fluids.

If the external force is written as $\mathbf{f} = -\nabla\chi$ (i.e. conservative) where $\chi = gz$ with g the constant acceleration of gravity, then the

Table 4.1. Dynamic viscosity μ and kinematic viscosity $\nu = \mu/\rho_0$ at temperature 15°C (1 atm).

	$\mu\,[\text{g}/(\text{cm}\cdot\text{s})]$	$\nu\,[\text{cm}^2/\text{s}]$
Air	$0.18\cdot 10^{-3}$	0.15
Water	$0.11\cdot 10^{-1}$	0.011
Glycerin	8.5	6.8
Olive oil	0.99	1.08
Mercury	$0.16\cdot 10^{-1}$	0.0012

---

[4]The Navier–Stokes equation was discovered and rederived at least five times, by Navier in 1821, by Cauchy in 1823, by Poisson in 1829, by Saint-Venent in 1837, and by Stokes in 1845. Each new discoverer either ignored or denigrated his predecessors' contribution. Each had his own way to justify the equation ... [Dar05].

Navier–Stokes equation takes the form,

$$\partial_t \mathbf{v} + (\mathbf{v} \cdot \nabla)\mathbf{v} = -\frac{1}{\rho_0}\nabla p_{\mathrm{m}} + \nu\nabla^2\mathbf{v}, \tag{4.10}$$

where p_{m} is the *modified* pressure defined by

$$p_{\mathrm{m}} = p + \rho_0\chi = p + \rho_0 gz. \tag{4.11}$$

Using the vector identity (3.16), Eq. (4.10) is transformed to

$$\partial_t\mathbf{v} + \boldsymbol{\omega}\times\mathbf{v} = -\nabla\left(\frac{\mathbf{v}^2}{2} + \frac{1}{\rho_0}p + \chi\right) + \nu\nabla^2\mathbf{v}. \tag{4.12}$$

This corresponds to (3.18) for an ideal fluid. Taking curl of this equation, we obtain the *vorticity equation* for a viscous fluid:

$$\partial_t\boldsymbol{\omega} + \mathrm{curl}\,(\boldsymbol{\omega}\times\mathbf{v}) = \nu\nabla^2\boldsymbol{\omega}, \tag{4.13}$$

where $\boldsymbol{\omega} = \mathrm{curl}\,\mathbf{v}$. Using (A.24), this is rewritten as

$$\partial_t\boldsymbol{\omega} + (\mathbf{v}\cdot\mathrm{grad})\boldsymbol{\omega} = (\boldsymbol{\omega}\cdot\mathrm{grad})\mathbf{v} + \nu\nabla^2\boldsymbol{\omega}, \tag{4.14}$$

since $\mathrm{div}\,\boldsymbol{\omega} = 0$ identically, and $\mathrm{div}\,\mathbf{v} = 0$ is assumed.

4.2. Energy equation and entropy equation

Equation of energy conservation in a fluid when there is viscosity and thermal conduction is given by

$$\partial_t\left[\rho\left(\frac{1}{2}v^2 + e\right)\right] + \partial_j\left[\rho v_j\left(\frac{1}{2}v^2 + h\right) - v_i\sigma_{ij}^{(v)} - k\partial_j T\right] = W, \tag{4.15}$$

where the heat flux $\mathbf{q} = -k\,\mathrm{grad}\,T = (-k\partial_j T)$ is added to the energy flux in addition to the energy transfer $-v_i\sigma_{ik}^{(v)}$ due to microscopic processes of internal friction.

For consistency with thermodynamic relations as well as the continuity equation and equation of motion (Navier–Stokes equation), the following equation for the entropy s per unit mass is required to hold [LL87, Sec. 49]:

$$\rho T(\partial_t s + (\mathbf{v} \cdot \nabla)s) = \sigma_{ik}^{(v)} \partial_k v_i + \mathrm{div}(k \,\mathrm{grad}\, T). \tag{4.16}$$

This is a *general equation of heat transfer* (see (9.36) for the equation of thermal conduction). In the entropy equation (3.25) of an ideal fluid, there was no term on the right-hand side, resulting in the conservation of entropy. The expression on the left $\rho T(\mathrm{D}s/\mathrm{D}t)$ is the quantity of heat gained per unit volume in unit time. Therefore, the expressions on the right denote the heat production due to viscous dissipation of energy and thermal conduction.

A deeper insight into the entropy equation will be gained if we consider the rate of change of total entropy $\int \rho s \, dV$ in a volume V_0. In fact, using Eq. (4.16) and the continuity equation, one can derive the following equation:

$$\frac{\mathrm{d}}{\mathrm{d}t} \int_{V_0} \rho s \mathrm{d}V = \int \frac{\mu}{2T} \left(\partial_k v_i + \partial_i v_k - (2/3)(\mathrm{div}\, \mathbf{v})\delta_{ik} \right)^2 \mathrm{d}V$$
$$+ \int \frac{\zeta}{T} (\mathrm{div}\, \mathbf{v})^2 \mathrm{d}V + \int k \, \frac{(\mathrm{grad}\, T)^2}{T^2} \, \mathrm{d}V \tag{4.17}$$

[LL87, Sec. 49]. The entropy can only increase. Namely, each integral term on the right must be positive. The first two terms are the rate of entropy production due to internal friction, while the last term is that owing to thermal conduction. Hence it follows that the viscosity coefficients μ and ζ must be positive as well as the thermal conduction coefficient k.

4.3. Energy dissipation in an incompressible fluid

In an incompressible viscous fluid of uniform density ρ_0 where $\mathrm{div}\, \mathbf{v} = 0$, the energy equation becomes simpler and clearer. Multiplying v_i to the Navier–Stokes equation (4.5), one has

$$\rho_0(v_i \partial_t v_i + v_i v_k \partial_k v_i) = -v_i \partial_i p + v_i \partial_k \sigma_{ik}^{(v)}. \tag{4.18}$$

On the left, we can rewrite as $v_i \partial_t v_i = \partial_t \left(\frac{1}{2} v^2 \right)$ where $v^2 = v_i^2$, and

$$v_i v_k \partial_k v_i = v_k \partial_k \left(\frac{1}{2} v^2 \right) = \partial_k \left(v_k \frac{1}{2} v^2 \right)$$

since $\partial_k v_k = \text{div } \mathbf{v} = 0$. Similarly, we have $v_i \partial_i p = \partial_i (v_i p)$. Using the viscous stress defined by (4.9), we obtain

$$v_i \partial_k \sigma_{ik}^{(v)} = \mu v_i (\partial_k \partial_k v_i + \partial_k \partial_i v_k) = \mu \partial_k (v_i \partial_k v_i) - \mu (\partial_k v_i)^2, \quad (4.19)$$

where the second term in the middle vanishes because $\partial_k \partial_i v_k = \partial_i \partial_k v_k = 0$. Thus Eq. (4.18) is rewritten as

$$\partial_t \left(\frac{1}{2} v^2 \right) + \nu (\partial_k v_i)^2 = \partial_k \left(-v_k \frac{1}{2} v^2 - v_k p + \nu v_i \partial_k v_i \right), \quad (4.20)$$

where $\nu = \mu / \rho_0$.

Suppose that the flow space V_∞ is unbounded and the velocity decays sufficiently rapidly at infinity. Taking the integral of Eq. (4.20) over a sufficiently large bounded space V and transforming the volume integral on the right to an integral over the surface S_∞ at large distance (which recedes to infinity later), we obtain

$$\partial_t \int_V \frac{1}{2} v^2 \mathrm{d}V + \nu \int_V (\partial_k v_i)^2 \mathrm{d}V$$
$$= \oint_{S_\infty} \left(-v_k \frac{1}{2} v^2 - v_k p + \nu v_i \partial_k v_i \right) n_k \mathrm{d}S.$$

It is assumed that the velocity v_k decays sufficiently rapidly and the surface integral vanishes in the limit when it recedes to infinity.[5] Thus we obtain the equation of energy decay of flows of a viscous fluid:

$$\frac{\mathrm{d}K}{\mathrm{d}t} = -\nu \int_{V_\infty} (\partial_k v_i)^2 \mathrm{d}V, \quad (4.21)$$

where K is the total kinetic energy (divided by ρ_0) of fluid motion in V:

$$K = \int_{V_\infty} \frac{1}{2} v^2 \mathrm{d}V. \quad (4.22)$$

The right-hand side of (4.21) represents the rate of energy dissipation by viscosity, which is denoted by $\varepsilon(V)$ where $V = V_\infty$. Using the rate-of-strain tensor e_{ik} defined by (1.18), the following expression of the

---

[5] If the fluid motion is driven by a solid body, the fluid velocity decays like r^{-3} as the distance r from the body increases (see (5.77)), while the surface S_∞ increases like r^2.

rate of energy dissipation can be verified:

$$\varepsilon(V) = \nu \int_V (\partial_k v_i)^2 \mathrm{d}V = 2\nu \int_V e_{ik}\partial_k v_i \mathrm{d}V = 2\nu \int_V \|e_{ik}\|^2 \mathrm{d}V,$$
$$(4.23)$$

where $\|e_{ik}\|^2 \equiv \sum_{i=1}^{3}\sum_{k=1}^{3} e_{ik}e_{ik}$, and the following relations are used,

$$(\partial_k v_i)^2 = \partial_k v_i \left(2e_{ik} - \partial_i v_k\right) = 2e_{ik}\partial_k v_i - \partial_k(v_i\partial_i v_k) + v_i\partial_i(\partial_k v_k).$$

The last term vanishes since $\partial_k v_k = 0$, and the term $\partial_k(\cdot)$ is transformed to vanishing surface integral. For the last equality, we used the definition of e_{ik}: $e_{ik} = \frac{1}{2}(\partial_k v_i + \partial_i v_k) = e_{ki}$, and the following relation:

$$e_{ik}\partial_k v_i = e_{ik}\partial_i v_k = e_{ik}e_{ik}.$$

We have another expression. Using $\omega_j = \varepsilon_{jki}\partial_k v_i$, we obtain

$$|\boldsymbol{\omega}|^2 = (\varepsilon_{jki}\partial_k v_i)^2 = (\partial_k v_i)^2 - \partial_i(v_k\partial_k v_i),$$

because $\varepsilon_{jki}\varepsilon_{jlm} = \delta_{kl}\delta_{im} - \delta_{km}\delta_{il}$ and $\partial_i v_i = 0$. Therefore, we find

$$\varepsilon(V) = \nu \int_V |\boldsymbol{\omega}|^2 \mathrm{d}V, \qquad (4.24)$$

since the integrated surface terms vanish.

4.4. Reynolds similarity law

In studying the motions of a viscous fluid, the flow fields are often characterized with a representative velocity U and length L. For example (Fig. 4.1), one can consider a spherical ball of diameter L moving with a velocity U. Another case is a viscous flow in a circular pipe of diameter L with a maximum flow speed U.

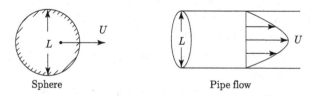

Sphere   Pipe flow

Fig. 4.1.   Representative scales.

When a flow of an incompressible viscous fluid of density ρ_0 has a single representative velocity U and a single representative length L, the state of flow is characterized by a single dimensionless number, defined by

$$R_e = \frac{UL}{\nu} = \frac{\rho_0 UL}{\mu}, \tag{4.25}$$

which is called the **Reynolds number**. This is formulated as follows.

Suppose that the space coordinates $\mathbf{x} = (x, y, z)$, time t, velocity $\mathbf{v}$ and pressure p are normalized to dimensionless variables. Denoting the dimensionless variables by primes, we define

$$\mathbf{x}' = \frac{\mathbf{x}}{L}, \quad t' = \frac{t}{\tau}, \quad \mathbf{v}' = \frac{\mathbf{v}}{U}, \quad p' = \frac{p - p_0}{\rho_0 U^2},$$

where the normalization is done using L as the dimension of *length*, U for the dimension of *velocity*, $\tau = L/U$ for *time*, and $U/\tau = U^2/L$ for *acceleration*. The dimension of *pressure* is the same as that of

[pressure] = [force]/[area] = ([mass] · [acceleration])/[area].

The representative value of pressure variation is of the order of $[(\rho_0 L^3)(U^2/L)/L^2] = [\rho_0 U^2]$, with the reference pressure denoted by p_0.

Substituting these into (4.8) (without the external force $\mathbf{f}$) and rewriting it with primed variables, we obtain

$$\frac{\partial}{\partial t'}\mathbf{v}' + (\mathbf{v}' \cdot \nabla')\mathbf{v}' = -\nabla' p' + \frac{1}{R_e}(\nabla')^2 \mathbf{v}' \tag{4.26}$$

($\nabla' \cdot \mathbf{v}' = 0$). Equation (4.26) thus derived is a dimensionless equation including a single dimensionless constant R_e (Reynolds number). If the values of R_e are the same between two flows under the same boundary condition, the flow fields represented by the two solutions are equivalent, even though the sets of values U, L, ν are different. This is called the *Reynolds similarity law*, and R_e is termed the *similarity parameter*.

Stating it in another way, if the values of R_e are different, the corresponding flow fields are different, even though the boundary conditions are the same. From this point of view, R_e is also termed as the *control parameter*.

Consider a *steady flow* in which all the field variables do not depend on time and hence the first term of (4.8) vanishes. Let us estimate relative magnitude of the second *inertia* term. Estimating the magnitude of velocity gradient $\nabla \mathbf{v}$ to be of the order of U/L, the magnitude of the second term of (4.8) is $|(\mathbf{v} \cdot \nabla)\mathbf{v}| = O(U(U/L)) = O(U^2/L)$. Regarding the last viscous term, analogous estimation leads to $|\nu \nabla^2 \mathbf{v}| = O(\nu U/L^2)$. Taking the ratio of the two terms, we obtain

$$\frac{O(|(\mathbf{v} \cdot \nabla)\mathbf{v}|)}{O(|\nu \nabla^2 \mathbf{v}|)} = \frac{U^2/L}{\nu U/L^2} = \frac{UL}{\nu} = R_e,$$

which is nothing but the Reynolds number. Note that the dimension of ν is the same as that of $UL = [L]^2[T]^{-1}$ with $[T]$ the dimension of time.

If the fluid viscosity is high and its kinematic viscosity ν is large enough, the viscous term $|\nu \nabla^2 \mathbf{v}|$ will be larger than the inertia term $|(\mathbf{v} \cdot \nabla)\mathbf{v}|$, resulting in $R_e < 1$. Such a flow of *low* Reynolds number is called a *viscous flow*, including the case of $R_e \simeq 1$ (see Sec. 4.8). The flow of $R_e < 1$ is said to be a *slow motion*, because a flow of very small velocity U makes $R_e < 1$ regardless of the magnitude of viscosity. The motion of a microscopic particle becomes inevitably a flow of low Reynolds number because the length L is sufficiently small.

On the contrary, we will have $R_e \gg 1$ for flows of large U, large L, or small ν. Such a flow is said to be a flow of *high* Reynolds number. In the flows of high Reynolds numbers, the magnitude of inertia term $|(\mathbf{v} \cdot \nabla)\mathbf{v}|$ is much larger than the viscous term $|\nu \nabla^2 \mathbf{v}|$, and the flow is mainly governed by the fluid inertia. However, it will be found in the next section that the role of viscosity is still important in the flows of high Reynolds numbers too. Most flows at high Reynolds numbers become **turbulent**.

Smooth flows observed at low Reynolds numbers are said to be **laminar**. Consider a sequence of states as the value of Reynolds number R_e is increased gradually from a low value at the state of a laminar flow. According to the increase in Reynolds number, the flow varies its state, and at sufficiently large values of R_e, it will change over to a turbulent state. This *transition* to turbulence occurs at a fixed value of Reynolds number, which is termed as the *critical* Reynolds number, often written as R_c.

4.5.   Boundary layer

As the Reynolds number R_e is increased, the viscous term takes small values over most part of the space of flow field, and the viscous action tends to be localized in space. But, however large the value of R_e, the viscosity effect can never disappear. This fact means that the flow in the limit of vanishing viscosity ν (i.e. $\nu \to 0$) does not necessarily coincide with the flow of inviscid flow ($\nu = 0$). As an example, we consider a plane boundary layer flow.[6]

Suppose that there is a flow of an incompressible viscous fluid over a plane wall AB and the flow tends to a uniform flow of velocity U far from the wall, and that the flow is steady (Fig. 4.2). We take the x axis in the direction of flow along the wall AB and the y axis perpendicular to AB. The flow can be described in the two-dimensional (x, y) space. The velocity becomes zero on the wall $y = 0$ by the *no-slip* condition. Velocity distributions with respect to the coordinate y are schematically represented in the figure. In this situation, the change of velocity occurs in a thin layer of thickness δ (say) adjacent to the wall, and a boundary layer is formed. The boundary

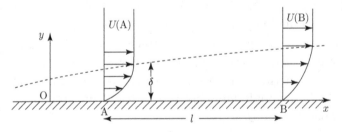

Fig. 4.2.   Plane boundary layer.

---

[6] *Plane* in fluid mechanics means "two-dimensional".

conditions are summarized as follows:

$$(u, v) \rightarrow (U, 0) \quad \text{as } y/\delta \rightarrow \infty, \tag{4.27}$$

$$(u, v) = (0, 0) \quad \text{at } y = 0. \tag{4.28}$$

Provided that the velocity field is expressed as (u, v) with the kinematic viscosity ν and the uniform density ρ, the Navier–Stokes equation (4.8) in the two-dimensional (x, y) space reduces to

$$u_t + uu_x + vu_y = -(1/\rho)\, p_x + \nu(u_{xx} + u_{yy}), \tag{4.29}$$

$$v_t + uv_x + vv_y = -(1/\rho)\, p_y + \nu(v_{xx} + v_{yy}), \tag{4.30}$$

$$u_x + v_y = 0, \tag{4.31}$$

where the first time derivative terms of (4.29) and (4.30) vanish in the steady problem under consideration. Within the boundary layer, the change in the y direction is more rapid compared with the change in the x direction. Hence, the term νu_{xx} may be much smaller than the term νu_{yy} in Eq. (4.29), and hence the term νu_{xx} may be neglected. But, there is a change in the x direction however slight. The two terms u_x and v_y in the continuity equation (4.31) balance each other.

On the basis of the above estimates and the estimates below (4.33), the *steady* flow in a boundary layer can be well described asymptotically by the following system of equations in the limit of small viscosity ν (Prandtl (1904); see Problem 4.4 for Blasius flow and Problem 4.5):

$$uu_x + vu_y = -(1/\rho)\, p_x + \nu u_{yy}, \tag{4.32}$$

$$u_x + v_y = 0. \tag{4.33}$$

These are consistent with the estimate of the order of magnitude just below. It is useful to recognize that the scales of variation are different in the two directions x and y. Suppose that the representative scales in the directions x and y are denoted by l and δ, respectively where $l \gg \delta$, and the magnitude of u is given by U. The viscous term on the right-hand side of Eq. (4.32) can be estimated as the order $O(\nu U/\delta^2)$. Both the first and second terms on the left-hand side are estimated as the order $O(U^2/l)$. Equating the above two estimates in the order of magnitude, we have

$$\frac{U^2}{l} = \frac{\nu U}{\delta^2}, \quad \text{or} \quad \frac{\delta}{l} = \sqrt{\frac{\nu}{Ul}} = \frac{1}{\sqrt{R_e}}, \tag{4.34}$$

where $R_e = Ul/\nu$. It is seen that the thickness δ of the boundary layer becomes smaller, as the Reynolds number R_e increases. From the above expression, one finds the behavior $\delta \propto \sqrt{l}$, a *parabolic* growth of the thickness δ along the wall. The boundary layer does not disappear, however small the viscosity ν is.

More importantly, in the boundary layer there exists nonzero vorticity ω. This is seen from the fact that $\omega = v_x - u_y \approx -u_y$ (nonzero), where v_x is very small. Far from the wall, the flow velocity tends to a uniform value and therefore $\omega \to 0$ as $y \to \infty$. This implies that the wall contributes to the generation of vorticity. An important difference of the flow at high Reynolds numbers from the flow of an inviscid fluid is the existence of such a rotational layer at the boundary. In fact, this is an essential difference from the inviscid flow.

4.6. Parallel shear flows

Consider a simple class of flows of a viscous fluid having only x component u of velocity $\mathbf{v}$:

$$\mathbf{v} = (u(y, z, t),\, 0,\, 0). \tag{4.35}$$

The continuity equation reduces to the simple form $\partial_x u = 0$, stating that u is independent of x, and consistent with (4.35). This is called the *parallel shear flow*, or *unidirectional flow*. In this type of flows, the convection term vanishes identically since $(\mathbf{v} \cdot \nabla)u = u\partial_x u = 0$. Without the external force $\mathbf{f}$, the x, y, z components of the Navier–Stokes equation (4.8) reduce to

$$\partial_t u - \nu(\partial_y^2 + \partial_z^2)u = -\frac{1}{\rho_0}\partial_x p \tag{4.36}$$

$$\partial_y p = 0, \quad \partial_z p = 0. \tag{4.37}$$

From the last equations, the pressure should be a function of x and t only: $p = p(x, t)$. The left-hand side of (4.36) depends on y, z, t, while the right-hand side depends on x, t. Therefore, the equality states that both sides should be a function of t only for the consistency of the equation. Writing it as $P(t)/\rho_0$, we have

$$- \operatorname{grad} p = (P, 0, 0), \qquad (4.38)$$

in which the fluid is driven to the positive x direction when P is positive.

4.6.1. *Steady flows*

In steady flows, we have $\partial_t u = 0$ and $P = \text{const}$. The above equation (4.36) reduces to

$$\partial_y^2 u + \partial_z^2 u = -\frac{P}{\mu}, \qquad (4.39)$$

where μ is the dynamic viscosity. The acceleration of a fluid particle vanishes identically in steady unidirectional flows, hence the density does not make its appearance explicitly.

A solution $u = u(y, z)$ satisfying (4.39) is in fact an exact solution of the incompressible Navier–Stokes equation. Some of such solutions are as follows :

$$u_C(y) = \frac{U}{2b}(y + b), \quad P = 0, \ (|y| < b) : \text{Couette flow}, \qquad (4.40)$$

$$u_P(y) = \frac{P}{2\mu}(b^2 - y^2), \quad P \neq 0, \ (|y| < b) : \text{2D Poiseuille flow.} \qquad (4.41)$$

Obviously, these satisfy Eq. (4.39). We have another axisymmetric solution, called the *Hagen–Poiseuille* flow which is considered as Problem 4.1 at the end of this chapter.

The first Couette flow represents a flow between the fixed plate at $y = -b$ and the plate at $y = b$ moving with velocity U in the x direction when there is no pressure gradient [Fig. 4.3(a)].

The second is the Poiseuille flow [Fig. 4.3(b)] between two parallel walls at $y = \pm b$ under a constant pressure gradient P ($d = 2b$ in (2.22)). The maximum velocity U is attained at the center $y = 0$,

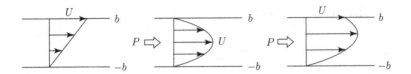

Fig. 4.3.   (a) Couette flow, (b) Poiseuille flow, and (c) combined flow.

given by

$$U = Pb^2/(2\mu).\tag{4.42}$$

The total rate of flow Q per unit time is

$$Q = \int_{-b}^{b} u_P(y)\,\mathrm{d}y = \frac{2Pb^3}{3\mu} = \frac{4}{3}bU.$$

In view of (4.39) which is an inhomogeneous linear differential equation for u, a linear combination of the above two solutions [Fig. 4.3(c)] is also another exact solution (for constants A, B):

$$u_*(y) = Au_C(y) + Bu_P(y) = A\frac{U}{h}y + B\frac{P}{2\mu}(b^2 - y^2).\tag{4.43}$$

4.6.2.   *Unsteady flow*

We consider unsteady parallel flows evolving with time t without pressure gradient, so that $P = 0$. It is assumed to be uniform in the z-direction so that $\partial_z u = 0$. Then, Eq. (4.36) reduces to

$$\partial_t u = \nu \partial_y^2 u.\tag{4.44}$$

(a) *Rayleigh's problem*
A simplest unsteady problem would be the flow caused by unsteady motion of a flat plate moving parallel to itself. Suppose that the fluid is bounded by a plane wall at $y = 0$, occupying the half space $y > 0$, and the wall $y = 0$ is moving. There are two well-known problems: an impulsive start of wall and an oscillating boundary layer. Below is given a detailed account of the first problem. Regarding the second problem, see Problem 4.2.

Rayleigh's problem: Let us consider a problem of impulsive start of the plane wall at $y = 0$. Suppose that the wall was at rest for $t < 0$, and has been driven impulsively to the positive x direction with velocity U at $t = 0$ and thereafter. The conditions to be satisfied are as follows:

$$u(y, t) = 0, \qquad\qquad\qquad \text{for } t < 0,$$
$$u(0, 0) = U, \quad u(y > 0, 0) = 0, \quad \text{at } t = 0, \qquad (4.45)$$
$$u(0, t) = U, \quad u(y \to \infty, t) \to 0, \quad \text{for } t > 0. \qquad (4.46)$$

This is called the *Rayleigh's problem*. We try to solve Eq. (4.44) in the following form of the *similarity solution*,

$$u(y, t) = Uf(\eta), \quad \eta = \frac{y}{2\sqrt{\nu t}}, \qquad (4.47)$$

as a function of a dimensionless similarity variable η. Differentiating this,

$$\partial_t u = Uf'(\eta)\, \partial_t \eta = -Uf'(\eta)\, \frac{\eta}{2t}, \quad \partial_y u = Uf'(\eta)\, \partial_y \eta = Uf'(\eta)\, \frac{1}{2\sqrt{\nu t}}.$$

Then, Eq. (4.44) becomes the following ordinary differential equation:

$$f'' + 2\eta f' = 0. \qquad (4.48)$$

The boundary conditions of (4.46) become

$$f(0) = 1, \quad f(\infty) = 0.$$

Integrating (4.48) once results in $f'(\eta) = Ce^{-\eta^2}$ with C a constant. Integrating once again and applying the above boundary conditions, we obtain

$$f(\eta) = 1 - \frac{2}{\sqrt{\pi}} \int_0^\eta e^{-\eta^2} d\eta,$$

where the integration formula $\int_0^\infty e^{-\eta^2} d\eta = \sqrt{\pi}/2$ is used. Thus, we have found the solution:

$$u(y, t) = U \left(1 - \frac{2}{\sqrt{\pi}} \int_0^\eta e^{-\eta^2} d\eta \right). \qquad (4.49)$$

This solution represents diffusion of a vorticity layer, which was localized initially in a thin boundary layer of thickness $\sqrt{\nu t}$ adjacent to

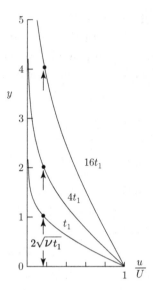

Fig. 4.4.   Rayleigh flow.

the wall $y = 0$ (Fig. 4.4). The velocity profile at a time t can be obtained by rescaling the vertical coordinate by the ratio $\sqrt{t/t_1}$, once the profile at t_1 is given.

(b) *Diffusive spreading of a shear layer*

Let us consider a time-evolving transition layer of velocity which was initially a zero thickness plane surface. Hence, it was a vortex sheet (see Problem 5.11) initially, coinciding with the plane $y = 0$. Suppose that

$$u(y,0) = U, \quad \text{for } y > 0; \qquad -U, \quad \text{for } y < 0$$

[Fig. 4.5(a)]. The x-velocity $u(y,t)$ is governed by Eq. (4.44): $\partial_t u = \nu \partial_y^2 u$, together with the boundary condition: $u \to \pm U$ as $y \to \pm\infty$. The symmetry implies $u(0,t) = 0$.

In this problem, there is no characteristic length. So that, we seek a solution $u(y,t)$ depending on a similarity variable composed of three parameters y, t and ν of the present problem. From these, one can form only one dimensionless variable, $y/\sqrt{\nu t}$,[7] since the dimension of ν is $[L]^2[T]^{-1}$.

---

[7]The same reasoning can be applied to the previous Rayleigh problem.

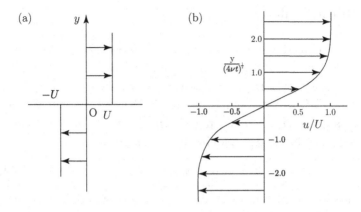

Fig. 4.5. (a) Vortex sheet, (b) a transition layer (a shear layer).

Assuming the *similarity* form $u = Uf(\eta)$ where $\eta = y/(2\sqrt{\nu t})$ as before, we can deduce Eq. (4.48) again for $f(\eta)$, while the present boundary conditions are

$$f = 0 \quad \text{at } \eta = 0, \qquad f \to 1 \quad \text{as } \eta \to \infty. \tag{4.50}$$

One can solve this problem immediately, and obtain

$$f(\eta) = \frac{2}{\sqrt{\pi}} \int_0^\eta e^{-\eta^2} \mathrm{d}\eta.$$

Thus, the time-evolving velocity profile is given by

$$u(y, t) = \frac{U}{\sqrt{\pi \nu t}} \int_0^y \exp\left(-\frac{y'^2}{4\nu t}\right) \mathrm{d}y'. \tag{4.51}$$

Note that the condition at $-\infty$ is satisfied as well : $u \to -U$ as $y \to -\infty$.

This solution represents a diffusive spreading of a transition layer of the velocity from $u = -U$ to $+U$. The similarity variable $\eta = y/(2\sqrt{\nu t})$ implies that the layer thickness grows in proportion to $\sqrt{\nu t}$. This kind of transition layer is called a *shear layer* [Fig. 4.5(b)].

4.7. Rotating flows

Next, let us consider rotating flows of a viscous incompressible fluid
in which all the stream lines are circular with a common axis. The
differential operators in the cylindrical coordinates (r, θ, z), such as
∇, div and ∇^2, are given in Appendix D.1, where the velocity is
written as $\mathbf{v} = (v_r, v_\theta, v_z)$. In the present situation of concentric
circular stream lines, the velocity is expressed as

$$\mathbf{v} = (0,\ v_\theta(r, t),\ 0). \tag{4.52}$$

The condition of incompressibility div $\mathbf{v} = 0$ is satisfied identically
by (4.52) and (D.9). Using the formulae of Appendix D.1, the (r, θ, z)
components of the Navier–Stokes equation (4.8) read as follows:

$$-\frac{v_\theta^2}{r} = -\frac{1}{\rho_0}\partial_r p,$$

$$\partial_t v_\theta = -\frac{1}{\rho_0 r}\partial_\theta p + \nu\left(\partial_r^2 v_\theta + \frac{1}{r}\partial_r v_\theta - \frac{v_\theta}{r^2}\right), \tag{4.53}$$

$$0 = -\frac{1}{\rho_0}\partial_z p.$$

From the last equation, we have $p = p(r, \theta, t)$. However, the pressure
p must be of the form $p = p(r, t)$. Otherwise, $\partial_\theta p$ is a function of r and
t from the second equation of (4.53), and the pressure p becomes a
multivalued function in general with respect to the azimuthal angle θ,
i.e. $p(\theta, t) \neq p(\theta + 2\pi, t)$, which is not permissible.

Assuming $p = p(r, t)$, the second equation of (4.53) becomes

$$\partial_t v_\theta = \nu\left(\partial_r^2 v_\theta + \frac{1}{r}\partial_r v_\theta - \frac{v_\theta}{r^2}\right). \tag{4.54}$$

Integrating the first of (4.53), we obtain

$$p(r, t) = p(r_0, t) + \rho_0 \int_{r_0}^{r} \frac{v_\theta^2}{r}\, dr. \tag{4.55}$$

It is seen that the pressure becomes higher for outer positions. Thus,
the force of the pressure gradient $-\mathrm{grad}\ p$ is directed inward. This is
necessary as a centripetal force of fluid particles in circular motion.

The circulating velocity $v_\theta(r)$ of *steady* rotation must satisfy the following ordinary differential equation from (4.54):

$$\frac{\mathrm{d}^2 v_\theta}{\mathrm{d}r^2} + \frac{1}{r}\frac{\mathrm{d}v_\theta}{\mathrm{d}r} - \frac{v_\theta}{r^2} = 0. \tag{4.56}$$

This can be readily solved to give a general solution of the following form,

$$v_\theta(r) = ar + \frac{b}{r} \qquad (a,\, b: \text{ constants}). \tag{4.57}$$

A typical example is the Taylor–Couette flow (see Problem 4.3).

4.8.   Low Reynolds number flows

4.8.1.   *Stokes equation*

When we studied the boundary layer in Sec. 4.5, the Reynolds number was assumed to be very high. Here, we consider another limit of very low Reynolds number with respect to a representative length L and a representative velocity U:

$$R_e = \frac{UL}{\nu} \ll 1.$$

This means that U is sufficiently small, or L is sufficiently short, or the fluid is very sticky such that ν is sufficiently large.

We consider *steady* flows. Then, the equation governing steady flows of a viscous fluid is given by

$$(\mathbf{v} \cdot \nabla)\mathbf{v} = -\frac{1}{\rho_0}\nabla p + \mathbf{f} + \nu\nabla^2\mathbf{v} \tag{4.58}$$

from (4.8). The fluid is assumed to be incompressible,

$$\mathrm{div}\,\mathbf{v} = 0. \tag{4.59}$$

The assumption of very low R_e permits us to omit the nonlinear convection term on the left (we consider such cases below), and we have

$$\nabla p - \mu \nabla^2 \mathbf{v} = \mathbf{F}, \quad \text{div } \mathbf{v} = 0, \tag{4.60}$$

where $\mathbf{F} = \rho_0 \mathbf{f}$, $\mu = \rho_0 \nu$. The external force $\mathbf{F}$ plays an essential role in the flows considered here.

The system of equations (4.60) is called the equations of *Stokes approximation*, or *Stokes equation*. The Stokes equation is useful for describing very slow *creeping* flows, or motion of a very tiny particle.

Note that any irrotational incompressible flow $\mathbf{v}_*$ is a solution of the Stokes equation. In fact, according to the vector identity (A.27) in Appendix A.5, we have

$$\nabla^2 \mathbf{v}_* = -\nabla \times (\nabla \times \mathbf{v}_*) + \nabla (\nabla \cdot \mathbf{v}_*) = 0, \quad \text{if } \text{curl } \mathbf{v}_* = 0, \ \text{div } \mathbf{v}_* = 0.$$

Therefore, this $\mathbf{v}_*$ satisfies Eq. (4.60) together with $p = \text{const.}$ and $\mathbf{F} = 0$.

4.8.2. *Stokeslet*

Consider an axisymmetric slow motion of a viscous incompressible fluid. Such a flow can be described by the Stokes' stream function $\Psi(r, \theta)$ with the spherical polar coordinates (r, θ, ϕ) (see Appendices B.3 and D.2). In terms of the Stokes's stream function Ψ, the velocity field is given by $\mathbf{v} = (v_r, v_\theta, 0)$, where

$$v_r = \frac{1}{r^2 \sin \theta} \frac{\partial \Psi}{\partial \theta}, \quad v_\theta = -\frac{1}{r \sin \theta} \frac{\partial \Psi}{\partial r}. \tag{4.61}$$

These satisfy the continuity equation (4.59) in the spherical polar coordinates always (see (D.20)).[8]

One of the most important flows in the low Reynolds-number hydrodynamics is the *Stokeslet*, whose stream function is given by

$$\Psi_S = r \sin^2 \theta. \tag{4.62}$$

This gives the following velocities according to (4.61):

$$v_r = \frac{2}{r} \cos \theta, \quad v_\theta = -\frac{1}{r} \sin \theta, \quad v_\phi = 0. \tag{4.63}$$

---

[8]See Sec. 5.7 and Appendix B.2 for the stream function of 2D problems.

This is also expressed in the following vector form:

$$\mathbf{v_S} = -\nabla(\cos\theta) + \frac{2}{r}\mathbf{i}, \quad \nabla = \left(\partial_r, \frac{1}{r}\partial_\theta, \frac{1}{r\sin\theta}\partial_\phi\right), \quad (4.64)$$

where $\mathbf{i} = (\cos\theta, -\sin\theta, 0)$ is the unit vector in the direction of polar axis x (coinciding with $\theta = 0$) and $\cos\theta = x/r$.

Expressing $\nabla^2\mathbf{v}$ with the spherical polar coordinates by using (D.24) and $\mathbf{v} = (v_r, v_\theta, 0)$ and substituting (4.63), Eq. (4.60) reduces to

$$\partial_r p = -\frac{4\mu}{r^3}\cos\theta + F_r, \quad \frac{1}{r}\partial_\theta p = -\frac{2\mu}{r^3}\sin\theta + F_\theta,$$

where $\mathbf{F} = (F_r, F_\theta, 0)$. This implies the following:

$$p = p_S = \frac{2\mu}{r^2}\cos\theta, \quad \mathbf{F} = 0 \quad (\text{if } r \neq 0). \quad (4.65)$$

In fact, in Problem 4.7, one is asked to verify the expression:

$$\mathbf{F} = (F_x, F_y, F_z) = (8\pi\mu, 0, 0)\,\delta(x)\,\delta(y)\,\delta(z). \quad (4.66)$$

Namely, the Stokeslet is a particular solution of the Stokes equation and represents a creeping flow of a viscous fluid subject to a concentrated external force of magnitude $8\pi\mu$ acting at the origin in the positive x direction.

4.8.3. *Slow motion of a sphere*

Consider a steady slow motion of a very small spherical particle of radius a in a viscous fluid with a constant velocity $-\mathbf{U}$. Let us observe this motion relative to the reference frame F_C fixed to the center of the sphere. Fluid motion observed in the frame F_C would be a steady flow around a sphere of radius a, tending to a uniform flow of velocity $\mathbf{U}$ at infinity. Then, the flow field is axisymmetric and described by the Stokes's stream function Ψ (Appendix B.3). Such a flow can be represented as

$$\Psi_{Us}(r,\theta) = \frac{1}{2} U r^2 \sin^2\theta + \frac{1}{4} U \frac{a^3}{r} \sin^2\theta - \frac{3}{4} U a \Psi_S(r,\theta). \qquad (4.67)$$

Each of the three terms is a solution of the Stokes equation (4.60). In fact, the third term is the Stokeslet considered in the previous subsection and a solution of (4.60) with a singularity at the origin. The first term represents a uniform flow of velocity U in the x direction.[9] The second term is a dipole in the positive x direction. These are equivalent to the stream functions given by (5.43) and (5.44) in Chapter 5. As described in detail there, the first two terms represent irrotational incompressible flows. Hence they are solutions of the Stokes equation as remarked at the end of Sec. 4.8.1.

Thus, the above stream function Ψ_{Us} is a solution of the Stokes equation as a whole as well because of the linearity of Eq. (4.60).[10] It is to be remarked that the stream function is of the form $U f(r) \sin^2\theta$. This manifests that the stream lines are rotationally symmetric with respect to the polar axis x, and in addition, has a symmetry $\sin^2\theta = \sin^2(\pi - \theta)$, i.e. a *left–right symmetry* or *fore–aft symmetry*.

Substituting (4.67) into (4.61), we obtain

$$v_r = U \left(1 - \frac{3}{2}\frac{a}{r} + \frac{1}{2}\frac{a^3}{r^3} \right) \cos\theta, \qquad (4.68)$$

$$v_\theta = -U \left(1 - \frac{3}{4}\frac{a}{r} - \frac{1}{4}\frac{a^3}{r^3} \right) \sin\theta. \qquad (4.69)$$

At $r = a$, we have $v_r = 0$, $v_\theta = 0$. Hence the no-slip condition is satisfied on the surface of the sphere of radius a. In addition, at infinity the velocity tends to the uniform flow:

$$(v_r, v_\theta) \to U(\cos\theta, \ -\sin\theta).$$

Thus, it has been found that the stream function of (4.67) represents a uniform flow of velocity U around a sphere of radius a (Fig. 4.6).

---

[9]The formula (4.61) for $\Psi = \frac{1}{2} U r^2 \sin^2\theta$ leads to $(v_r, v_\theta) = U(\cos\theta, -\sin\theta) = U\mathbf{i}$.
[10]Equation (4.60) has an inhomogeneous term $\mathbf{F}$. However, because of the forces associated with each Ψ_{Us} described below, the linear combination is still valid.

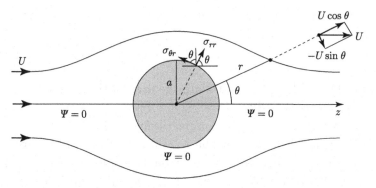

Fig. 4.6. Low Reynolds number flow around a sphere: $\sigma_{\theta r} = 2\mu e_{\theta r}$, $\sigma_{rr} = -p + 2\mu e_{rr}$.

Next, let us consider the force acting on the sphere of radius a (i.e. the drag (D_i)), which is given by the integral of stress over the sphere surface S_a:

$$D_i = \oint_{S_r} \sigma_{ij} n_j \mathrm{d}S, \quad (n_r, n_\theta, n_\phi) = (1, 0, 0), \qquad (4.70)$$

where σ_{ij} is the stress tensor defined by (4.4). In the present case, σ_{ij} takes the form,

$$\sigma_{ij} = -p\delta_{ij} + 2\mu e_{ij}, \qquad (4.71)$$

where the Reynolds stress term $\rho_0 v_i v_j$ is absent in the Stokes equation. The rate-of-strain tensors e_{ij} in the spherical polar coordinates are defined by (D.28) and (D.29) in Appendix D.2. Using (4.68) and (4.69) and setting $r = a$, we obtain the rate-of-strain tensors in the frame (r, θ, ϕ) as follows:

$$e_{\theta r} = -\frac{3}{4} \frac{U}{a} \sin\theta, \quad e_{rr} = e_{\theta\theta} = e_{\phi\phi} = e_{\theta\phi} = e_{\phi r} = 0. \qquad (4.72)$$

The pressure p of the Stokeslet is given by p_S of (4.65), while the pressure of the first two terms of (4.67) is a constant p_0, as shown in

Sec. 4.8.1, because they represent irrotational flows. Thus, we have

$$p = -\frac{3}{4} Uap_S + p_0 = -\frac{3}{2} \frac{\mu U}{a} \cos\theta + p_0. \qquad (4.73)$$

From (4.71)–(4.73), we can estimate the drag of the sphere by (4.70), where $dS = 2\pi a^2 \sin\theta\, d\theta$.

However, there is a more compact way to find the drag on the sphere. It can be shown that the irrotational flows of the first two terms do not give rise to any force on the sphere (Problem 4.8). Therefore, the force is due to the Stokeslet only, which is given by $F = -\frac{3}{4} Ua\, 8\pi\mu = -6\pi\mu aU$ (toward the negative x direction). By the law of reaction, the drag D experienced by the sphere toward the positive x direction is given by

$$D = -F = 6\pi\mu aU. \qquad (4.74)$$

This is called the **Stokes' law of resistance** for a moving sphere.

Stokes (1850) applied the formula (4.74) to a tiny cloud droplet of radius a of a mass m falling with a steady velocity U in air under the gravity force mg, and explained the suspension of clouds. Then we have the equality,

$$6\pi\mu aU = mg.$$

Since the drop mass is proportional to a^3, the falling velocity U is proportional to a^2 and μ^{-1}, becoming negligibly small as $a \to 0$.

4.9. Flows around a circular cylinder

When a long rod of circular cross-section is placed in a uniform stream of water, it is often observed that vortices are formed on the back of the body. The vortices sometime form a regular queue, and sometime fluctuate irregularly. On the other hand, if a rod is moved with some speed with respect to water at rest, then similar vortex formations are observed too. The hydrodynamic problem of motion of a rod and associated generation of vortices are studied as *Flow of a viscous fluid around a circular cylinder*.

When a flow has the same velocity vector **U** at all points in the flow without presence of a rod, the flow is called a **uniform flow**. If a circular cylinder of diameter D is placed in such a uniform flow, a characteristic flow pattern is observed downstream of the body. Such a flow pattern is called the **wake**. Particularly well-known wake is the *Kármán vortex street*, observed at some values of the Reynolds number $R_e = UD/\nu$ (observed when $70 \lesssim R_e \lesssim 200$ experimentally and computationally). The Kármán vortex street consists of two lines of vortices with one line having a common sense of rotation but the other line having its opposite and the vortices being located at staggered positions. At much higher values of the Reynolds number, the wake becomes irregular and turbulent, but there is still some periodic component immersed in the turbulent wake.

However, far downstream from a body in the uniform flow in the x direction, the velocity decays and the stream lines become nearly parallel (but not exactly parallel). There the flow may be regarded as a certain kind of boundary layer with its breadth being relatively thin compared with the distance from the body. The pressure variation across the wake may be very small. The limiting form of the wake is considered in Problem 4.6.

Flow around a circular cylinder is the most typical example of flows around a body, and regarded as exhibiting the Reynolds similarity in an idealistic way, where the Reynolds number R_e plays the role of control parameter exclusively. Various types of flows are summarized in Table 4.2. Figure 4.7 shows stream-lines of such flows obtained by computer simulations. Figure 1.2 is a photograph of an air flow around a circular cylinder at $R_e = 350$.

4.10.   Drag coefficient and lift coefficient

Suppose that a body is fixed in a uniform stream of velocity U, and subjected to a drag D D and a lift L. The drag is defined as a force on the body in the direction of the stream, while the lift is defined as a force on the body perpendicular to the stream direction.

In fluid mechanics, these forces are normalized, and they are represented as dimensionless coefficients of drag and lift. They are defined

Table 4.2. A sequence of flows around a circular cylinder (diameter D) with respect to the Reynolds number $R_e = UD/\nu$ with uniform velocity U in the direction of x axis (horizontal), the z axis coinciding with the cylinder axis, and the y axis taken vertically.

$R_e \lesssim 5$	:	*Slow viscous flows*, with the stream-lines symmetric *w.r.t.* x axis (*up–down symmetry*), and symmetric *w.r.t.* y axis (*fore–aft symmetry*).
$5 \lesssim R_e \lesssim 40$	:	A pair of steady *separation eddies* are formed on the rear side of the cylinder. The *fore–aft symmetry is broken*.
$40 \lesssim R_e \lesssim 70$	:	Wake *oscillations* are observed: *Hopf bifurcation*. The *up–down symmetry is broken*.
$70 \lesssim R_e \lesssim 500$	:	Kármán vortex streets are observed in the wake. *Nonlinear* modulation to the Hopf bifurcation.
$500 \lesssim R_e$	:	Wake is in turbulent state on which Kármán vortex street is superimposed. *Transition to turbulent wake*.

as follows.

$$\text{Drag coefficient}: \quad C_{\mathrm{D}} = \frac{D}{\frac{1}{2}\rho U^2 S}, \qquad (4.75)$$

$$\text{Lift coefficient}: \quad C_{\mathrm{L}} = \frac{L}{\frac{1}{2}\rho U^2 S}, \qquad (4.76)$$

where S is a specific reference surface area (cross-section or others), regarded as appropriate for the nomalization.

4.11. Problems

Problem 4.1 Hagen–Poiseuille flow

(i) Using the cylindrical coordinates (x, r, θ) (Appendix D.2, by rearranging from (r, θ, z) to (x, r, θ)), show that Eq. (4.39) can be written as

$$\frac{1}{r}\frac{\partial}{\partial r}\left(r\frac{\partial}{\partial r}\right)u + \frac{1}{r^2}\frac{\partial^2}{\partial \theta^2}u = -\frac{P}{\mu}. \qquad (4.77)$$

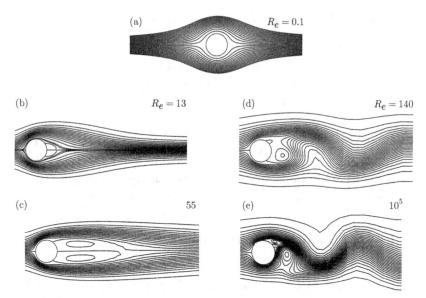

Fig. 4.7. Stream-lines from computer simulations of flows around a circular cylinder, by courtesy of K. Kuwahara and M. Hashiguchi [1994]. The flows of (c)–(e) are time-dependent so that the stream-lines are snapshots.

(ii) Suppose that a steady parallel viscous flow is maintained in a circular pipe of radius a under a constant pressure gradient $\mathrm{grad}\, p = (-P, 0, 0)$ with $P > 0$ in the x-direction, so that Eq. (4.77) is valid. Assuming that $u = u(r)$, derive the following velocity profile,

$$u_{\mathrm{HP}}(y) = \frac{P}{4\mu}(a^2 - r^2), \quad (r < a) \qquad (4.78)$$

(Hagen, 1839 and Poiseuille, 1840).

(iii) Show that the total rate of flow Q per unit time is given by

$$Q = \frac{\pi a^4 P}{8\mu} = \frac{\pi a^4}{8\mu}\frac{\Delta p}{L},$$

where Δp is the magnitude of pressure drop along the length $L : P = \Delta p/L$. [Note that $Q \propto a^4/\mu$.]

Problem 4.2 Oscillating boundary layer

Suppose that the plane wall $y = 0$ is oscillating with velocity $U_w(t)$ and angular frequency ω in its own plane with: $U_w = U \cos \omega t$, and that velocity u of a viscous fluid in space $y > 0$ is subject to Eq. (4.44). Show that the solution decaying at infinity is given by

$$u(y,t) = U \exp\left[-\sqrt{\frac{\omega}{2\nu}}\, y\right] \cos\left(\omega t - \sqrt{\frac{\omega}{2\nu}}\, y\right) \qquad (4.79)$$

(Fig. 4.8). In addition, express a representative thickness δ of the oscillating boundary layer in terms of ω and viscosity ν.

Problem 4.3 Taylor–Couette flow

Suppose that there is a viscous fluid between two concentric cylinders of radius r_1 and r_2 $(> r_1)$ and the cylinders are rotating steadily with angular velocity Ω_1 and Ω_2, respectively (Fig. 4.9). Determine the steady velocity distribution $v(r)$, governed by (4.56).

Problem 4.4 Blasius flow

(i) By applying the estimates of order of magnitude described in Sec. 4.5, show the consistency of the approximate Eq. (4.32)

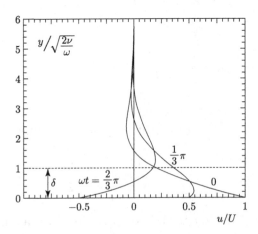

Fig. 4.8. Oscillating boundary layer.

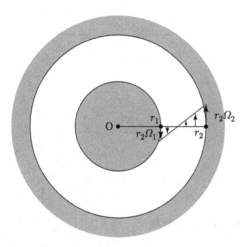

Fig. 4.9. Taylor–Couette flow.

and describe what is the order of the neglected term νu_{xx}. In addition, show that Eq. (4.30) can be replaced by

$$\partial p / \partial y = 0. \tag{4.80}$$

(ii) The velocity (u, v) in the boundary layer along a flat plate is governed by Eq. (4.32) with $p_x = 0$. By assuming that the stream function takes the following similarity form,

$$\psi(x, y) = \left(\nu U x\right)^{\frac{1}{2}} f(\eta), \quad \eta = y \left(\frac{U}{\nu x}\right)^{\frac{1}{2}}, \tag{4.81}$$

under the boundary conditions (4.27) and (4.28), show that the equation for $f(\eta)$ and the boundary conditions are written as

$$f''' + \frac{1}{2} f f'' = 0 \quad (0 < \eta < \infty), \tag{4.82}$$

$$f = f' = 0 \quad \text{at } \eta = 0, \quad f' \to 1 \quad \text{as } \eta \to \infty. \tag{4.83}$$

[The boundary layer flow determined by the third-order ordinary differential equation (4.82) is called the Blasius flow. Its velocity $u = U f'(\eta)$ can be determined numerically (Fig. 4.10).]

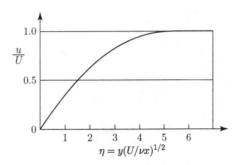

Fig. 4.10. Blasius flow.

Problem 4.5 Jet flow (2D)

If a jet from a two-dimensional slit became sufficiently thin in the far down stream, the boundary layer equations (4.32) (with $p_x = 0$) and (4.33) may be applied together with the boundary conditions (Fig. 4.11):

$$\partial_y u = 0, \quad v = 0 \quad \text{at } y = 0; \quad \text{and} \quad u \to 0 \quad \text{as } y \to \pm\infty. \quad (4.84)$$

(i) Assume that the stream function takes the following similarity form,

$$\psi(x, y) = Ax^{\frac{1}{3}}f(\zeta), \qquad \zeta = \frac{y}{x^{2/3}}. \quad (4.85)$$

By choosing the constant A appropriately, show that the function $f(\zeta)$ satisfies the following ordinary differential equation

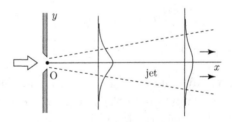

Fig. 4.11. Jet flow.

and boundary conditions:

$$f''' + ff'' + (f')^2 = 0 \qquad (0 < \eta < \infty), \qquad (4.86)$$
$$f = f'' = 0 \quad \text{at } \zeta = 0, \quad f, f' \to 0 \quad \text{as } \zeta \to \pm\infty. \qquad (4.87)$$

(ii) Show that the following integral M is independent of x:

$$M = \rho \int_{-\infty}^{\infty} u^2 dy,$$

where M is the momentum-flux.

(iii) Determine the function $f(\zeta)$ and the jet velocity profile $u(x, y)$. This is called the *Bickley jet*.

Problem 4.6 Wake flow (2D)

We consider a two-dimensional steady flow around a cylindrical body. Suppose that the cylindrical body is fixed in a uniform stream of velocity U in the x direction and the (x, y)-plane is taken perpendicularly to the cylindrical axis with the origin inside the cross-section.

In the wake far downstream from the body, the velocity decays significantly and the stream-lines are nearly parallel, and the pressure variation across the wake would be very small (Fig. 4.12). The velocity in the steady wake may be written as $(U - u, v)$, where $u > 0$ is assumed, and Eq. (4.32) could be linearized with respect to small components u and v, as follows:

$$U\partial_x u = \nu \partial_y^2 u, \qquad (4.88)$$

with the boundary condition for $u > 0$,

$$u \to 0, \qquad \text{as } y \to \pm\infty. \qquad (4.89)$$

(i) Determine the velocity $u(x, y)$ of the steady wake satisfying the above Eq. (4.88) and the boundary condition. Furthemore, show the following integral (volume flux of inflow) is independent of x:

$$\int_{-\infty}^{\infty} u(x, y) \, dy = Q \quad \text{(indep. of } x\text{)}.$$

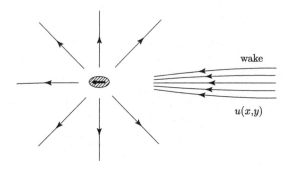

Fig. 4.12.   Wake (right) and a source-like flow away from the body (left).

(ii) Show that the drag D on the cylindrical body giving rise to the wake is given by

$$D = \rho U \int_{-\infty}^{\infty} u(x,y)\,\mathrm{d}y = \rho U Q. \qquad (4.90)$$

[**Hint**: The *Momentum Theorem* may be applied. Namely, the drag is related to a momentum sink in the flow, which can be estimated by the momentum conservation integral (3.23) over a large volume including the body. This result shows $D \propto \rho Q$ (mass flux of *dragging* fluid).]

Problem 4.7 Stokeslet

Derive the expression (4.66) of the force **F** of the Stokeslet (4.62), according to the drag formula (4.70), by using (4.63) and (4.65).

Problem 4.8 Force by an irrotational flow

Show that an irrotational flow outside of a closed surface S does not give rise to any force on S if there is no singularity out of S, according to the drag formula (4.70) and Stokes equation (4.60).

Chapter 5

Flows of ideal fluids

Theory of flows of ideal fluids not only provides us the basis of study of fluid flows, but also gives us fundamental physical ideas of continuous fields in Newtonian mechanics. Extending the ideas of ideal fluid flows, the physical concepts can be applied to extensive areas in physics and mathematics.

Governing equations of flows of an ideal fluid derived in Chapter 3 are summarized as follows:

$$\partial_t \rho + \mathrm{div}(\rho \, \mathbf{v}) = 0, \tag{5.1}$$

$$\partial_t \mathbf{v} + (\mathbf{v} \cdot \nabla)\mathbf{v} = -\frac{1}{\rho}\,\mathrm{grad}\,p + \mathbf{f}, \tag{5.2}$$

$$\partial_t s + (\mathbf{v} \cdot \nabla)s = 0. \tag{5.3}$$

Boundary condition for an ideal fluid flow is that the normal component of the velocity vanishes on a solid boundary surface at rest:

$$v_n = \mathbf{v} \cdot \mathbf{n} = 0, \tag{5.4}$$

where $\mathbf{n}$ is the unit normal to the boundary surface. If the boundary is in motion, the normal component v_n of the fluid velocity should coincide with the normal velocity component V_n of the boundary

$$v_n = V_n. \tag{5.5}$$

Tangential components of both of the fluid and moving boundary do not necessarily coincide with each other in an ideal fluid.

5.1. Bernoulli's equation

One of the basic theorems of flows of an ideal fluid is Bernoulli's theorem, which can be derived as follows. Suppose that the fluid's entropy is uniform, i.e. the entropy s per unit mass of fluid is constant everywhere, and the external force has a potential χ represented by $\mathbf{f} = -\text{grad}\,\chi$. Then, Eq. (5.2) reduces to (3.30), which is rewritten here again:

$$\partial_t \mathbf{v} + \boldsymbol{\omega} \times \mathbf{v} = -\text{grad}\left(\frac{|\mathbf{v}|^2}{2} + h + \chi\right), \qquad (5.6)$$

where h is the enthalpy ($h = e + p/\rho$) and e the internal energy. From this equation, one can derive two important equations. One is the case of *irrotational* flows for which $\boldsymbol{\omega} = 0$. The other is the case of *steady* flows. Here, we consider the latter case.

In *steady flows*, the field variables like the velocity is independent of time t, and their time derivatives vanish identically. In this case, we have

$$\boldsymbol{\omega} \times \mathbf{v} = -\text{grad}\left(\frac{v^2}{2} + h + \chi\right), \quad v^2 = |\mathbf{v}|^2. \qquad (5.7)$$

Since the vector product $\boldsymbol{\omega} \times \mathbf{v}$ is perpendicular to the vecor $\mathbf{v}$ (also perpendicular to $\boldsymbol{\omega}$), the inner product of $\mathbf{v}$ and $\boldsymbol{\omega} \times \mathbf{v}$ vanishes. Taking the inner product of $\mathbf{v}$ with Eq. (5.7), we have

$$\mathbf{v} \cdot \text{grad}\left(\frac{v^2}{2} + h + \chi\right) = \frac{\partial}{\partial s}\left(\frac{v^2}{2} + h + \chi\right) = 0, \qquad (5.8)$$

where $v = |\mathbf{v}|$. The derivative $\partial/\partial s$ in the middle is the differentiation along a stream-line parameterized with a variable s.[1] Thus, it is

---

[1] According to the relation (1.9) along a stream-line, $\mathbf{v} \cdot \text{grad} = (\partial x/\partial s)\partial_x + (\partial y/\partial s)\partial_y + (\partial z/\partial s)\partial_z = \partial/\partial s$.

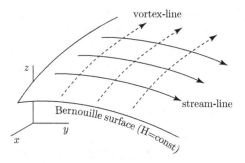

Fig. 5.1. Bernoulli surface.

found that

$$H := \frac{v^2}{2} + h + \chi = \text{const.} \quad \text{along a stream-line}. \tag{5.9}$$

This is called the **Bernoulli theorem**. The function H is constant along a stream-line. Its value may be different for different stream-lines.

In view of the property that the inner product of $\boldsymbol{\omega}$ and $\boldsymbol{\omega} \times \mathbf{v}$ vanishes as well, the function H is constant along a vortex-line[2] (defined in Sec. 5.3) analogously. This implies that the surface given by $H(x,y,z) = \text{const.}$ is covered with a family of stream-lines and the vortex-lines crossing with the stream-lines. In most flows, they are not parallel.[3] The surface defined by $H = \text{const.}$ is called the *Bernoulli surface* (Fig. 5.1).

The Bernoulli equation (5.9) is often applied to such a case where the fluid density is a constant everywhere, denoted by ρ_0, and $\chi = gz$ (g is the constant gravity acceleration). Then Eq. (5.9) can be written as[4]

$$\rho_0 H = \rho_0 \frac{v^2}{2} + p + \rho_0 gz = \text{const.}, \tag{5.10}$$

---

[2] A vortex-line is defined by $dx/\omega_x = dy/\omega_y = dz/\omega_z$ where $\boldsymbol{\omega} = (\omega_x, \omega_y, \omega_z)$.

[3] The flow in which $\mathbf{v}$ is parallel to $\boldsymbol{\omega}$ is said to a *Beltrami flow*.

[4] Daniel Bernoulli (1738) derived this Eq. (5.10), actually a form equivalent to it, in the hydraulic problem of efflux. He coined the word *Hydrodynamica* for the title of this dissertation.

where $h = p/\rho_0 + e$ is used, and the internal energy e is included in the "const." on the right since e is constant in a fluid of constant density and constant entropy (see (3.27)).

When the gravity is negligible, the Bernoulli equation reduces to

$$\frac{1}{2}\rho_0 v^2 + p = p_0, \quad p_0 : \text{a constant.} \tag{5.11}$$

We consider an example of its application. Suppose that a two-dimensional cylindrical body is immersed in a uniform flow of velocity U in the x-direction, and a steady flow around the body is maintained in the (x, y)-plane. At distances far from the body, the velocity $\mathbf{v} = (u, v)$ tends to $(U, 0)$, and the pressure tends to a constant value p_∞ (Fig. 5.2). The Bernoulli equation (5.11) holds for each stream-line. However at infinity, the value on the left is given by the same value $\frac{1}{2}\rho_0 U^2 + p_\infty$ for all the stream-lines. Hence, the constant on the right is the same for all the stream-lines. Thus, we have the following equations valid for all the stream-lines:

$$\frac{1}{2}\rho_0 v^2 + p = \frac{1}{2}\rho_0 U^2 + p_\infty. \tag{5.12}$$

There exists always a dividing stream-line coming from the upstream which divides the flow into one going around the upper side B_+ of the cylindrical body and the one going around its lower side B_- (Fig. 5.2). The dividing stream-line terminates at a point on the body surface, where the fluid velocity necessarily vanishes. This point is termed the *stagnation point* (see Problem 5.1), where the pressure p_s

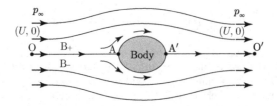

Fig. 5.2.   Steady uniform flow around a body.

is given by

$$p_s = p_\infty + \frac{1}{2}\rho_0 U^2.$$

Namely, the stagnation pressure is higher than the pressure at infinity by the value $\frac{1}{2}\rho_0 U^2$. This value $\frac{1}{2}\rho_0 U^2$ is called the *dynamic pressure*, whereas the pressure p on the left of (5.12) is called the *static pressure*.

5.2. Kelvin's circulation theorem

A fundamental theorem of flows for an ideal fluid is the circulation theorem. Given a velocity field $\mathbf{v}(\mathbf{x}, t) = (u, v, w)$, the *circulation* along a closed curve C in a flow field is defined by

$$\Gamma(C) := \oint_C \mathbf{v} \cdot d\mathbf{l} = \oint_C (u\,dx + v\,dy + w\,dz), \qquad (5.13)$$

where $\oint$ denotes an integral along a *closed* curve C, and $d\mathbf{l} = (dx, dy, dz)$ denotes a line element along C.

An important property of fluid motion can be derived if the closed curve C is a *material* curve, i.e. each point on the curve moves with the velocity $\mathbf{v}$ of fluid (material) particle $\mathbf{a}$. In this sense, the material curve and material line element are denoted as C_a and $d\mathbf{l}_a$, respectively.

In the flow governed by Eq. (5.6), the **Kelvin's circulation theorem** (Kelvin, 1869) reads as follows. The circulation Γ_a along the material closed curve C_a is invariant with respect to time (Fig. 5.3),

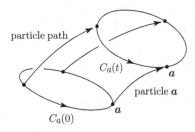

Fig. 5.3.   Motion of a material closed curve.

namely,

$$\Gamma_a = \oint_{C_a} \mathbf{v} \cdot \mathrm{d}\mathbf{l}_a \quad \text{is independent of } t, \tag{5.14}$$

or

$$\frac{\mathrm{D}}{\mathrm{D}t} \oint_{C_a} \mathbf{v} \cdot \mathrm{d}\mathbf{l}_a = 0. \tag{5.15}$$

Since each line element $\mathrm{d}\mathbf{l}_a$ of C_a moves with the fluid material, the time derivative must be the Lagrange derivative $\mathrm{D}/\mathrm{D}t$. The theorem is verified as follows.

Euler's equation of motion (3.29) is rewritten as

$$\frac{\mathrm{D}\mathbf{v}}{\mathrm{D}t} = -\mathrm{grad}(h + \chi). \tag{5.16}$$

In addition, it is noted that, denoting the velocities of two neighboring fluid particles at $\mathbf{x}_a$ and $(\mathbf{x} + \delta\mathbf{x})_a$ by $\mathbf{v}$ and $\mathbf{v} + \delta\mathbf{v}$ respectively, we have the relation

$$\frac{\mathrm{D}}{\mathrm{D}t}\delta\mathbf{x}_a = \delta\mathbf{v}, \tag{5.17}$$

since $\mathrm{D}/\mathrm{D}t(\mathbf{x} + \delta\mathbf{x})_a = \mathrm{D}\mathbf{x}_a/\mathrm{D}t + \mathrm{D}\delta\mathbf{x}_a/\mathrm{D}t$ (see (1.32)).

Now, the time derivative $\mathrm{D}\Gamma_a/\mathrm{D}t$ is given by

$$\frac{\mathrm{D}}{\mathrm{D}t} \oint_{C_a} \mathbf{v} \cdot \mathrm{d}\mathbf{l}_a = \lim \sum_n \frac{\mathrm{D}}{\mathrm{D}t}(\mathbf{v}^{(n)} \cdot \mathrm{d}\mathbf{l}_a^{(n)}) = \oint_{C_a} \frac{\mathrm{D}}{\mathrm{D}t}(\mathbf{v} \cdot \mathrm{d}\mathbf{l}_a),$$

where the curve C_a is divided into a number of small line elements with n denoting its nth element and $\sum_n$ denotes the summation with respect to all the elements. "lim" denotes taking a limit in such a way that the total number N of elements is made infinite by keeping each line element infinitesimally small. Thus we have

$$\frac{\mathrm{D}}{\mathrm{D}t}\Gamma_a = \oint_{C_a} \frac{\mathrm{D}\mathbf{v}}{\mathrm{D}t} \cdot \mathrm{d}\mathbf{l}_a + \oint_{C_a} \mathbf{v} \cdot \frac{\mathrm{D}}{\mathrm{D}t}\mathrm{d}\mathbf{l}_a. \tag{5.18}$$

Due to (5.17), the last term vanishes:

$$\oint_{C_a} \mathbf{v} \cdot \frac{\mathrm{D}}{\mathrm{D}t}\mathrm{d}\mathbf{l}_a = \oint_{C_a} \mathbf{v} \cdot \mathrm{d}\mathbf{v} = \oint_{C_a} \mathrm{d}\left(\frac{1}{2}v^2\right) = 0,$$

since $\mathbf{v} \cdot \mathrm{d}\mathbf{v} = v_i \mathrm{d}v_i = \mathrm{d}(\frac{1}{2}|\mathbf{v}|^2)$. Vanishing of the integral is obtained because the velocity field is single-valued and $|\mathbf{v}|^2$ returns to its original value after one circuit of integration. The first integral of (5.18) vanishes too because of (5.16), i.e.

$$\oint_{C_a} \frac{D\mathbf{v}}{Dt} \cdot \mathrm{d}\mathbf{l}_a = - \oint_{C_a} \mathrm{d}\mathbf{l}_a \cdot \mathrm{grad}(h + \chi) = - \oint_{C_a} \mathrm{d}(h + \chi) = 0,$$

since $\mathrm{d}\mathbf{x} \cdot \mathrm{grad}\, f = \mathrm{d}f$ for a scalar function f in general. This is due to single-valuedness of h and χ. Thus Eq. (5.15) has been verified.

In summary, the *circulation theorem* has been verified under the condition of Euler equation (5.16) together with the homentropy of the fluid and the conservative external force, but the fluid density is not necessarily assumed constant.

The same circulation theorem can be proved as well for the case of a uniform density (ρ is constant), or the case of a *barotropic* fluid (in which the pressure p is a function of density ρ only, i.e. $p = p(\rho)$), instead of the homentropic property of uniform value of entropy s.

5.3. Flux of vortex-lines

Given a velocity field $\mathbf{v}(\mathbf{x})$, the vorticity $\boldsymbol{\omega} = (\omega_x, \omega_y, \omega_z)$ is defined by $\nabla \times \mathbf{v}$. In an analogous way to the stream-line, a vortex-line is defined by

$$\frac{\mathrm{d}x}{\omega_x(x, y, z)} = \frac{\mathrm{d}y}{\omega_y(x, y, z)} = \frac{\mathrm{d}z}{\omega_z(x, y, z)} = \frac{\mathrm{d}l}{|\boldsymbol{\omega}|}, \tag{5.19}$$

where the vector $\boldsymbol{\omega}$ is tangent to the vortex-line and l is the arc-length along it (Fig. 5.4).

The circulation Γ of (5.13) is also represented with a surface integral in terms of $\boldsymbol{\omega}$ by using the Stokes theorem (A.35) as

$$\Gamma[C] = \oint_C \mathbf{v} \cdot \mathrm{d}\mathbf{l} = \int_S (\nabla \times \mathbf{v}) \cdot \mathbf{n}\mathrm{d}S = \int_S \boldsymbol{\omega} \cdot \mathbf{n}\mathrm{d}S, \tag{5.20}$$

where S is an open surface immersed in a fluid and bounded by the closed curve C with $\mathbf{n}$ being a unit normal to the surface element $\mathrm{d}S$ (Fig. 5.5). The integrand of the surface integral on the right

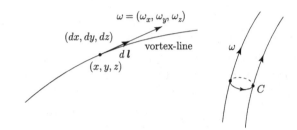

Fig. 5.4.   Vortex-line and tube.

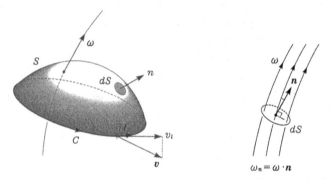

Fig. 5.5.   Closed curve C and surface S.

represents the magnitude of flux of vortex-lines passing through the surface element:

$$\omega_n \mathrm{d}S. \tag{5.21}$$

Hence, the circulation $\Gamma[C]$ is equal to the total magnitude of flux of vortex-lines passing through the closed curve C.

Thus, the Kelvin's circulation theorem is rephrased in the following way. *The total magnitude of flux of vortex-lines passing through a material closed curve C_a, i.e. circulation $\Gamma[C_a]$, is invariant with respect to the time for the motion of an ideal fluid of homentropy (or uniform density, or barotropy) under the conservative external force.*

A *vortex tube* is defined as a tube formed by the set of all vortex-lines passing through the closed curve C. The strength k, i.e. circulation, of a small vortex tube of an infinitesimal cross-section δS is given by $k = \omega_n \delta S$, which is invariant along the tube itself. This is verified by using the property (a vector identity (A.26)),

$$\operatorname{div} \boldsymbol{\omega} = \nabla \cdot (\nabla \times \mathbf{v}) = 0, \qquad (5.22)$$

because there is no flux out of the tube surface.

5.4.   Potential flows

From the Kelvin circulation theorem, one can deduce an important property. Suppose that the flow field is steady, hence stream-lines are unchanged. In addition, suppose that the vorticity curl $\mathbf{v}$ is zero in a domain D including a point P. Consider a small closed curve C within D encircling a stream-line L passing through the point P. Then the circulation $\Gamma[C]$ vanishes by (5.20). The circulation theorem assures that the zero-value of the circulation $\Gamma[C]$ is kept invariant during the translational motion of the closed curve C along the line L carried by the steady fluid flow. Namely, once the vorticity was zero in a domain D$'$ including a point on a stream-line L, then the vorticity is zero (curl $\mathbf{v} = 0$) in an extended domain D$''$ including the stream-line L which is swept by D$'$ carried along by the flow. If the velocity field is unsteady, the property curl $\mathbf{v} = 0$ is *maintained* in a domain carried along by the moving fluid.

The field where the vorticity is zero (curl $\mathbf{v} = 0$) is termed *irrotational*. Any vector field $\mathbf{v}(\mathbf{x})$ whose curl vanishes in a simply-connected domain D is represented in terms of a potential function $\Phi(\mathbf{x})$ as

$$\mathbf{v} = \operatorname{grad} \Phi = (\partial_x \Phi, \, \partial_y \Phi, \, \partial_z \Phi) = \nabla \Phi \qquad (5.23)$$

(see Appendix B.1). It is obvious by the vector identity (A.25) that this is sufficient.[5] In this sense, an irrotational flow is also called a *potential flow*. The scalar function $\Phi(\mathbf{x})$ is called the *velocity potential*. The flow in which the vorticity is not zero is said to be *rotational*.

Suppose that a body is immersed in an otherwise uniform flow and that the newly formed flow is steady. At upstream points of infinity,

---

[5]It can be verified that this is also necessary, as verified in Appendix B.1.

the vorticty is zero because the velocity field is uniform there. Thus, we find that this is a potential flow.

For a potential flow, setting $\boldsymbol{\omega} = 0$ in the equation of motion (5.6), we have

$$\partial_t \mathbf{v} + \mathrm{grad}\left(\frac{|\mathbf{v}|^2}{2} + h + \chi \right) = 0. \tag{5.24}$$

We can integrate this Euler's equation of motion. Substituting the expression (5.23) for $\mathbf{v}$ and writing as $v = |\mathbf{v}|$, we obtain

$$\mathrm{grad}\left(\frac{\partial \Phi}{\partial t} + \frac{v^2}{2} + h + \chi \right) = 0. \tag{5.25}$$

This means that the expression in ( ) is a function of time t only,

$$\partial_t \Phi + \frac{1}{2} v^2 + h + \chi = f(t), \tag{5.26}$$

where $f(t)$ is an arbitrary function of time. However, if we introduce a function Φ' by $\Phi' = \Phi - \int^t f(t')\mathrm{d}t'$, we obtain

$$\mathrm{grad}\, \Phi' = \mathrm{grad}\, \Phi = \mathbf{v}, \quad \partial_t \Phi' = \partial_t \Phi - f(t).$$

Then, Eq. (5.26) reduces to

$$\partial_t \Phi' + \frac{1}{2} v^2 + h + \chi = \mathrm{const.}$$

This (or (5.26)) is called the *integral of the motion*, where

$$v^2 = (\partial_x \Phi)^2 + (\partial_x \Phi)^2 + (\partial_x \Phi)^2. \tag{5.27}$$

Henceforce, in the above integral, the function Φ is used instead of Φ'[6]:

$$\partial_t \Phi + \frac{1}{2} v^2 + h + \chi = \mathrm{const.} \tag{5.28}$$

When the flow is steady, the first term $\partial_t \Phi$ vanishes. Then we obtain the same expression as Bernoulli's equation (5.9). However

---

[6]In this case, the function Φ is to be supplemented with an arbitrary function of time. The invariance with the transformation $\Phi \to \Phi'$ may be termed as *gauge invariance*, which is analogous to that of electromagnetism. See Chapter 12.

in the present potential flow, the constant on the right holds at all points (not restricted to a single stream-line).

If the gravitational potential is written as $\chi = gz$, the integral of motion (5.28) for a fluid of uniform density ρ_0 is given as

$$\partial_t \Phi + \frac{1}{2}v^2 + (p/\rho_0) + gz = \text{const.} \qquad (5.29)$$

When the velocity potential Φ is known, this equation gives the pressure p since v^2 is also known from (5.27).

5.5. Irrotational incompressible flows (3D)[7]

An irrotational flow has a velocity potential Φ, and the velocity is represented as $\mathbf{v} = \text{grad}\,\Phi$. In addition, if the fluid is incompressible, i.e. $\text{div}\,\mathbf{v} = 0$, then we have the following equation,

$$\text{div}\,\mathbf{v} = \nabla \cdot \nabla \Phi = \nabla^2 \Phi = 0, \qquad (5.30)$$

where $\nabla^2 = \partial_x^2 + \partial_y^2 + \partial_z^2$. This is also written as

$$0 = \nabla^2 \Phi = \Phi_{xx} + \Phi_{yy} + \Phi_{zz} = \Delta \Phi. \qquad (5.31)$$

This is the *Laplace equation* for Φ where Δ is the Laplacian. In general, the functions $\Phi(x, y, z)$ satisfying the Laplace equation is called *harmonic functions*. Thus, it is found that irrotational incompressible flows are described by harmonic functions.

In other words, any harmonic function Φ (satisfying $\Delta \Phi = 0$) represents a certain irrotational incompressible flow. To see what kind of flow is represented by Φ, we have to examine what kind of boundary conditions are satisfied by the function Φ.

When a body is fixed in an irrotational flow, the normal component v_n of the velocity must vanish on the surface S_b of the body:

$$v_n = \mathbf{n} \cdot \text{grad}\,\Phi = \partial \Phi / \partial n = 0, \qquad (5.32)$$

where $\mathbf{n}$ is a unit normal to S_b.

---

[7]3D: three-dimensional, 2D: two-dimensional.

It is remarkable that Eq. (5.31) is linear with respect to the velocity potential Φ (as well as the boundary condition (5.32) is so). If two functions Φ_1 and Φ_2 satisfy (5.31), then their linear combination $\Phi_1 + \Phi_2$ does satisfy it too. Often the combined potential $\Phi_1 + \Phi_2$ satisfies a required boundary condition even if each potential does not. Then we get a required solution $\Phi_1 + \Phi_2$.

A *Neumann problem* for harmonic functions is a boundary value problem in the theory of partial differential equations, which is to find a function Φ satisfying the Laplace equation $\Delta\Phi = 0$ together with the boundary condition: $\partial\Phi/\partial n$ *given* on a surface S. In the context of fluid dynamics, if $\partial\Phi/\partial n = 0$ on the boundary S and $\operatorname{grad}\Phi \to \mathbf{U}$ at infinity (with $\mathbf{U}$ a constant vector), this is a problem to determine the velocity potential Φ of a flow around a solid body S at rest in a uniform stream $\mathbf{U}$.

5.6. Examples of irrotational incompressible flows (3D)

A simplest example is given by a linear function, $\Phi = Ax + By + Cz$ (with A, B, C: constants), which satisfies the Laplace equation $\Delta\Phi = 0$ identically. The velocity $\mathbf{v} = \operatorname{grad}\Phi = (A, B, C)$ is a constant vector at every point, and hence this expresses a *uniform flow*.

In particular, $\Phi_U = Ux$ is the velocity potential of a uniform flow of velocity U in the x direction.

5.6.1. *Source (or sink)*

Consider a point $\mathbf{x} = (x, y, z)$ in the cartesian (x, y, z) coordinates. In the spherical polar coordinates (r, θ, ϕ) (see Appendix D.3), the radial coordinate r denotes its distance from the origin. A flow due to a source (or a sink) located at the origin is represented by the velocity potential Φ_s:

$$\Phi_s = -\frac{m}{r}, \tag{5.33}$$

where $r = \sqrt{x^2 + y^2 + z^2}$.

Applying the Laplacian (A.33), or (D.22) in the spherical coordinates, we obtain

$$\nabla^2 \Phi_s = -\frac{1}{r^2}\frac{\partial}{\partial r}\left(r^2 \frac{\partial}{\partial r}\frac{m}{r}\right) = -\frac{1}{r^2}\frac{\partial}{\partial r}(-m) = 0, \quad \text{for } r \neq 0,$$

where a singular point $r = 0$ (the origin) is excluded. This verifies that the function Φ_s is a harmonic function except when $r = 0$. The property that the function Φ_s is the velocity potential describing a flow due to a source can be shown by its velocity field, which is given by

$$\mathbf{v} = \text{grad}\left(-\frac{m}{r}\right) = \frac{m}{r^2}\left(\frac{x}{r}, \frac{y}{r}, \frac{z}{r}\right),$$

since $\partial r/\partial x = x/r$, etc. Taking a sphere S_R of an arbitrary radius R with its center at the origin, the volume flux of fluid flow out of S_R per unit time is given by

$$\oint_{S_R} \mathbf{v} \cdot \mathbf{n} dS_R = \oint_{S_R} \frac{m}{r^2}\left(\frac{x}{r}\frac{x}{r} + \frac{y}{r}\frac{y}{r} + \frac{z}{r}\frac{z}{r}\right) dS_R = \frac{m}{R^2} 4\pi R^2 = 4\pi m,$$

since $\mathbf{n} = (x/r, y/r, z/r)$. Equivalently, we have the relation,

$$\oint_{S_R} \mathbf{v} \cdot \mathbf{n} dS_R = \int \text{div } \mathbf{v} dV = \int_{V_R} \nabla^2 \Phi_s dV = 4\pi m,$$

where V_R is the spherical volume of radius R with its center at the origin. This implies $\int \nabla^2(1/r)dV = -4\pi$, namely

$$\nabla^2 \frac{1}{r} = -4\pi\delta(x)\,\delta(y)\,\delta(z), \tag{5.34}$$

(see Appendix Sec. A.7 for the definition of δ-function).[8] Thus, it is found that the flux out of the spherical surface is independent of its radius and equal to a constant $4\pi m$. Therefore, the velocity potential $\Phi_s = -m/r$ describes a source of fluid flow from the origin if $m > 0$, or a sink if $m < 0$. Figure 5.6 shows such a source flow (left), and a sink flow (right) in the upper half space.

---

[8] The function $1/(4\pi r)$ is called a fundamental solution of the Laplace equation due to the property (5.34).

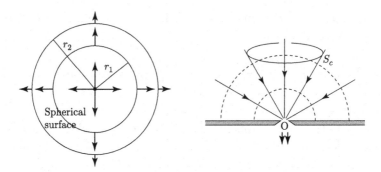

Fig. 5.6.   (Left) *source* $(m > 0)$; (right) *sink* for a half space $(m < 0)$.

5.6.2.   *A source in a uniform flow*

One can consider a linear combination of two potentials:

$$\Phi_{Us} := \Phi_U + \Phi_s = Ux - \frac{m}{r}. \tag{5.35}$$

What kind of flow is represented by Φ_{Us}? The velocity field is given by

$$\mathbf{v} = (U,\, 0,\, 0) + \frac{m}{r^2}\left(\frac{x}{r},\, \frac{y}{r},\, \frac{z}{r}\right),$$

in the cartesian (x, y, z) space. At large distances as $r \to \infty$, the velocity $\mathbf{v}$ approaches the first uniform flow. On the x axis $(y = 0$, $z = 0)$, the x component velocity is given by $u = U + mx/r^3$, whereas the other two components are zero. There is a *stagnation point* where the velocity vanishes. The x component u vanishes at a point $x = -r_s$ $(r_s > 0,\, y = 0,\, z = 0)$ on the negative x axis, where r_s is given by

$$r_s = \sqrt{m/U}.$$

Evidently the flow field has a rotational symmetry with respect to the x axis. The stream-lines emanating from the stagnation point forms a surface of rotation symmetry S_* around the x axis, which divides the fluid coming out of the origin (the source) from the fluid of the on-coming uniform flow from the left (Fig. 5.7). At sufficiently downstream, this surface S_* tends to the surface of a circular cylinder of radius a (say). The rate of outflow per unit time is $4\pi m$ as given in the previous case. This amount is equal to the uniform flow of

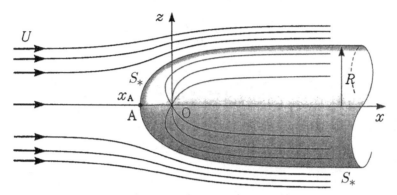

Fig. 5.7. A source at O in a uniform flow of velocity U. The external flow is equivalent to a flow around a semi-infinite cylindrical body B_*.

velocity U passing through the circular cross-section of radius a. Thus, we have

$$4\pi m = \pi a^2 U, \quad \therefore \ a = 2\sqrt{m/U}$$

One can replace the surface S_* (and its inside space) by a solid body B_*, which is a semi-infinite cylinder with a rounded head. The flow around the solid body B_* is given by the same velocity potential of (5.35) (if the body axis coincides with the uniform flow) with the restriction that only the flow external to B_* is considered.

5.6.3. *Dipole*

Differentiating the source potential Φ_s of (5.33) with respect to x, we obtain $\partial_x \Phi_s = mx/r^3$. This form x/r^3 is called a dipole potential. The reason is as follows.

Suppose that there is a source $-mf(x, y, z)$ at the origin $(0, 0, 0)$ and a sink $mf(x + \epsilon, y, z)$ at $(-\epsilon, 0, 0)$, where $f(x, y, z)$ is defined by $1/\sqrt{x^2 + y^2 + z^2}$ (where $m, \epsilon > 0$). The velocity potential of the combined flow is given by

$$\Phi = m(f(x + \epsilon, y, z) - f(x, y, z)) = m\epsilon \frac{\partial}{\partial x} f(x, y, z) + O(\epsilon^2)$$
$$= -m\epsilon \frac{x}{r^3} + O(\epsilon^2).$$

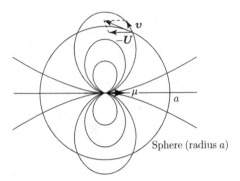

Fig. 5.8.   Stream-lines of a dipole.

Furthermore, consider the limit that the pair of the source and sink becomes closer indefinitely as $\epsilon \to 0$, but that the product $m\epsilon$ tends to a nonzero constant μ. Then, we obtain the following dipole potential (Fig. 5.8),

$$\Phi_d = -\mu \frac{x}{r^3}, \tag{5.36}$$

where μ is termed the strength of a dipole in the x direction. The function $\Phi_d = \mu \partial_x (1/r)$ is a harmonic function except at the origin because the derivative of a harmonic function $1/r$ is another harmonic function. In a mathematical expression, if $\Delta f = 0$ is satisfied, we have $\Delta(\partial_x f) = 0$.

5.6.4.   *A sphere in a uniform flow*

Let us consider a flow represented by the linear combination of a uniform flow Φ_U and a dipole $-\Phi_d$:

$$\Phi_{Ud} := \Phi_U - \Phi_d = Ux + \mu \frac{x}{r^3}. \tag{5.37}$$

Using the spherical coordinates (r, θ, ϕ) where the x axis coincides with the polar axis, this is written as

$$\Phi_{Ud} = U\left(r + \frac{\mu}{Ur^2}\right)\cos\theta, \tag{5.38}$$

since $x = r\cos\theta$. This is an axisymmetric flow. In fact, using (A.30), or (D.17) for the ∇ operator of the spherical coordinates (r, θ, ϕ) in

Appendix D.3, the velocity in the spherical frame is given by

$$\mathbf{v} = \left(U\left(1 - \frac{2\mu}{Ur^3}\right)\cos\theta, -U\left(1 + \frac{\mu}{Ur^2}\right)\sin\theta, 0 \right). \tag{5.39}$$

At infinity as $r \to \infty$, the velocity tends to the uniform flow $(U\cos\theta, -U\sin\theta, 0)$ in the spherical frame, which is equivalent to $(U, 0, 0)$ in the (x, y, z) frame.

The radial component v_r is given by $v_r = U(1 - (2\mu/Ur^3))\cos\theta$. From this, it is found that there is a sphere of radius $a = (2\mu/U)^{1/3}$ on which the normal velocity component v_r vanishes for all θ (Fig. 5.9). This implies that the expression (5.39) represents the velocity of a flow around a sphere of radius a. A dipole placed in a uniform flow can represent a flow field around a sphere placed in a uniform stream. If the radius a is given in advance, the dipole strength μ should be $\frac{1}{2}Ua^3$.

Thus, it is found that the velocity potential of the uniform flow of velocity U around a sphere of radius a is given by

$$\Phi_{U\text{sph}} = Ux + \frac{1}{2}Ua^3\frac{x}{r^3} = U\left(r + \frac{a^3}{2r^2}\right)\cos\theta. \tag{5.40}$$

For this axisymmetric flow, we can introduce the Stokes' stream function defined by (B.11) in Appendix B.3 as

$$v_r = \frac{1}{r^2\sin\theta}\frac{\partial\Psi}{\partial\theta}, \quad v_\theta = -\frac{1}{r\sin\theta}\frac{\partial\Psi}{\partial r}. \tag{5.41}$$

Using (5.39), we obtain the stream function $\Psi_{U\text{sph}}$ of the uniform flow of velocity U around a sphere of radius a:

$$\Psi_{U\text{sph}} = \frac{1}{2}Ur^2\left(1 - \frac{a^3}{r^3}\right)\sin^2\theta. \tag{5.42}$$

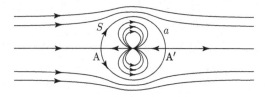

Fig. 5.9. Flow around a sphere, represented by a dipole and a uniform flow.

Obviously, this shows that the sphere $r = a$ coincides with a stream-line, more precisely a surface of stream-lines.

The first term represents the stream function of a *uniform flow*:

$$\Psi_U = \frac{1}{2} U r^2 \sin^2 \theta. \tag{5.43}$$

The second term represents the stream function of a *dipole* toward the negative x-direction:

$$\Psi_d = -\frac{1}{2} U \frac{a^3}{r} \sin^2 \theta. \tag{5.44}$$

5.6.5.   *A vortex line*

In the cylindrical coordinates (r, θ, z) with z as the polar axis, the following function Φ_v is a harmonic function:

$$\Phi_v = \frac{k}{2\pi} \theta, \quad (r \neq 0, \quad k : \text{a constant}). \tag{5.45}$$

In the cartesian (x, y, z) coordinates, the variable θ is an angle defined by

$$\theta = \tan^{-1}(y/x) = \tan^{-1}\xi, \quad \xi = \frac{y}{x}. \tag{5.46}$$

It can be readily verified that Φ_v, in fact, satisfies the Laplace equation, $\nabla^2 \Phi_v = 0$, by using the formula (D.10). The function Φ_v is actually the velocity potential of a *vortex-line* of strength k coinciding with the z axis (Fig. 5.10). This can be shown as follows.

First, the function Φ_v is independent of z by the above definition. Hence the z component of velocity is zero: $w = \partial_z \Phi_v = 0$. Next, the x component is given by

$$u = \partial_x \Phi_v = \frac{\partial}{\partial x}\left(\frac{k}{2\pi}\theta\right) = \frac{k}{2\pi}\frac{\partial \xi}{\partial x}\frac{\partial}{\partial \xi}\tan^{-1}\xi$$

$$= \frac{k}{2\pi}\left(-\frac{y}{x^2}\right)\frac{1}{1+\xi^2} = -\frac{k}{2\pi}\frac{y}{x^2+y^2}. \tag{5.47}$$

Similarly, the y component is

$$v = \partial_y \Phi_v = \frac{k}{2\pi}\frac{x}{x^2+y^2}. \tag{5.48}$$

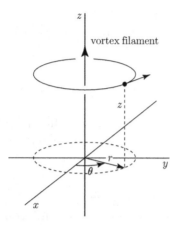

Fig. 5.10. A vortex-line coinciding with the z axis.

Thus, the velocity is given by $\mathbf{v} = (u, v, 0)$, which is rewritten as

$$\mathbf{v} = \frac{k}{2\pi} \frac{1}{x^2 + y^2} (-y,\, x,\, 0) = (0, 0,\, \Omega) \times (x,\, y,\, z),$$

where $\Omega = (1/2\pi)\,k/(x^2 + y^2)$ is the angular velocity of rotating motion around the z axis depending on $r = \sqrt{x^2 + y^2}$ only. The velocity circulation along a closed curve C around the z axis (surrounding it once), called the strength of a vortex, is

$$\oint_C \mathbf{v} \cdot d\mathbf{l} = \oint_C (\partial_x \Phi_v dx + \partial_y \Phi_v dy) = \oint_C d\Phi_v = [\Phi_v]_C = \frac{k}{2\pi}\, 2\pi = k.$$

Thus, it is found that the velocity potential Φ_v represents the flow field due to a concentrated vortex of strength k coinciding with the z axis.

5.7. Irrotational incompressible flows (2D)

Let us consider the velocity at a point $\mathbf{x} = (x, y)$ in the two-dimensional cartesian space, given by

$$\mathbf{v}(\mathbf{x}) = (u(x, y), v(x, y)).$$

Any two-dimensional irrotational flow of an incompressible fluid can be represented by a complex function of the complex variable z defined as

$$z = x + iy = r\cos\theta + ir\sin\theta = re^{i\theta}, \qquad (5.49)$$

where $i = \sqrt{-1}$ is the imaginary unit. The magnitude $r = \sqrt{x^2 + y^2}$ and the angle $\theta = \tan^{-1}(y/x)$ are written by the modulus $|z| = r$ and the argument, $\arg(z) = \theta$, where $z = re^{i\theta}$ is called polar representation.

Conversely, any analytic complex function $f(z)$ represents a certain two-dimensional irrotational flow of an incompressible fluid. The present section is devoted to such a powerful method.

An irrotational flow can be represented by a *velocity potential* $\Phi(x, y)$ (Appendix B.1). In the two-dimensional space, velocity components are expressed as

$$u(x, y) = \partial_x \Phi(x, y), \quad v(x, y) = \partial_y \Phi(x, y). \qquad (5.50)$$

On the other hand, a two-dimensional incompressible flow can be represented in terms of a *stream function* $\Psi(x, y)$ as well (Appendix B.2), such that

$$u = \partial_y \Psi(x, y), \quad v = -\partial_x \Psi(x, y), \qquad (5.51)$$

which satisfies $\partial_x u + \partial_y v = 0$ identically. A contour curve is defined by $\Psi(x, y) = C$ (a constant), which coincides with a stream-line because

$$0 = d\Psi = \Psi_x dx + \Psi_y dy = -v dx + u dy.$$

This is equivalent to the equation of a stream-line: $dx/u = dy/v$.

From (5.50) and (5.51), we have the relation,

$$\partial_x \Phi = \partial_y \Psi, \quad \partial_y \Phi = -\partial_x \Psi. \qquad (5.52)$$

This is nothing but the *Cauchy–Riemann's relations* [CKP66] which are the fundamental property to be satisfied by a complex *analytic* function defined by

$$F(z) = \Phi(x, y) + i\Psi(x, y), \quad z = x + iy. \tag{5.53}$$

From (5.52), it can be readily shown that both the velocity potential $\Phi(x, y)$ and the stream function $\Psi(x, y)$ satisfy the Laplace equation:

$$\Phi_{xx} + \Phi_{yy} = 0, \quad \Psi_{xx} + \Psi_{yy} = 0. \tag{5.54}$$

The *analyticity condition* (5.52) is another word for *differentiability*. In fact, the Cauchy–Riemann's conditions assure that the complex function $F(z)$ has a well-defined derivative,[9]

$$\frac{dF}{dz} = \frac{\partial \Phi}{\partial x} + i\frac{\partial \Psi}{\partial x} = \frac{\partial \Psi}{\partial y} - i\frac{\partial \Phi}{\partial y} = u - iv. \tag{5.55}$$

(See Problem 5.1.) In addition, it is assured by (5.52) that the complex function $F(z)$ is differentiable any number of times [CKP66].

The function $F(z)$ is called the *complex potential*, and dF/dz the *complex velocity*. The complex velocity is written as $w = u - iv$. The magnitude $|w|$ gives the magnitude of velocity $\mathbf{v} = (u, v)$, while the argument of w is $-\phi$, where ϕ is defined by the counter-clockwise angle of $\mathbf{v}$ from the x axis:

$$w = u - iv = \frac{dF}{dz} = q\,e^{-i\phi}, \quad u + iv = q\,e^{i\phi}. \tag{5.56}$$

Since $u^2 + v^2 \equiv q^2 = |dF/dz|^2$, the Bernoulli equation (5.11) can be rewritten in the form for the pressure p as[10]

$$p = p_0 - \frac{1}{2}\rho_0 q^2, \quad q^2 = \left|\frac{dF}{dz}\right|^2 = \frac{dF}{dz}\overline{\left(\frac{dF}{dz}\right)}, \tag{5.57}$$

where the overline denotes complex conjugate.

The definition of the complex function $F(z) = \Phi + i\Psi$ by (5.53) is regarded as a map from a complex variable $z = x + iy$ to another

---

[9]The form $d(\Phi + i\Psi) = (u - iv)(dx + idy)$ was first given by d'Alembert (1761). The second expression of (5.55) is the derivative with $\Delta z = \Delta x$, whereas the third one is that with $\Delta z = i\Delta y$. Namely, *differentiability* means that the derivative does not depend on the direction of differentiation.

[10]$\overline{A} \equiv a - ib$ is the complex conjugate of $A = a + ib$ (a, b: real).

complex variable $F = \Phi + i\Psi$, or an inverse map from F to z. This is called the *conformal map*. In fact, in the (x, y)-plane, we have two families of curves defined by

$$\Phi(x, y) = \text{const.}, \quad \Psi(x, y) = \text{const.} \tag{5.58}$$

The second describes a family of stream-lines as explained in Appendix B.2, while the first may be called a family of equi-potential curves. The two families of curves form an *orthogonal net* in the (x, y)-plane (see Problem 5.3). In the (Φ, Ψ)-plane, the family of curves $\Phi = \text{const.}$ are straight lines parallel to the imaginary axis, while the family of curves $\Psi = \text{const.}$ are parallel to the real axis. Thus the two sets of orthogonal parallel straight lines in the (Φ, Ψ) plane are mapped to the orthogonal net in the (x, y)-plane by the complex function $F(z)$. The property of angle-preservation by a complex analytic function $F(z)$ is valid for any intersecting angle (see Problem 5.4).

From (5.55), we have

$$\mathrm{d}F = w\mathrm{d}z = (u - iv)(\mathrm{d}x + i\mathrm{d}y) = (u\mathrm{d}x + v\mathrm{d}y) + i(u\mathrm{d}y - v\mathrm{d}x).$$

Integrating this along a closed curve C, we obtain

$$\oint_C w\mathrm{d}z = \oint_C (u\mathrm{d}x + v\mathrm{d}y) + i \oint_C (u\mathrm{d}y - v\mathrm{d}x) = \Gamma(C) + iQ(C). \tag{5.59}$$

The first real part is just the velocity circulation $\Gamma(C)$ along C, according to (5.13). The second imaginary part, if multiplied by ρ, is the mass flux across C. In fact, we can write

$$Q(C) = \oint_C (u\mathrm{d}y - v\mathrm{d}x) = \oint_C (un_x + vn_y)\,\mathrm{d}l, \tag{5.60}$$

where $(n_x, n_y) = \mathbf{n}$ is the unit normal to the closed curve C.[11] The right-hand side clearly means the outward flux across the counterclockwise closed curve C.

---

[11]Using the line element $\mathrm{d}\mathbf{l} = (\mathrm{d}x, \mathrm{d}y)$ and its magnitude $\mathrm{d}l = |\mathrm{d}\mathbf{l}|$, the unit tangent is defined by $\mathbf{t} = (t_x, t_y) = (\mathrm{d}x/\mathrm{d}l, \mathrm{d}y/\mathrm{d}l)$, while the unit normal is defined by $\mathbf{n} = (n_x, n_y) = (\mathrm{d}y/\mathrm{d}l, -\mathrm{d}x/\mathrm{d}l)$, directed to the right of $\mathbf{t}$. Obviously, $\mathbf{n} \cdot \mathbf{t} = 0$.

The contour integral of a complex function $w(z)$ round a counter-clockwise closed contour C is given by the sum of *residues* within C by the theory of complex function [CPC66]:

$$\oint_C w(z)\,\mathrm{d}z = 2\pi i \sum_k R_k, \tag{5.61}$$

where R_k is the residue of $w(z)$ at the simple pole z_k within C (Problem 5.6).

5.8. Examples of 2D flows represented by complex potentials

Monomials like z^n (for $n = 1, 2, \ldots$) represent typical flows:

(i) $F = Cz$ (where $C = Ue^{-i\theta}$ with U and θ real constants):
This represents a uniform flow of magnitude U of velocity in the direction inclined at an angle θ counter-clockwise with respect to the x axis because $\mathrm{d}F/\mathrm{d}z = Ue^{-i\theta}$.

(ii) $F = \frac{1}{2}Az^2$ (A is a real constant):
$w = \mathrm{d}F/\mathrm{d}z = Az = Ax + iAy \, (= u - iv)$. This represents a stagnation-point flow where $(u, v) = (Ax, -Ay)$ (Problem 5.2).

(iii) $F = (A/(n+1))\, z^{n+1}$ ($A > 0$ and $n \geq -1/2$):
$\mathrm{d}F/\mathrm{d}z = Az^n = Ar^n(\cos n\theta + i \sin n\theta)$ where $z = re^{i\theta}$. This represents a corner flow in a sector of angle π/n. If $n > 0$, the velocity at the origin $z = 0$ vanishes, while the velocity at $z = 0$ diverges for $0 > n \geq -1/2$. The case $n = 0$ corresponds to a uniform flow. The case $n = -1/2$ corresponds to the maximum sector angle 2π and describes the flow around a semi-infinite plate from the lower to the upper side (or *vice versa*).

5.8.1. *Source (or sink)*

A source (or sink) at the origin $z = 0$ is represented by

$$F_s(z) = m \log z \quad (m : \text{real}). \tag{5.62}$$

Substituting the polar representation $z = re^{i\theta}$,

$$F_s \equiv \Phi + i\Psi = m \log r + im\theta.$$

Stream-lines, expressed by $\Psi = m\theta = $ const., are all *radial*. The velocity field is given by $dF_s/dz = w = |\mathbf{v}|\,e^{-i\phi}$:

$$w = \frac{dF_s}{dz} = \frac{m}{z} = \frac{m}{r}\,e^{-i\theta}. \tag{5.63}$$

Volume flux of outflow per unit time from a domain bounded by a closed curve C is given by $Q(C)$ of (5.59), which is the imaginary part of the contour integral of the complex velocity w round C. Such a contour integral is given by (5.61).

The complex velocity $w(z)$ has only one simple pole at $z = 0$. Hence, if C encloses the origin, we have

$$Q(C) = \Im\left[\oint_C w\,dz\right] = \Im\left[\oint_C \frac{m}{z}\,dz\right] = \Im[2\pi i m] = 2\pi m. \tag{5.64}$$

If C does not enclose the origin, we have $Q(C) = 0$. The circulation $\Gamma(C)$ along C is zero since the contour integral $\oint_C w\,dz = 2\pi i m$ is purely imaginary with a real m.

Thus it is found that the complex potential F_s describes a source of fluid flow from the origin if $m > 0$, or a sink if $m < 0$.

5.8.2. *A source in a uniform flow*

Consider the following complex potential:

$$F_{Us} := Uz + F_s = Uz + m\log z. \tag{5.65}$$

The complex velocity is given by

$$w = \frac{dF_{Us}}{dz} = U + \frac{m}{z}.$$

At large distances as $|z| \to \infty$, the velocity w approaches the first uniform flow $w = U$. There is a *stagnation-point* on the real axis (x axis) where the velocity w vanishes (see Problem 5.2 for a stagnation-point flow). On the real axis, we have $z = x$, and the x component is given by $u = U + m/x$, whereas the y component vanishes. u vanishes at a point x_s on the negative x axis, where $x_s = -m/U$. The flow field has a symmetry with respect to the x axis. The stream-lines emanating from the stagnation point forms a curve C_* of mirror-symmetry with respect to the x axis, which divides the fluid emerging out of the source at $z = 0$ from the fluid of oncoming uniform flow from the left. At sufficiently downstream, this curve C_* tends to two parallel lines of distance d (say). The rate of outflow per unit time is $2\pi m$ as given in the previous example. This amount is equal to the uniform flow of velocity U between the two parallel lines of width d. Thus, we have

$$2\pi m = Ud, \quad \therefore \ d = 2\pi m/U$$

One can replace the curve C_* (and its inside space) by a solid body C_*, which is a semi-infinite plate of thickness d with a rounded edge.

5.8.3.   Dipole

A derivative of the source potential F_s of (5.62) with respect to x becomes a dipole potential. The reason is as follows.

Suppose that there is a source $m \log z$ at the origin $(0,0)$ and a sink $-m \log(z + \epsilon)$ at $(-\epsilon, 0)$, where $m, \epsilon > 0$. The complex potential of the combined flow is given by

$$\begin{aligned}
F &= m \log z - m \log(z + \epsilon) \\
&= -m[\log(z + \epsilon) - \log z] \\
&= -m\epsilon \frac{\partial}{\partial z} \log z + O(\epsilon^2) = -m\epsilon \frac{1}{z} + O(\epsilon^2).
\end{aligned}$$

Furthermore, consider the limit that the pair of source and sink becomes closer indefinitely as $\epsilon \to 0$, but that the product $m\epsilon$ tends to a nonzero constant μ. Then, we obtain the following dipole potential,

$$F_d = -\frac{\mu}{z}. \tag{5.66}$$

where μ is termed as the strength of dipole in the x-direction. The dipole in the y-direction is obtained by replacing ϵ by $i\epsilon$, and hence replacing μ by $i\mu$.

The function $F_d = -\mu \, (\mathrm{d}/\mathrm{d}z) \log z$ is an analytic function except at the origin because the derivative of an analytic function $\log z$ $(z \neq 0)$ is another analytic function. According to the theory of complex functions, if $f(z)$ is an analytic function in a domain, then the derivative $\mathrm{d}^n f / \mathrm{d} z^n$ is another analytic function for any integer n in the same domain.

5.8.4.   *A circular cylinder in a uniform flow*

The dipole $-F_d$ in a uniform stream Uz,

$$F_{Ud} := Uz - F_d = Uz + \frac{\mu}{z} \qquad (5.67)$$

describes a uniform flow around a circle.

The complex velocity is given by $\mathrm{d}F_{Ud}/\mathrm{d}z$, which is

$$w = u - iv = U - \frac{\mu}{z^2} = \left(U - \frac{\mu}{r^2} \cos 2\theta \right) + i \frac{\mu}{r^2} \sin 2\theta, \qquad (5.68)$$

where $z^2 = r^2 e^{i2\theta}$. As $z \to \infty$, the velocity tends to the uniform flow $(U, 0)$.

Consider a circle C_r of radius r centered at $z = 0$. The radial component of velocity at the angular position θ on C_r is given by

$$v_r = u \cos \theta + v \sin \theta$$

$$= \left(U - \frac{\mu}{r^2} \cos 2\theta \right) \cos \theta - \frac{\mu}{r^2} \sin 2\theta \sin \theta$$

$$= \left(U - \frac{\mu}{r^2} \right) \cos \theta.$$

From this, it is found that there is a circle of radius $a = \sqrt{\mu/U}$ on which the normal velocity component v_r vanishes for all θ. This implies that the expression (5.68) represents the velocity of a flow around a circle of radius a. A dipole (in the negative x-direction) placed in a uniform flow (in the positive x-direction) can represent a flow field around a circle placed in a uniform stream. If the radius a is given in advance, the dipole strength μ is given by Ua^2.

Thus, it is found that the complex potential of the uniform flow of velocity U around a circle of radius a is given by

$$F_{U\,\text{circ}} = U\left(z + \frac{a^2}{z}\right) \tag{5.69}$$

$$= U\left(r + \frac{a^2}{r}\right)\cos\theta + i\left(r - \frac{a^2}{r}\right)\sin\theta. \tag{5.70}$$

5.8.5.   Point vortex (a line vortex)

A vortex at the origin $z = 0$ is represented by the complex potential,

$$F_v(z) = \frac{k}{2\pi i}\log z \quad (k : \text{real}). \tag{5.71}$$

Substituting the polar representation $z = re^{i\theta}$,

$$F_v \equiv \Phi + i\Psi = \frac{k}{2\pi}\theta - i\frac{k}{2\pi}\log r.$$

The velocity field is given by $dF_v/dz = w = u - iv$:

$$w(z) = \frac{k}{2\pi i}\frac{1}{z} = \frac{k}{2\pi r}(-\sin\theta - i\cos\theta), \tag{5.72}$$

which has only one simple pole at $z = 0$. We obtain

$$u = -\frac{k}{2\pi r}\sin\theta = -\frac{k}{2\pi}\frac{y}{r^2}, \quad v = \frac{k}{2\pi r}\cos\theta = \frac{k}{2\pi}\frac{x}{r^2}, \tag{5.73}$$

where $r^2 = x^2 + y^2$. It is seen that these are equivalent to (5.47) and (5.48).

Circulation $\Gamma(C)$ and volume flux $Q(C)$ with respect to a contour C is given by the contour integral of (5.59). If C encloses the origin, we have

$$\oint_C w\,dz(= \Gamma(C) + iQ(C)) = \oint_C \frac{k}{2\pi i}\frac{dz}{z} = 2\pi i\frac{k}{2\pi i} = k. \tag{5.74}$$

Therefore, the velocity circulation round C enclosing $z = 0$ is given by $\Gamma(C) = k$, whereas $Q(C) = 0$. If C does not enclose the origin, we have $\Gamma(C) = 0$ and $Q(C) = 0$.

Thus it is found that the complex potential $F_v = (k/2\pi i)\log z$ represents a concentrated vortex located at $z = 0$ with the circulation k, counter-clockwise if $k > 0$, or clockwise if $k < 0$.

5.9. Induced mass

5.9.1. *Kinetic energy induced by a moving body*

Let us consider an ingenious analysis of potential flow induced by a solid body moving through an invisid fluid otherwise at rest in the three-dimensional space (x, y, z). This is a formulation in which the total momentum and energy of fluid motion thus induced can be represented in terms of a constant tensor depending only on the asymptotic behavior of velocity potential $\Phi(\mathbf{r})$ at infinity, where $\mathbf{r} = (x, y, z)$. [LL87, Sec. 11]

Potential flow of an incompressible fluid satisfies the Laplace equation $\nabla^2\Phi = 0$. We investigate the fluid velocity at great distances from the moving body. Since the fluid is at rest at infinity, the velocity grad Φ must vanish at infinity. We take the origin inside the moving body at a particular instant of time. We already know (Sec. 5.6) that the function $1/r$ is a particular solution of the Laplace equation ($r \neq 0$), where r is the distance from the origin. In addition, its space derivatives of any order are also solutions. Their linear combination is also a solution. All these solutions vanish at infinity. Thus, the general form of solutions of Laplace equation is represented by

$$\Phi = \frac{a}{r} + \mathbf{A} \cdot \mathrm{grad}\, \frac{1}{r} + [\text{higher order derivatives of } 1/r], \qquad (5.75)$$

at large distances from the body, where a and $\mathbf{A}$ are constants independent of coordinates. The constant a must be zero in incompressible flows since the first term represents a flow due to a fluid source at the origin (Sec. 5.6.1). The second term represents a flow due to a dipole at the origin (Sec. 5.6.3). The series form (5.75) is called the *multipole expansion*.

We concentrate on the dipole term including $\mathbf{A}$, since the terms of higher order could be neglected at large distances in the following analysis. Then, we have

$$\Phi = \mathbf{A} \cdot \operatorname{grad} \frac{1}{r} = -\frac{\mathbf{A} \cdot \mathbf{e}}{r^2}, \quad \mathbf{e} \equiv \frac{\mathbf{r}}{|\mathbf{r}|}, \tag{5.76}$$

where $\mathbf{r} = (x, y, z)$ is the radial vector. The velocity is given by

$$\mathbf{v} = \operatorname{grad} \Phi = (\mathbf{A} \cdot \operatorname{grad}) \operatorname{grad} \frac{1}{r} = \frac{3(\mathbf{A} \cdot \mathbf{e})\mathbf{e} - \mathbf{A}}{r^3}. \tag{5.77}$$

It is remarkable that the velocity diminishes as r^{-3} at large distances, which is a characteristic property of dipoles. The constant vector $\mathbf{A}$ depends on the detailed shape of the body.[12]

The vector $\mathbf{A}$ appearing in (5.76) can be related to total momentum and energy of fluid motion induced by the moving body. The total kinetic energy is[13]

$$E = \frac{1}{2}\rho \int_{V_*} |\mathbf{v}|^2 dV,$$

where the integration is to be taken over all space outside the body, denoted by V_*. The integral in the unbounded space is dealt with in the following way. First, we choose a region of space V bounded by a sphere S_R of a large radius R with its center at the origin and integrate over V. Next, we let R tend to infinity. We denote the volume of the body by V_b and its surface by S_b.

Denoting the velocity of the body by $\mathbf{U}$, we note the following identity:

$$\int_{V_*} |\mathbf{v}|^2 dV = \int_{V_*} |\mathbf{U}|^2 dV + \int_{V_*} (\mathbf{v} - \mathbf{U}) \cdot (\mathbf{v} + \mathbf{U}) \, dV.$$

The first integral on the right is simply $U^2(V - V_b)$, since $U \equiv |\mathbf{U}|$ is a constant. In the second integral, we use the expressions: $\mathbf{v} = \operatorname{grad} \Phi$,

---

[12]The vector $\mathbf{A}$ can be determined by solving the equation $\nabla^2 \Phi = 0$ in the complete domain, taking into account the boundary condition at the surface of the moving body.

[13]The internal energy of an incompressible ideal fluid is constant.

and $\mathbf{U} = \mathrm{grad}(\mathbf{U} \cdot \mathbf{r})$ for the second factor. Then, we have

$$(\mathbf{v} - \mathbf{U}) \cdot (\mathbf{v} + \mathbf{U}) = (\mathbf{v} - \mathbf{U}) \cdot \mathrm{grad}(\Phi + \mathbf{U} \cdot \mathbf{r}) = \mathrm{div}[(\Phi + \mathbf{U} \cdot \mathbf{r})(\mathbf{v} - \mathbf{U})],$$

since $\mathrm{div}\,\mathbf{v} = 0$ and $\mathrm{div}\,\mathbf{U} = 0$. The second integral is now transformed into a surface integral over S_R and S_b. Thus, we have

$$\int_{V_*} |\mathbf{v}|^2 \, dV = U^2(V - V_b) + \oint_{S_R + S_b} (\Phi + \mathbf{U} \cdot \mathbf{r})(\mathbf{v} - \mathbf{U}) \cdot \mathbf{n} dS,$$

where $\mathbf{n}$ is the unit outward normal to the bounding surface. On the body's surface S_b, the normal component $\mathbf{v} \cdot \mathbf{n}$ must be equal to $\mathbf{U} \cdot \mathbf{n}$. Hence the surface integral over S_b vanishes identically. On the remote surface S_R, we can use (5.76) for Φ and (5.77) for $\mathbf{v}$. Then, we obtain

$$\int |\mathbf{v}|^2 dV = U^2 \left(\frac{4}{3}\pi R^3 - V_b \right) + \int \left[3(\mathbf{A} \cdot \mathbf{e})(\mathbf{U} \cdot \mathbf{e}) \right.$$
$$\left. - R^3(\mathbf{U} \cdot \mathbf{e})^2 - 2R^{-3}(\mathbf{A} \cdot \mathbf{e})^2 \right] d\Omega, \qquad (5.78)$$

where, on the spherical surface S_R, we used the expressions: $\mathbf{n} = \mathbf{e}$ and $dS = R^2 \, d\Omega$ with $d\Omega$ an element of solid angle ($d\Omega = \sin\theta \, d\theta d\phi$).

Here, we apply the following formula for the integral of $(\mathbf{A} \cdot \mathbf{e}) \times (\mathbf{B} \cdot \mathbf{e})$ over $d\Omega$ divided by 4π (equivalent to an average over a unit sphere):

$$\frac{1}{4\pi} \int (\mathbf{A} \cdot \mathbf{e})(\mathbf{B} \cdot \mathbf{e}) \, d\Omega = \frac{1}{3} \mathbf{A} \cdot \mathbf{B}, \qquad (5.79)$$

for constant vectors $\mathbf{A}$ and $\mathbf{B}$ (see Problem 5.9 for its proof).

Carrying out the integration of (5.78) by using (5.79), neglecting the last term ($\propto R^{-3}$) since it vanishes as $R \to \infty$ and dropping the cancelling terms of $(4/3)\pi R^3 U^2$, we finally obtain the following expression for the total energy of fluid motion (multiplying by $\frac{1}{2}\rho$):

$$E = \frac{1}{2}\rho \int |\mathbf{v}|^2 \, dV = \frac{1}{2}\rho \left(4\pi \mathbf{A} \cdot \mathbf{U} - V_b U^2 \right). \qquad (5.80)$$

In order to obtain the exact expression of $\mathbf{A}$, we need a complete solution of the Neumann problem of the Laplace equation,

$$\nabla^2 \Phi = 0, \quad \text{for } \mathbf{r} \in V_*; \quad \mathbf{n} \cdot \nabla\Phi = \mathbf{n} \cdot \mathbf{U} \quad \text{on } S_b. \qquad (5.81)$$

From this, we can make a deductive reasoning about the general nature of the dependence of **A** on velocity **U** of the body. In view of the properties that the governing equation is linear with respect to Φ, and that the boundary condition for the normal derivative of Φ is linear in **U**, it follows that the complete solution can be represented by a linear combination of potentials in the form of multipole expansion (5.75), and that the coefficient **A** of the dipole term must be a linear function of the components of **U**, i.e. $A_i = c_{ij}U_j$.[14]

5.9.2. *Induced mass*

Thus, the energy E of (5.80) is represented by a quadratic function of U_i, and can be written in the following form:

$$E = \frac{1}{2}\, m_{ij}\, U_i U_j, \tag{5.82}$$

$$m_{ij} = \rho\big(4\pi c_{ij} - V_b\, \delta_{ij}\big), \tag{5.83}$$

where m_{ij} is a certain symmetrical tensor (the expression (5.82) enables symmetrization of m_{ij} even if the original m_{ij} is not so). m_{ij} is called the *induced-mass tensor*, or the *added mass*. The latter meaning will become clear by the reasoning given next.

Knowing the energy E, we can obtain an expression of the total momentum **P** of fluid motion. Suppose that the body is subject to an external force **F**. The momentum of fluid will thereby be increased by d**P** (say) during a short time dt. This increase is related to the force by d**P** = **F**dt. Scalar multiplication with the velocity **U** leads to **U** · d**P** = **F** · **U**dt, i.e. the work done by the force **F** through the distance **U**dt. This must be equal to the increase of energy dE of

---

[14]The dipole potential Φ of (5.76) are of the form $A_i\phi_i$ by using three dipole components ϕ_i. On the other hand, with $\Phi = U_j\psi_j$, the Neumann problem (5.81) can be transformed to $\nabla^2\psi_j = 0$ with the boundary condition $\mathbf{n} \cdot \nabla\psi_j = n_j$. The three solutions ψ_j thus determined depend on the body shape, but are independent of U_i. ψ_j may include a dipole term $\psi_j^{(d)} = c_{ij}\phi_i$ in the asymptotic multipole expansion at large distances. Therefore, we have $A_i\phi_i = U_j\psi_j^{(d)} = U_jc_{ij}\phi_i$. From this, we obtain $A_i = c_{ij}U_j$. For the problem in Sec. 5.6.4 of a sphere of radius a moving with $(-U,0,0)$, we found $\mathbf{A} = (A_x, A_y, A_z) = \big(\frac{1}{2}a^3(-U),0,0\big)$.

the fluid. We already have an expression of E from (5.82). Hence, we obtain two expressions of dE:

$$dE = \mathbf{U} \cdot d\mathbf{P}, \quad dE = m_{ij}U_j dU_i,$$

where symmetry of the mass tensor m_{ij} is used. From the equivalence of the two expressions, we find that

$$P_i = m_{ij}U_j, \quad \text{or} \quad \mathbf{P} = 4\pi\rho\mathbf{A} - \rho V_b\mathbf{U}. \tag{5.84}$$

It is remarkable that the total fluid momentum is given by a *finite* quantity, although the fluid velocity is distributed over unbounded space.

5.9.3. *d'Alembert's paradox and virtual mass*

The momentum transmitted to the fluid by the body is $d\mathbf{P}/dt$ per unit time, the reaction force $\mathbf{F}'$ from the fluid on the body is given by

$$\mathbf{F}' = -d\mathbf{P}/dt.$$

Suppose that the body is moving steadily in an ideal fluid at a velocity $\mathbf{U}$ and inducing an irrotational flow around it. Then, we should have $\mathbf{P} = \text{const.}$, since $\mathbf{U}$ is constant, so we obtain $\mathbf{F}' = 0$. Hence, there would be no force. This is the result known as the *d'Alembert paradox*. This paradox is clearly seen by considering the drag. The presence of a drag in uniform motion of a body would mean that work is continually done on the fluid by the external force (required to maintain the steady motion) and that the fluid will gain energy continually. However, there is no dissipation of energy in an ideal fluid by definition, and the velocity field decays so rapidly with increasing distance from the body that there can be no energy flow to infinity (out of the space). Thus, no force is possible for uniform motion of a body in an ideal fluid.[15] If there was a nonzero force in such a case, that would be a real *paradox*.

Suppose that a body of mass m is moving with acceleration under the action of an external force $\mathbf{f}$. This force must be equated to

---

[15]Except for the lift $\mathbf{L}$ in the two-dimensional problem of uniform flow around a cylindrical body with a circulation (Problem 5.8). We have $\mathbf{L} \cdot \mathbf{U} = 0$.

the time derivative of the total momentum of the system. The total momentum is the sum of the momentum $m\mathbf{U}$ of the body and the momentum $\mathbf{P}$ of the fluid. Thus,

$$m\frac{d\mathbf{U}}{dt} + \frac{d\mathbf{P}}{dt} = \mathbf{f}. \tag{5.85}$$

This can also be written as

$$(m\delta_{ij} + m_{ij})\frac{dU_j}{dt} = f_i. \tag{5.86}$$

This is the equation of motion of a body immersed in an ideal fluid. This clearly shows that m_{ij} is an added mass (tensor). The factor $(m\delta_{ij} + m_{ij})$ in front of the acceleration dU_j/dt is the *virtual mass* (tensor).

The added mass can be estimated from $\Phi_{U\,\mathrm{sph}}$ of Sec. 5.6.4 (and from the remark of the footnote in Sec. 5.9.1), which implied $c_{ij} = \frac{1}{2}a^3\delta_{ij}$ for the motion of a sphere of radius a. Hence, the formula (5.83) with $V_b = (4\pi/3)a^3$ yields

$$m_{ij} = \frac{1}{2}\left(\frac{4\pi}{3}a^3\rho\right)\delta_{ij} = \frac{1}{2}(m_{\mathrm{fluid}})\,\delta_{ij}. \tag{5.87}$$

Namely, the added mass for the motion of a sphere is a half of the fluid mass displaced by the sphere.

For the motion of a circular cylinder, we will obtain $m_{ij} = (m_{\mathrm{fluid}})\,\delta_{ij}$ from Sec. 5.8.4.

5.10.   Problems

Problem 5.1 *Complex velocity*

Taking differential of a complex function $F(z)$: $dF = \partial_x F\,dx + \partial_y F\,dy$, show the equalities of (5.55).

Problem 5.2 *Stagnation-point flow*

Determine the stream-lines of two-dimensional velocity field given by $\mathbf{v} = (u, v) = (Ax, -Ay)$ with A a real constant (Fig. 5.11). [The origin $(0, 0)$ is called the *stagnation-point* because $\mathbf{v}$ vanishes there.]

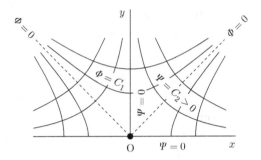

Fig. 5.11. Stagnation-point flow (Ψ: stream function, Φ: velocity potential).

Problem 5.3 Orthogonal net of a complex analytic function

The expressions (5.58) define two families of equi-potential lines and stream-lines. Show mutual orthogonal-intersection of the two families.

Problem 5.4 Conformal property

Suppose that we are given two complex planes $z = x + iy$ and $Z = X + iY$, and that there is a point z_0 in the z plane and two points z_1 and z_2 infinitesimally close to z_0. The two planes are related by a complex analytic function $Z = F(z)$, and the three points z_0, z_1, z_2 are mapped to Z_0, Z_1, Z_2 in the Z plane by $F(z)$, respectively. Show that the intersecting angle θ between two infinitesimal segments (from $z_1 - z_0$ to $z_2 - z_0$) is invariant by the map $F(z)$, namely the intersecting angle between two infinitesimal segments (from $Z_1 - Z_0$ to $Z_2 - Z_0$) is θ.

Problem 5.5 Joukowski transformation

Suppose that we have two complex planes $z = x + iy$ and $\zeta = \xi + i\eta$, and that we are given the following transformation between z and ζ:

$$z = f(\zeta), \quad f(\zeta) = \zeta + \frac{a^2}{\zeta} \quad (a : \text{real positive}), \tag{5.88}$$

called the *Joukowski transformation*.

(i) Using the polar representation $\zeta = \sigma e^{i\phi}$, show that the circle $\sigma = a$ in the ζ plane is mapped to a segment L on the real axis of z plane: $x \in [-2a, +2a]$, $y = 0$. In addition, show that the exterior of the circle $\sigma = a$ of the ζ plane is mapped to the whole z plane except L (a cut along the real axis x), and that the interior of the circle $\sigma = a$ is mapped to the entire z plane except L, too.

(ii) Show that the following potential $F_\alpha(\zeta)$ describes an inclined uniform flow around a circle of radius a in the ζ plane:

$$F_\alpha(\zeta) = U\left(\zeta e^{-i\alpha} + \frac{a^2 e^{i\alpha}}{\zeta}\right) \quad (\alpha : \text{ real}). \tag{5.89}$$

In addition, determine what is the angle of inclination with respect to the real axis ξ [Fig. 5.12(a)].

(iii) Show that the corresponding flow in the z plane is an inclined uniform flow around a flat plate of length $4a$ [Fig. 5.12(b)].

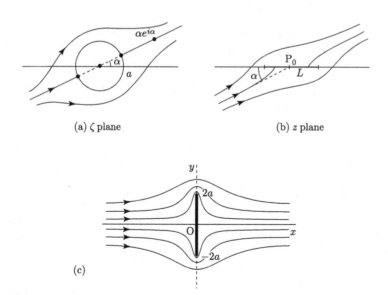

(a) ζ plane   (b) z plane

(c)

Fig. 5.12.

(iv) Show that the potential $F_\perp(\zeta)$ represents a flow impinging on a vertical flat plate (on the imaginary axis) in the Z plane at right angles from left (from the negative real axis of Z) [Fig. 5.12(c)]:

$$F_\perp(\zeta) = -iU\left(\zeta - \frac{a^2}{\zeta}\right), \quad Z = -i\left(\zeta + \frac{a^2}{\zeta}\right). \tag{5.90}$$

Problem 5.6 Residue

Show that the following contour integral of a complex function $F(z)$ along C around $z = 0$:

$$\oint_C F(z)\, dz = 2\pi i\, A_{-1},$$

$$F(z) = \sum_{n=-\infty}^{\infty} A_n z^n = \cdots + \frac{A_{-1}}{z} + \cdots, \tag{5.91}$$

where C is a counter-clockwise closed contour around the origin $z = 0$. The coefficient A_{-1} is called the *residue* at the simple pole $z = 0$.

Problem 5.7 Blasius formula

Suppose that a body B is fixed within a two-dimensional incompressible irrotational flow represented by the complex potential $F(z)$ on the (x, y) plane, and that the force acting on B is given by (X, Y). By using the expression (5.57) for the pressure p on the body surface (given by a closed curve C), show the following *Blasius's formula*:

$$X - iY = \frac{1}{2}i\rho_0 \oint_C \left(\frac{dF}{dz}\right)^2 dz. \tag{5.92}$$

Problem 5.8 Kutta–Joukowski's theorem

(i) Show that the following complex potential $F_\gamma(z)$ represents a uniform flow (of velocity U) around a circular cylinder of radius a (Fig. 5.13):

$$F_\gamma(z) = U\left(z + \frac{a^2}{z}\right) - \frac{\gamma}{2\pi i}\log z, \quad \gamma = \text{const.} \ (> 0). \tag{5.93}$$

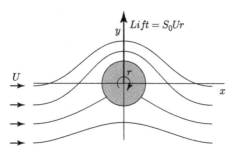

Fig. 5.13. Flow of $F_\gamma(z)$.

In addition, compute the velocity circulation around the cylinder.

(ii) Determine the force (X, Y) acting on the cylinder C in the flow $F_\gamma(z)$ by using the Blasius formula (5.92).

[The force formula to be obtained is the *Kutta–Joukowski's formula* for the *lift*, $Y = \rho_0 U \gamma$ (with $X = 0$), acting on the cylinder.]

Problem 5.9 Integral formula

Verify the integral formula (5.79) for any constant vectors **A** and **B**, where **e** is the unit vector in the radial direction, expressed as

$$\mathbf{e} = (e_i) = (e_x,\, e_y,\, e_z) = (\sin\theta\cos\phi,\, \sin\theta\sin\phi,\, \cos\theta) \qquad (5.94)$$

in the (x, y, z) cartesian frame in terms of the polar angle θ and azimuthal angle ϕ of the spherical polar system, and $d\Omega = \sin\theta\, d\theta d\phi$.

Problem 5.10 Motion of a sphere in a fluid

According to the formulation in Sec. 5.9, determine the equation of motion for a *sphere* of mass m_s, subject to a force **f** in an incompressible inviscid fluid, assuming that the flow is irrotational. In addition, apply it to the motion of a spherical bubble (assuming its density zero) in water in the gravitational field of acceleration g.

Chapter 6

Water waves and sound waves

Fluid motions are characterized by two different elements, i.e. *vortices* and *waves*. In this chapter, we consider water waves and sound waves. There exists a fundamental difference in character between the two waves from a physical point of view. The vortices will be considered in the next chapter.

A liquid at rest in a gravitational field is in general bounded above by a free surface. Once this free surface experiences some disturbance, it is deformed from its equilibrium state, generating fluid motion. Then, the deformation propagates over the surface as a wave. Waves are observed on water almost at any time and are called *water waves* which are sometimes called a *surface wave*. The surface wave is a kind of dispersive waves whose phase velocity depends on its wave length.

On the other hand, a sound wave is *nondispersive* and the phase velocity of different wave lengths take the same value. This results in a remarkable consequence, i.e. invariance of wave form during propagation.

6.1. Hydrostatic pressure

Suppose that water of uniform density ρ is at rest in a uniform field of gravity. Then, setting $\mathbf{v} = 0$ and $\mathbf{f} = \mathbf{g} = (0, 0, -g)$ in (5.2), with respect to the (x, y, z)-cartesian frame, the z axis taken vertically upward (where g is a constant of gravitational acceleration), the Euler equation of motion reduces to

$$\operatorname{grad} p = \rho \mathbf{g}, \quad = \rho \, (0, 0, -g). \tag{6.1}$$

Horizontal x and y components and vertical z component of this equation are

$$\frac{\partial p}{\partial x} = 0, \quad \frac{\partial p}{\partial y} = 0, \quad \frac{\partial p}{\partial z} = -\rho g.$$

Since the density ρ is a constant, we obtain

$$p = p(z) := -\rho g z + \text{const.}$$

Provided that the pressure on the surface is equal to the uniform value p_0 (the atmospheric pressure) at every point, the surface is given by $z = \text{const.}$, called the *horizontal plane*. Since the surface is determined by the pressure solely without any other constraint, it is also called a *free surface*.

Let the horizontal free surface be at $z = 0$, where $p = p_0$. Then we have from the above equation,

$$p = p_0 - \rho g z \quad (z < 0). \tag{6.2}$$

This pressure distribution is called the **hydrostatic pressure** (Fig. 6.1).

Our problem is as follows. Initially, it is assumed that the water is at rest. Next, a small external disturbance pressure acts on the surface of water and deforms it. Needless to say, the state of rest is irrotational since $\mathbf{v} = 0$. Kelvin's circulation theorem tells us that the

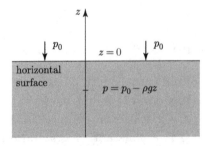

Fig. 6.1.   Hydrostatic pressure.

irrotationality maintains itself in the fluid motion thereafter because the gravity force is conservative. The deformation propagates as a wave, and the water motion can be represented by a velocity potential Φ. It will be found that the wave motion is limited to a neighborhood of the surface, and the water motion decays rapidly with increasing depth from the surface. Hence, it is called a **surface wave**. The surface wave is characterized as a *dispersive wave* which will be clarified in the analysis below.

6.2.   Surface waves on deep water

It is assumed that the water is inviscid and that its motion is incompressible and irrotational. Then according to Sec. 5.5, the velocity potential is governed by the Laplace equation. Furthermore for simplification, the motion is assumed to be two-dimensional in the x (horizontal) and z (vertical) plane, uniform in the y (horizontal) direction. In this case, the velocity is represented as

$$\mathbf{v} = \operatorname{grad} \Phi(x, z, t) = (u, 0, w) \tag{6.3}$$

(with t the time), and the velocity potential satisfies the following equation:

$$\Delta \Phi = \Phi_{xx} + \Phi_{zz} = 0. \tag{6.4}$$

In addition, it is assumed that the depth is infinite. The surface deformation is expressed as

$$z = \zeta(x, t). \tag{6.5}$$

Under the surface, there is an irrotational fluid motion. The Laplace equation (6.4) is investigated in the domain, $-\infty < z < \zeta(x, t)$. The wave is determined by two boundary conditions imposed on the surface $z = \zeta$, which are now considered (Fig. 6.2).

6.2.1.   *Pressure condition at the free surface*

This is the condition that the surface $z = \zeta$ is acted on by the uniform atmospheric pressure p_0. The fact that the fluid motion is irrotational

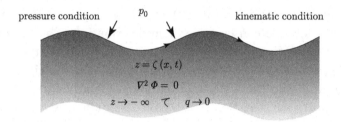

Fig. 6.2. Surface wave on deep water.

enables us to use the integral (5.29), which can be written as

$$\Phi_t + \frac{1}{2}q^2 + (p/\rho) + gz = C,$$

$$q^2 = u^2 + w^2 = \Phi_x^2 + \Phi_z^2. \tag{6.6}$$

In Eq. (6.6), p is set equal to p_0 on the surface $z = \zeta$ and the constant C is equal to p_0/ρ at the initial instant when the water is at rest: namely $p = p_0$ at $z = 0$, $\Phi_t = 0$, and $q^2 = 0$. Thus, we obtain the first condition,

$$\text{BC1}: \ \Phi_t + \frac{1}{2}q^2 + g\zeta = 0 \quad \text{at } z = \zeta(x, t). \tag{6.7}$$

This is called the **dynamic** condition.

From (6.6) with $C = p_0/\rho$, the pressure within the fluid is given by

$$p = p_0 - \rho\left(\Phi_t + \frac{1}{2}q^2 + gz\right) \quad (z < \zeta) \tag{6.8}$$

This is an extension of the formula (6.2) taking account of additional terms of $\rho\Phi_t$ and $\frac{1}{2}\rho q^2$.

6.2.2. *Condition of surface motion*

Surface deformation is caused by the motion of fluid particles moving with velocity (u, w) on the surface. Its mathematical representation is given as follows. Using a function $\zeta(x, t)$ of the surface, we define a new function f by

$$f(x, z, t) := z - \zeta(x, t). \tag{6.9}$$

Every point (x, z) on the free surface satisfies the equation $f(x, z, t) = 0$ at any time t. At a time $t + \Delta t$ infinitesimally after t, a particle located at (x, z) at t displaces to $(x + u\Delta t, z + w\Delta t)$, and still satisfies $f(x + u\Delta t, z + w\Delta t, t + \Delta t) = 0$. This means that the following equation holds:

$$\frac{Df}{Dt} = 0, \quad \text{or equivalently } f_t + uf_x + wf_z = 0.$$

Here, we have $f_t = -\zeta_t$, $f_x = -\zeta_x$ and $f_z = 1$ from (6.9). Substituting these in the above equation, we obtain the second condition,

$$\text{BC2}: \ \zeta_t + u\zeta_x = w \quad \text{at } z = \zeta(x, t). \tag{6.10}$$

This is called the **kinematic** condition.

In *steady* problem, one has the form $\zeta = \zeta(x)$ and hence $\zeta_t = 0$. Then, the above equation (6.10) reduces to $d\zeta/dx = w/u$. This describes simply that the surface coincides with a stream-line, in other words, fluid particles move over the surface as its stream-line.

Another condition is for the bottom. The fluid motion decays at an infinite depth:

$$\text{BC3}: \ q = |\text{grad } \Phi| = \sqrt{u^2 + w^2} \to 0 \quad \text{as } z \to -\infty. \tag{6.11}$$

The three conditions BC1–BC3 are the boundary conditions for the motion in deep water.

Note that the governing equation (6.4) is linear with respect to Φ. However, the boundary conditions (6.7) and (6.10) on the free surface are nonlinear with respect to the velocity and surface deformation.

6.3. Small amplitude waves of deep water

6.3.1. *Boundary conditions*

It is assumed that the displacement ζ of the free surface and the fluid velocities u, w are small in the sense described in Sec. 6.3.3 below. Then one may linearize the problem by neglecting terms of second or higher orders with respect to those small quantities. This is called a *linearization*. The magnitudes of velocity potential $|\Phi|$ is assumed to be infinitesimally small as well because the water was at rest initially. Equation (6.4) of Φ is linear originally, hence no term is neglected.

The surface boundary condition (6.7) has a quadratic term q^2. Neglecting this, we have

$$\Phi_t + g\zeta = 0 \quad \text{at } z = \zeta(x, t).$$

Taylor exansion of the first term with respect to z is

$$\Phi_t(x, \zeta, t) = \Phi_t(x, 0, t) + \zeta\partial_z\Phi_t(x, 0, t) + O(\zeta^2).$$

The second term consists of two first order terms ζ and $\partial_z\Phi_t$, i.e. of second order. Thus, keeping the first term only, we have

$$\Phi_t + g\zeta = 0 \quad \text{at } z = 0. \tag{6.12}$$

Likewise, the boundary condition (6.10) is linearized to

$$\zeta_t = w \quad (w = \Phi_z) \quad \text{at } z = 0. \tag{6.13}$$

Eliminating ζ from (6.12) and (6.13), we obtain

$$\Phi_{tt} + g\Phi_z = 0 \quad \text{at } z = 0. \tag{6.14}$$

Thus, we have arrived at the following mathematical problem including Φ only:

$$\Delta\Phi = \Phi_{xx} + \Phi_{zz} = 0 \quad (0 > z > -\infty), \tag{6.15}$$

$$|\text{grad }\Phi| = \sqrt{(\Phi_x)^2 + (\Phi_z)^2} \to 0 \quad \text{as } z \to -\infty, \tag{6.16}$$

together with the surface condition (6.14).

This is a system of equations (6.14)–(6.16) which determine water waves of small amplitude. Once $\Phi(x, z, t)$ is found, the wave form $\zeta(x, t)$ is determined by (6.12).

6.3.2. *Traveling waves*

Let us try to find a solution of the following form,

$$\Phi = f(z) \sin(kx - \omega t), \qquad (6.17)$$

(k, ω: constants, assumed positive for simplicity), which represents a sinusoidal wave of the free surface traveling in the x direction with phase velocity ω/k. The constant k is called the **wavenumber** and ω the **angular frequency**, and $\lambda = 2\pi/k$ is the **wavelength** (Fig. 6.3). The amplitude $f(z)$ is a function of z to be determined.

Substituting (6.17) in the Laplace equation (6.15), and dropping off the common factor $\sin(kx - \omega t)$, we obtain

$$f''(z) - k^2 f(z) = 0.$$

Its general solution takes the following form,

$$f(z) = Be^{kz} + Ce^{-kz} \quad (B, C : \text{constants}). \qquad (6.18)$$

The boundary condition (6.16) requires $C = 0$. Therefore,

$$\Phi = Be^{kz} \sin(kx - \omega t). \qquad (6.19)$$

This must satisfy the condition (6.14). Using (6.19), we find

$$\omega^2 = gk, \qquad (6.20)$$

which represents a relation to be satisfied by the frequency and wavenumber. Such a relation is called the **dispersion relation**.

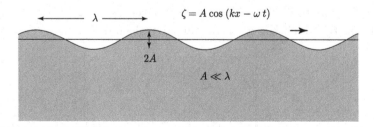

Fig. 6.3. Traveling wave.

The surface form is given by

$$\zeta = A\cos(kx - \omega t), \tag{6.21}$$

$$A = (k/\omega)B, \quad A, B > 0, \tag{6.22}$$

from (6.12) and (6.20). This is a surface wave generated by gravity and often called the **surface gravity wave**, or simply *gravity wave*.[1]

The velocity of a fluid particle is found from derivatives of Φ:

$$
\begin{aligned}
u &= \Phi_x = kBe^{kz}\cos(kx - \omega t), \\
w &= \Phi_z = kBe^{kz}\sin(kx - \omega t), \quad (\omega A = kB).
\end{aligned}
\tag{6.23}
$$

The velocities u and w decay rapidly owing to the factor e^{kz} as the depth $|z|$ ($z < 0$) increases. Thus the particle motion is practically limited in the neighborhood of surface, permitting the name *surface wave*.

Fixing (x, z) to a point (x_0, z_0), we obtain $u^2 + w^2 = (kBe^{kz_0})^2 =$ const. Namely, the velocity vector rotates along a circle of radius $|kB|\, e^{kz_0}$ in the (u, w) plane for a fixed point (x_0, z_0).

6.3.3.   *Meaning of small amplitude*

Let us estimate the linear approximation made in (Sec. 6.3.1) for the traveling wave. The term $\frac{1}{2}q^2$ at $z = 0$ is from (6.23)

$$\frac{1}{2}q^2 = \frac{1}{2}(\Phi_x^2 + \Phi_z^2) = \frac{1}{2}k^2B^2 = \frac{1}{2}\omega^2A^2,$$

which was neglected. On the other hand, the two terms in the boundary condition (6.7) are of the form $\mp gA\cos(kx - \omega t)$. Hence, the condition by which the q^2 term can be neglected is

$$\frac{\frac{1}{2}\omega^2A^2}{gA} = \frac{1}{2}kA = \pi\frac{A}{\lambda} \ll 1$$

---

[1]This must be discriminated from the gravity wave in the general relativity theory for gravitation.

from (6.20) where $\lambda = 2\pi/k$. Hence, the condition for the linear approximation to be valid is that the wave amplitude A is much less than the wavelength λ.

In the kinematic condition (6.10), a term of form $u\zeta_x$ was neglected. Its magnitude is of order $k^2 AB = k\omega A^2$ from (6.21) and (6.23). On the other hand, the linear terms ζ_t and w retained are of form $\omega A \sin(kx - \omega t)$. The condition for linear approximation to be valid is still the same as before, $kA \ll 1$, stating that the wave amplitude should be much smaller than the wavelength.

6.3.4. *Particle trajectory*

Let us consider the trajectories of fluid particles moving in the traveling wave (6.21). Suppose that a particle position is denoted by $(X(t), Z(t))$. Assuming that its deviation from the mean position (x_0, z_0) is small, the variables x and z in the expressions of u and w are replaced by x_0 and z_0, respectively. Then the equation of motion of the particle is given by

$$\frac{dX}{dt} = u(x_0, z_0, t), \quad \frac{dZ}{dt} = w(x_0, z_0, t).$$

Using (6.23) and $kB = \omega A$, these equations can be integrated, giving

$$X(t) - x_0 = -A e^{kz_0} \sin(kx_0 - \omega t),$$
$$Z(t) - z_0 = A e^{kz_0} \cos(kx_0 - \omega t).$$

$$(6.24)$$

It is found that the particle moves around a circle centered at (x_0, z_0) and that its radius $A e^{kz_0}$ decays rapidly as the depth $|z_0|(z_0 < 0)$ increases. Most part of the kinetic energy is included within the depth of a fourth of a wavelength under the surface ($|z_0|/\frac{1}{4}\lambda = k|z_0|/\frac{1}{2}\pi < 1$).

6.3.5. *Phase velocity and group velocity*

The surface elevation of the form $\zeta = A\cos(kx - \omega t)$ represents a traveling wave. This is understood from the phase $kx - \omega t$ of the cos-function. Requiring that the phase is equal to a fixed value ϕ_0, we have $kx - \omega t = \phi_0$, from which we obtain

$$\left(\frac{\mathrm{d}x}{\mathrm{d}t}\right)_{\phi_0=\text{const.}} = \frac{\omega}{k} =: c_\mathrm{p}. \tag{6.25}$$

This means that the same value of elevation $\zeta_0 = A\cos\phi_0$ moves with the speed $c_\mathrm{p} = \omega/k$ in the x direction. The velocity c_p is termed the **phase velocity**. In the case of water waves, we have another important velocity. That is, **group velocity** c_g is defined by the following:

$$c_\mathrm{g} := \frac{\mathrm{d}\omega}{\mathrm{d}k}. \tag{6.26}$$

In general, wave motion is characterized by a general *dispersion relation*,

$$\omega = \omega(k). \tag{6.27}$$

This functional relation immediately gives c_p and c_g defined above.

In water waves characterized by the dispersion relation (6.20), we have

$$\omega(k) = \sqrt{gk}. \tag{6.28}$$

This leads to (recall $\lambda = 2\pi/k$ the wavelength)

$$c_\mathrm{p}(k) = \omega/k = \sqrt{g/k} = \sqrt{g\lambda/2\pi}, \tag{6.29}$$

$$c_\mathrm{g}(k) = \mathrm{d}\omega/\mathrm{d}k = \frac{1}{2}\sqrt{g/k} = \frac{1}{2}c_\mathrm{p}. \tag{6.30}$$

The two velocities are interpreted as follows. The phase velocity denotes the moving speed of a phase ϕ_0 of a component with wavenumber k and frequency ω. On the other hand, the group velocity $c_\mathrm{g}(k)$ is given various interpretations of different physical significance:

(a) Speed of a *wave packet*, a group of waves having a wavenumber k_0 and wavenumbers of its immediate neighborhood (Problem 6.3).
(b) Speed of transport of *wave energy* of wavenumber k.

(c) Suppose that initial disturbance is localized spatially. Then, $d\omega/dk$ is the *traveling speed of k component* characterized by wavelength $\lambda_k = 2\pi/k$.

It is remarkable that waves of larger wavelength λ moves faster than shorter ones in deep water, according to (6.29) of phase velocity. Suppose that the initial wave is composed of wave components of various wavelengths. Because the phase velocities are different for different wavelengths, *each component of the wave moves with different speeds*. Therefore, the total wave form obtained by superposition of those components *deforms* in the course of time. This indicates that the waves disperse. Namely, the waves are **dispersive** if the phase velocity depends on the wavenumber k.

Water wave is a familiar example of dispersive waves. An example of nondispersive waves is the sound wave, which is investigated in Sec. 6.6.

6.4. Surface waves on water of a finite depth

Here we consider surface gravity waves at a finite depth of water, a more realistic problem. The bottom is assumed to be horizontal at a depth h (Fig. 6.4). The boundary condition at the bottom is given by vanishing normal velocity,

$$\text{BC3}' : \quad w = \Phi_z = 0 \quad \text{at } z = -h. \tag{6.31}$$

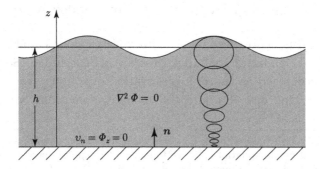

Fig. 6.4. Surface wave at a finite depth.

The general form of the velocity potential Φ is given by (6.17) and (6.18). After some calculations, we obtain the wave displacement ζ, velocity potential Φ, dispersion relation $\omega(k)$, and phase velocity c_p as follows:

$$\zeta = A \cos(kx - \omega t), \tag{6.32}$$

$$\Phi = B \cosh k(z + h) \sin(kx - \omega t), \tag{6.33}$$

$$\omega^2 = gk \tanh kh, \tag{6.34}$$

$$c_p^2 = (g/k) \tanh kh = (g\lambda/2\pi) \tanh(2\pi h/\lambda) \tag{6.35}$$

It is seen that the phase velocity c_p tends to $\sqrt{gh}$ as the wavelength λ increases since $\tanh(2\pi h/\lambda)$ is approximated with $2\pi h/\lambda$ for small $2\pi h/\lambda$.

If the water becomes shallower so that $kh = 2\pi h/\lambda \ll 1$ is satisfied, we obtain

$$c_p \approx \sqrt{gh} \left(1 - \frac{1}{3}(kh)^2 + \cdots \right)^{1/2} = \sqrt{gh} \left(1 - \frac{1}{6}(kh)^2 + \cdots \right), \tag{6.36}$$

since we have $\tanh kh = kh - \frac{1}{3}(kh)^3 + \cdots$ for small kh.

Namely, the wave speed is given approximately as $\sqrt{gh}$ in shallow water, and becomes slower as the depth decreases. This property is applied to explain the fact that crests of sea water wave near a coast become parallel to the coast line as the waves approach a coast, because the wave speed slows down according to $\sqrt{gh}$, as the depth decreases.[2]

6.5.   KdV equation for long waves on shallow water

John Scott Russel observed large solitary waves along canals between Glasgow and Edinburgh (or and Ardrossan) of Scotland in 1834, which are now recognized as the first observation of solitary waves, called the *soliton*. One day (*the happiest* to him), something unexpected

---

[2]Lagrange (1782) solved Eq. (6.15) for water waves at shallow water of depth h, and obtained the dispersion relation $\omega^2 = gk \tanh kh$ and the phase speed $c_p = \sqrt{gh}$ on the basis of the boundary conditions (6.12) and (6.14). Laplace also obtained $\omega^2 = gk \tanh kh$ earlier in 1776 by solving the *Laplace equation* (6.15) under a different formulation [Dar05].

happened. He was on a vessel moving at a high velocity in order to understand an anomalous decrease of resistance as a young engineer of naval architecture. He observed that a large wave was generated when it stopped suddenly. He immediately left the vessel and got on horseback. The wave propagated a long distance without change of its form. He then confirmed it was in fact a *large, solitary, progressive wave* [Dar05].

Later, both Boussinesq (1877) and Korteweg and de Vries (1895) (apparently unaware of the Boussinesq's study [Dar05]) succeeded in deriving an equation allowing stationary advancing waves without change of form, i.e. solutions which do not show breakdown at a finite time. In the problem of long waves in a shallow water channel of depth h_*, it is important to recognize that there are two dimensionless parameters which are small:

$$\alpha = \frac{a_*}{h_*}, \quad \beta = \left(\frac{h_*}{\lambda}\right)^2, \tag{6.37}$$

where a_* is a wave amplitude and λ is a representative horizontal scale characterizing the wave width.

In order to derive the equation allowing permanent waves (traveling without change of form), it is assumed that $\alpha \approx \beta \ll 1$. Performing a systematic estimation of order-of-magnitudes under such conditions, one can derive the following equation,

$$\partial_\tau u + \frac{3}{2} u \partial_\xi u + \frac{1}{6} \partial_\xi^3 u = 0 \tag{6.38}$$

(see [Ka04, Ch. 5 & App. G] for its derivation), where

$$\xi = \left(\frac{\alpha}{\beta}\right)^{1/2} \frac{x - c_* t}{\lambda}, \quad \tau = \left(\frac{\alpha^3}{\beta}\right)^{1/2} \frac{c_* t}{\lambda}. \tag{6.39}$$

The function $u(x, t)$ denotes not only the surface elevation normalized by a_*, but also the velocity (normalized by ga_*/c_*),

$$u = dx_p/dt, \tag{6.40}$$

of the water particle with its location at $x = x_p(t)$.

One of the characteristic features is the existence of the third order derivative term $\frac{1}{6}\partial_\xi^3 u$ in Eq. (6.38). A significance of this term is interpreted as follows. Linearizing Eq. (6.38) with respect to u, we obtain $\partial_\tau u + \alpha \partial_\xi^3 u = 0$ $\left(\text{where } \alpha = \frac{1}{6}\right)$. Assuming a wave form $u_w \propto \exp[i(\omega\tau - k\xi)]$ (the wavenumber k and frequency ω) and substituting it, we obtain a dispersion relation, $\omega = -\alpha k^3$. Phase velocity is defined as $c(k) := \omega/k = -\alpha k^2$. Namely, a small amplitude wave u_w propagates with the nonzero speed $c(k) = -\alpha k^2$, and the speed is different at different wavelengths $(= 2\pi/k)$. This effect was termed as *wave dispersion* in Sec. 6.3. What is important is that the new term takes into account the above *wave propagation*, in addition to the particle motion $\mathrm{d}x_p/\mathrm{d}t$ (a physically different concept).

Replacing u by $v = \frac{3}{2}u$, we obtain

$$\partial_\tau v + v\partial_\xi v + \frac{1}{6}\partial_\xi^3 v = 0. \tag{6.41}$$

This equation is now called the **KdV equation** after Korteweg and de Vries (1895). Equation (6.41) allows *steady* wave solutions, which they called the permanent wave. Setting $v = f(\xi - b\tau)$ (b: a constant) and substituting it into (6.41), we obtain $f''' + 6ff' - 6bf' = 0$. This can be integrated twice. Choosing two integration constants appropriately, one finds two wave solutions as follows:

$$v = A \operatorname{sech}^2 \left[\sqrt{\frac{A}{2}} \left(\xi - \frac{A}{3}\tau \right) \right] \quad \text{(solitary wave)}, \tag{6.42}$$

$$v = A \operatorname{cn}^2 \left[\sqrt{\frac{d}{2}} \left(\xi - \frac{c}{3}\tau \right) \right], \quad c = 2A - d, \tag{6.43}$$

where $\operatorname{sech} x \equiv 2/(e^x + e^{-x})$ and $\operatorname{cn} x \equiv \operatorname{cn}(\beta x, k)$ (Jacobi's elliptic function) with $\beta = \sqrt{d/2}$ and $k = \sqrt{a/d}$. The first solitary wave solution is obtained by setting two integration constants zero (and $b = A/3$) (Fig. 6.5). The second solution represents a periodic wave train called the *cnoidal wave*. These are called the **soliton** solutions.

6.6. Sound waves

In general, fluids are compressible, and the density changes with motion. So far, we have been interested in the flow velocity itself by assuming that density does not change. Here, we consider a sound wave which depends on the compressibility essentially.

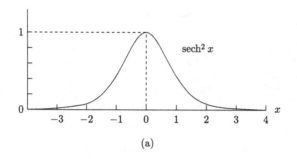

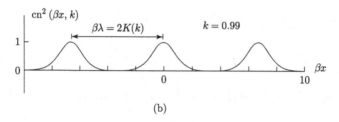

Fig. 6.5. (a) Solitary wave, and (b) cnoidal wave.

6.6.1. *One-dimensional flows*

In order to simplify the equations, we consider one-dimensional unsteady problems of an inviscid fluid. That is, the density ρ, pressure p, entropy s and velocity $\mathbf{v}$ are assumed to depend on the time t and a spatial coordinate x only. Furthermore, the velocity vector is assumed to have the x component u only: $\mathbf{v} = (u, 0, 0)$. Thus we have: $\rho(x,t)$, $p(x,t)$ and $u(x,t)$. In Problem 3.1, we considered three conservation equations of one-dimensional unsteady flows, which are reproduced here:

$$\text{Mass}: \ \rho_t + (\rho u)_x = 0, \tag{6.44}$$

$$\text{Momentum}: \ (\rho u)_t + (\rho u^2)_x + p_x = 0, \tag{6.45}$$

$$\text{Energy}: \ \left[\rho\left(\frac{1}{2}u^2 + e\right)\right]_t + \left[\rho u\left(\frac{1}{2}u^2 + h\right)\right]_x = 0. \tag{6.46}$$

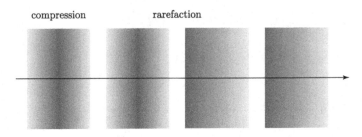

Fig. 6.6. Rarefaction and compression wave.

The continuity equation (6.44) is rewritten as

$$\partial_t \rho + \rho \partial_x u + u \partial_x \rho = 0 \,. \tag{6.47}$$

Euler's equation of motion (5.2) without external force ($\mathbf{f} = 0$) reduces to

$$\partial_t u + u \partial_x u = -\rho^{-1} \partial_x p \,, \tag{6.48}$$

which can be obtained also from (6.44) and (6.45). The entropy equation (5.3) becomes

$$\partial_t s + u \partial_x s = 0 \,. \tag{6.49}$$

This can be derived also from (6.46) with the help of (6.44), (6.48) and the thermodynamic relations.[3] These three are the governing equations of one-dimensional unsteady motions of an inviscid fluid.

6.6.2. *Equation of sound wave*

Sound is a wave of infinitesimal amplitude in a compressible fluid (Fig. 6.6). Suppose that the density, pressure and entropy of a uniform fluid at rest are denoted as ρ_0, p_0 and s_0 respectively, and that the uniform state is perturbed by a disturbance of infinitesimal

---

[3]$h = e + p\rho^{-1}$, with $dh = Tds + \rho^{-1}dp$ and $de = Tds - pd\rho^{-1}$.

amplitude. We denote the perturbed density and pressure by

$$\rho = \rho_0 + \rho_1, \quad p = p_0 + p_1,$$

where ρ_1 and p_1 are perturbations, satisfying $|\rho_1|/\rho_0 \ll 1$ and $|p_1|/p_0 \ll 1$.

The unperturbed velocity u_0 is zero, and $|u|$ is assumed so small that the nonlinear term $u\partial_x u$ in (6.48) can be neglected in the equation. Then, Eq. (6.48) can be approximated as

$$\partial_t u = -\frac{1}{\rho_0 + \rho_1}\,\partial_x p_1 \approx -\frac{1}{\rho_0}\,\partial_x p_1. \tag{6.50}$$

Likewise, the term ρu_x in the continuity equation is approximated as

$$\rho \partial_x u = (\rho_0 + \rho_1)\,\partial_x u \approx \rho_0 \partial_x u. \tag{6.51}$$

The third term of the continuity equation (6.47) becomes $u\partial_x \rho_1$, which is of the second order of smallness, therefore can be neglected. Thus, the continuity equation (6.47) becomes

$$\partial_t \rho_1 + \rho_0 \partial_x u = 0. \tag{6.52}$$

This sort of procedure is called *linearization*, since all the terms in the equations are linear with respect to the small terms ρ_1, p_1 or u. The third entropy equation becomes simply as

$$\partial_t s = 0. \tag{6.53}$$

That is, the entropy is unchanged with respect to time and takes the uniform value s_0 always.

Under the condition of constant entropy, we have a functional relation between ρ and p, because a thermodynamic state of a uniform gas of a single species is represented in terms of two thermodynamic variables and the entropy s is now constant. Namely, if the pressure p is expressed in general by ρ and s like $p(\rho, s)$, we have now an expression $p = p(\rho)$. In view of a differential relation $dp = (\partial p/\partial \rho)_s d\rho$, we have

$$p_1 = \left(\frac{\partial p}{\partial \rho}\right)_s \rho_1, \tag{6.54}$$

where $(\partial p/\partial\rho)_s$ denotes the partial differentiation of p with respect to ρ keeping s fixed. We define the value of $(\partial p/\partial\rho)_s$ at $\rho = \rho_0$ by c^2 which is a constant. Then we obtain

$$p_1 = c^2\rho_1, \tag{6.55}$$

$$c = \sqrt{(\partial p/\partial\rho)_s}. \tag{6.56}$$

Thus, the linearized equations (6.50) and (6.52) are written as

$$\rho_0\partial_t u + c^2\partial_x\rho_1 = 0, \tag{6.57}$$

$$\partial_t\rho_1 + \rho_0\partial_x u = 0. \tag{6.58}$$

Differentiating the first with respect to x and the second with t, and eliminating u between the two resulting equations, one obtains a second order differential equation,

$$\partial_t^2\rho_1 - c^2\partial_x^2\rho_1 = 0. \tag{6.59}$$

Similarly, interchanging x and t in the differentiations, and eliminating ρ_1 this time, one obtains the equation of u in the same form,

$$\partial_t^2 u - c^2\partial_x^2 u = 0. \tag{6.60}$$

Owing to (6.55), the pressure p_1 satisfies the same differential equation. This type of second order differential equation is called the **wave equation**.

There is a standard method to find the general solution to the wave equation. To that end, one introduces a pair of new variables by the following transformation,

$$\xi = x - ct, \quad \eta = x + ct.$$

Using the new variables ξ and η, the x derivatives are

$$\frac{\partial}{\partial x} = \frac{\partial\xi}{\partial x}\frac{\partial}{\partial\xi} + \frac{\partial\eta}{\partial x}\frac{\partial}{\partial\eta} = \frac{\partial}{\partial\xi} + \frac{\partial}{\partial\eta}$$

$$\frac{\partial^2}{\partial x^2} = \left(\frac{\partial}{\partial\xi} + \frac{\partial}{\partial\eta}\right)^2 = \frac{\partial^2}{\partial\xi^2} + 2\frac{\partial^2}{\partial\xi\partial\eta} + \frac{\partial^2}{\partial\eta^2},$$

where $\partial\xi/\partial x = 1$, $\partial\eta/\partial x = 1$ are used. Similarly, the t derivatives are

$$\frac{\partial}{\partial t} = \frac{\partial\xi}{\partial t}\frac{\partial}{\partial\xi} + \frac{\partial\eta}{\partial t}\frac{\partial}{\partial\eta} = -c\frac{\partial}{\partial\xi} + c\frac{\partial}{\partial\eta}$$

$$\frac{\partial^2}{\partial t^2} = c^2\left(-\frac{\partial}{\partial\xi} + \frac{\partial}{\partial\eta}\right)^2 = c^2\left(\frac{\partial^2}{\partial\xi^2} - 2\frac{\partial^2}{\partial\xi\partial\eta} + \frac{\partial^2}{\partial\eta^2}\right),$$

since $\partial\xi/\partial t = -c$, $\partial\eta/\partial t = c$. Substituting these into Eq. (6.59) for ρ_1, one obtains

$$\frac{\partial^2}{\partial\xi\partial\eta}\rho_1 = 0. \tag{6.61}$$

Integrating this equation with respect to η, we find $\partial\rho_1/\partial\xi = f(\xi)$, where $f(\xi)$ is an arbitrary function of ξ. Integrating again, we obtain $\rho_1 = f_1(\xi) + f_2(\eta)$. Thus, we have the d'Alembert's expression (1750):

$$\rho_1 = f_1(\xi) + f_2(\eta) = f_1(x - ct) + f_2(x + ct), \tag{6.62}$$

where $f_1(\xi) = \int^\xi f(\xi)d\xi$ is an arbitrary function of the variable ξ only, while $f_2(\eta)$ is an arbitrary function of η only. Both are to be determined by initial condition or boundary condition. It is readily verified that this satisfies Eq. (6.61). Same expressions can be given to u and p_1, too. This can be shown by noting that these are connected to ρ_1 with (6.55) and (6.57) (or (6.58)).

The solution just found can be easily understood as follows. Regarding the first term $f_1(x - ct)$, if the time t is fixed, f_1 is a function of x only, which represents a spatial wave form. Next, if the coordinate x is fixed, f_1 is a function of t only, which represents a time variation of ρ_1. Furthermore, it is important to recognize that the value of function f_1 is unchanged if the value $x - ct$ has the same value for different values of x and t.

For example, suppose that x is changed by $\Delta x = c\Delta t$ when the time t advanced by Δt, then we have $x + \Delta x - c(t + \Delta t) = x - ct$. Hence the value of f_1 is unchanged. This means that the value of f_1 is moving forward with the velocity $\Delta x/\Delta t = c$ in the positive direction of x. Namely, the wave form $f_1(x - ct)$ is moving with the velocity c at all points on x since c is a constant.

This is understood to mean that the function $f_1(x - ct)$ represents *propagation* of a wave of speed c, which is the **sound speed**.

Likewise, it is almost evident that the second term $f_2(x + ct)$ represents propagation of a wave in the negative x direction ($\Delta x = -c\Delta t$). Thus, it is found that Eq. (6.59) or (6.60) is a differential equation representing waves propagating in the positive or negative x directions with the speed c. The wave is called **longitudinal** because the variable u expresses fluid velocity in the same x direction as the wave propagation.

Provided that the propagation direction is chosen as that of positive x, the velocity, density and pressure are related by simple useful relations. Suppose that

$$\rho_1 = f(x - ct).$$

In view of $\partial \rho_1 / \partial x = f'(x - ct)$, Eq. (6.57) gives

$$\rho_0 \partial_t u = -c^2 f'(x - ct), \quad \text{hence } u = (c/\rho_0) f(x - ct).$$

Therefore, the velocity and density variation are related by

$$u = \frac{c}{\rho_0} \rho_1. \tag{6.63}$$

Furthermore, using (6.55), we have

$$u = \frac{1}{\rho_0 c} p_1. \tag{6.64}$$

In the case of an *ideal gas*, the sound speed c is proportional to the square root of temperature T. In fact, using the adiabatic relation where $s = $ const.,

$$p/p_0 = (\rho/\rho_0)^\gamma,$$

the definition of the sound speed (6.56) leads to

$$c^2 = \left(\frac{\partial p}{\partial \rho} \right)_s = \gamma p_0 \frac{\rho^{\gamma-1}}{\rho_0{}^\gamma} = \gamma \frac{p}{\rho} = \gamma \frac{RT}{\mu_m} \tag{6.65}$$

where the equation of state $p = \rho RT/\mu_m$ is used with T as the temperature, μ_m the molecular weight and R the gas constant (see the footnote in Sec. 1.2 and Appendix C).

Historically, in early times, Newton (1986) gave the expression $c^2 = p/\rho$ (an isothermal relation, obtained by an *ad hoc* model). Later, Laplace (1816) improved it by giving the above form, and found a good agreement with experiments. The sound speed is $c = 340.5\,\text{m/s}$ in air at $15°\text{C}$ and $1\,\text{atm}$, whereas it is about $1500\,\text{m/s}$ in water.

6.6.3. *Plane waves*

Let us write a solution of the wave equation (6.60) in a form of complex representation,[4]

$$u = \bar{u}(x)\,e^{-i\omega t} = \bar{u}(x)\,(\cos \omega t - i \sin \omega t), \qquad (6.66)$$

where i is the imaginary unit (i.e. $i^2 = -1$) and ω the angular frequency (assumed real). Substituting this into (6.60), using the relation $\partial_t^2 u = -\omega^2 u$, and omitting the common factor $e^{-i\omega t}$, we obtain an equation for the complex amplitude $\bar{u}(x)$,

$$\bar{u}_{xx} + \frac{\omega^2}{c^2}\,\bar{u} = 0.$$

A general solution to this differential equation is represented as

$$\bar{u}(x) = A \exp[i(\omega/c)x] + B \exp[-i(\omega/c)x], \qquad (6.67)$$

where A, B are arbitrary constants. The real part (Re) of the complex expressions is considered to represent the sound wave in the real world:

$$u(x,t) = \text{Re}\big[\bar{u}(x)\,e^{-i\omega t}\big]$$

The imaginary part also satisfies the wave equation.

Such a type of solution as (6.66) characterized by a single frequency ω is called a *monochromatic wave*, according to optics. Furthermore, if the solution (6.67) is substituted, it is found that u of (6.66) becomes

$$u(x,t) = A \exp[i(\omega/c)(x - ct)] + B \exp[-i(\omega/c)(x + ct)],$$

which has the form of (6.62).

---

[4]See (5.49) for the definition of a complex number z by $z = re^{i\theta}$.

Setting $A = ae^{i\alpha}$ (where a, α are real constants) and taking the real part of the first term of the above expression (assuming $B = 0$), we obtain

$$u_r(x, t) = a\,\cos(kx - \omega t + \alpha) = a\,\cos\phi, \qquad (6.68)$$

$$k = \omega/c, \quad \phi := kx - \omega t + \alpha \qquad (6.69)$$

where a is the wave amplitude, and ϕ the phase of the wave. In addition, $k = \omega/c$ is the wavenumber. This is due to the property that the phase ϕ of the above wave increases by $2\pi k$ when x increases by 2π. That is, the number of wave crests is k in the x-interval 2π. The wavelength is given by $\lambda = 2\pi/k$.

The wave we have investigated so far is regarded as a **plane wave** in the (x, y, z) three-dimensional space, because $x = $ const. describes a plane perpendicular to the x axis. The wave of the form (6.68) is called a *monochromatic plane wave*.

A complex monochromatic plane wave is given by

$$u_k(x, t) = A_k \exp[i(kx - \omega t)] = A_k \exp[ik(x - ct)]. \qquad (6.70)$$

The theory of *Fourier series* and *Fourier integrals* places a particular significance at this representation, because arbitrary waves are represented by superposition of monochromatic plane waves of different wavenumbers. So, the monochromatic plane waves are regarded as Fourier components, or spectral components.

The relation (6.69) between the wavenumber k and frequency ω is the dispersion relation considered in Sec. 6.3.5. Using the above dispersion relation $k = \omega/c$ of the sound wave, we can calculate the phase velocity c_{p} and the group velocity c_{g}:

$$c_{\mathrm{p}} = \frac{\omega}{k} = c, \quad c_{\mathrm{g}} = \frac{d\omega}{dk} = c.$$

Namely, in the sound wave, the phase velocity is the same at all wavelengths, and in addition, it is equal to the group velocity. Owing to this property, the wave form of sound waves is unchanged during propagation. This property is an advantage if it is used as a medium of communication. Interestingly, light has the same nondispersive property, and light speed is the same for all wavelengths in vacuum space.

6.7. Shock waves

Shock waves are formed around a body placed in a supersonic flow, or formed by strong impulsive pressure increases which act on a surface surrounding a fluid. Shocks are sometimes called *discontinuous surfaces* because velocity, pressure and density change discontinuously across the surface of a shock wave. So far, we have considered continuous fields which are represented by continuous differentiable functions of position $\mathbf{x}$ and time t, and the governing equations are described by partial differential equations. Let us investigate what is the circumstance when discontinuity is allowed.

Fluid motions are governed by three conservation laws of mass, momentum and energy considered in Chapter 3. These are basic constraints to be satisfied by fluid motions. Even when there exists a discontinuous surface, if these conservation laws are not violated, it should be allowed to exist physically. For example, mass conservation is satisfied if the rate of inflow of fluid into the discontinuous surface from one side is the same as the rate of outflow from the other side (Fig. 6.7).

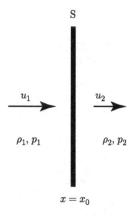

Fig. 6.7. Discontinuous surface.

In the one-dimensional problem considered in Sec. 6.6.1, the discontinuous surface should be a plane perpendicular to the x axis. Suppose that a discontinuous surface S, i.e. a shock wave, is at rest at position $x = x_0$ (Fig. 6.7), and that the fluid is flowing into the surface S from the left with velocity u_1 and flowing out of S to the right with velocity u_2. It is assumed that the states on both sides of S are steady and uniform, and that the density and pressure are ρ_1 and p_1 on the left and ρ_2 and p_2 on the right, respectively. The mass flux into the surface S from the left is $\rho_1 u_1$ per unit time and per unit area of S. The outflux is expressed by $\rho_2 u_2$. Therefore, the mass conservation is given by $\rho_1 u_1 = \rho_2 u_2$. Precisely speaking, the equation must be considered on the basis of the conservation equations.

Because the problem under consideration is *steady*, the time derivative term ∂_t in the equation of mass conservation (6.44) vanishes. Then we have $\partial_x(\rho u) = 0$, giving $\rho u = \text{const.}$ This leads to

$$\rho_1 u_1 = \rho_2 u_2. \tag{6.71}$$

This is equivalent to the relation obtained in the above consideration.

Similarly, neglecting the time derivative terms in the conservation equations of momentum (6.45) and energy (6.46), we obtain the following:

$$\rho_1 u_1^2 + p_1 = \rho_2 u_2^2 + p_2, \tag{6.72}$$

$$\frac{1}{2}u_1^2 + \frac{c_1^2}{\gamma - 1} = \frac{1}{2}u_2^2 + \frac{c_2^2}{\gamma - 1} \tag{6.73}$$

where the enthalpy h has been replaced by the expression of an ideal gas, $h = c^2/(\gamma - 1)$, given as (C.6)

Introducing J by

$$J = \rho_1 u_1 = \rho_2 u_2, \tag{6.74}$$

Eq. (6.72) leads to the following two expressions:

$$J^2 = (p_2 - p_1)\frac{\rho_1 \rho_2}{\rho_2 - \rho_1}, \tag{6.75}$$

$$u_1 - u_2 = \frac{p_2 - p_1}{J} = \sqrt{(p_2 - p_1)\left(\frac{1}{\rho_1} - \frac{1}{\rho_2}\right)}. \tag{6.76}$$

Furthermore, using (6.72) and (6.74), the third equation (6.73) results in

$$\frac{c_1^2}{\gamma - 1} - \frac{c_2^2}{\gamma - 1} = \frac{1}{2}(p_1 - p_2)\left(\frac{1}{\rho_1} + \frac{1}{\rho_2}\right). \tag{6.77}$$

These relations connecting the two states of upstream and downstream of a discontinuity surface are called *Rankin–Hugoniot*'s relation (Rankin (1870) and Hugoniot (1885)), known as the *shock adiabatic*.

For given ρ_1 and p_1, the relations (6.76) and (6.77) determine ρ_2 and p_2 with the help of the relation $c^2 = \gamma p/\rho$ of (C.2), depending on the parameters J and γ.

6.8. Problems

Problem 6.1 One-dimensional finite amplitude waves

We consider one-dimensional finite amplitude waves on the basis of the continuity equation (6.47) and the equation of motion (6.48).

(i) Using the isentropic relation (6.55) and rewriting (6.47) in terms of pressure p and velocity u, derive the following system of equations:

$$\partial_t u + \frac{1}{\rho c}\partial_t p + (u + c)\left(\partial_x u + \frac{1}{\rho c}\partial_x p\right) = 0, \tag{6.78}$$

$$\partial_t u - \frac{1}{\rho c}\partial_t p + (u - c)\left(\partial_x u - \frac{1}{\rho c}\partial_x p\right) = 0. \tag{6.79}$$

(ii) The two equations of (i) can be written as

$$[\partial_t + (u + c)\partial_x]J_+ = 0, \tag{6.80}$$
$$[\partial_t + (u - c)\partial_x]J_- = 0. \tag{6.81}$$

Determine the two functions J_+ and J_-.

(iii) The functions $J_\pm$ are called the *Riemann invariants*. Explain why and how $J_\pm$ are invariant.

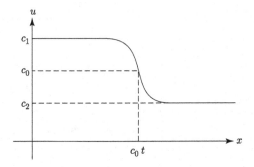

Fig. 6.8.   A weak shock wave.

Problem 6.2 *Burgers equation*

The following *Burgers equation* for $u(x, t)$ can represent a structure of weak shock wave (Fig. 6.8):

$$u_t + uu_x = \nu u_{xx}. \tag{6.82}$$

Assuming a solution of the steady progressing wave $u(x - c_0 t)$ with a constant c_0, determine a solution which satisfies the conditions: $u \to c_1$ as $x \to -\infty$ and $u \to c_2$ as $x \to +\infty$, where c_1, c_2 are constants and $c_1 > c_2 > 0$.

Problem 6.3 *Wave packet and group velocity* c_g

Suppose that we have a linear system of waves characterized by a dispersion relation $\omega = \omega(k)$ for the frequency ω and wavenumber k (see Sec. 6.3.5), and that there is a traveling wave solution of the form $\zeta(x, t) = Ae^{i(kx - \omega t)}$. In view of the assumed linearity of the wave system, a general solution is represented in the form of the following integral (a Fourier representation):

$$\zeta(x, t) = \int_{-\infty}^{\infty} A(k) e^{i(kx - \omega t)} \mathrm{d}k. \tag{6.83}$$

Consider a wavemaker which oscillates at a single frequency ω_0. Its amplitude first increased from zero to a maximum and then returned to zero again, slowly with a time scale much larger than the oscillation period $2\pi/\omega_0$. By this wave excitation, it is found that most of the wave energy is concentrated on a narrow band of wavenumbers around k_0. Hence, the dispersion relation is approximated by the following linear relation:

$$\omega(k) = \omega_0 + c_g(k - k_0), \quad \omega_0 = \omega(k_0), \quad c_g = d\omega/dk. \quad (6.84)$$

(The amplitude $A(k)$ is regarded as zero for such k-values in which the above linear relation loses its validity.)

(i) Show that the resulting wave would be given by the following form of a *wave packet*, with $\xi = x - c_g t$:

$$\zeta(x, t) = F(\xi) e^{i(k_0 x - \omega_0 t)}. \quad (6.85)$$

In addition, write down Fourier representation of the amplitude function $F(\xi)$.

(ii) When the Fourier amplitude is $A(k) = A_0 \exp[-a(k - k_0)^2]$, a Gaussian function around k_0, give an explicit form of the wave packet.

The function $F(\xi)$ is an envelope moving with the group velocity c_g and enclosing carrier waves $e^{i(k_0 x - \omega_0 t)}$ within it (Fig. 6.9).

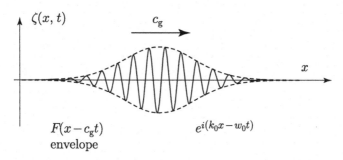

Fig. 6.9. Wave packet and group velocity c_g.

Chapter 7

Vortex motions

Vortex motions are vital elements of fluid flows. In fact, most dynamical aspects of fluid motions are featured by vortex motions. Already we considered some motions of vortices: vortex sheet and shear layer [Sec. 4.6.2(b)], vortex line and point vortex (Secs. 5.6.5 and 5.8.5), etc.

Analytical study of vortex motions is mostly based on vorticity which is defined as twice the angular velocity of local rotation (Sec. 1.4.3). The vorticity equation is derived in Sec. 4.1 from the equation of motion. It is to be remarked that the no-slip condition and the boundary layer are identified as where vorticity is generated (Sec. 4.5). In Chapter 12, it will be shown that vorticity is in fact a gauge field associated with *rotational symmetry* of the flow field, on the basis of the gauge theory of modern theoretical physics.

7.1. Equations for vorticity

7.1.1. *Vorticity equation*

Equation of the vorticity $\boldsymbol{\omega} = \mathrm{curl}\,\mathbf{v}$ was given for *compressible* flows in Sec. 3.4 in an inviscid fluid ($\nu = 0$) as

$$\partial_t\,\boldsymbol{\omega} + \mathrm{curl}\,(\boldsymbol{\omega} \times \mathbf{v}) = 0. \tag{7.1}$$

For *viscous incompressible* flows, two equivalent equations given in Sec. 4.1 are reproduced here:

$$\partial_t\boldsymbol{\omega} + \mathrm{curl}\,(\boldsymbol{\omega} \times \mathbf{v}) = \nu\nabla^2\boldsymbol{\omega}, \tag{7.2}$$

$$\partial_t\boldsymbol{\omega} + (\mathbf{v} \cdot \nabla)\boldsymbol{\omega} = (\boldsymbol{\omega} \cdot \nabla)\mathbf{v} + \nu\nabla^2\boldsymbol{\omega}. \tag{7.3}$$

7.1.2. *Biot–Savart's law for velocity*

Let us consider a *solenoidal* velocity field $\mathbf{v}(\mathbf{x})$ (defined by div $\mathbf{v} = 0$, i.e. incompressible) induced by a *compact vorticity field* $\boldsymbol{\omega}(\mathbf{x})$. That is, the vorticity vanishes out of a bounded open domain D:

$$\boldsymbol{\omega}(\mathbf{x}) \begin{cases} \neq 0, & \text{for } \mathbf{x} \in D \\ = 0, & \text{for } \mathbf{x} \text{ out of } D. \end{cases} \tag{7.4a}$$

It is to be noted in this case that the following space integral of each component $\omega_k(\mathbf{y})$ vanishes for $k = 1, 2, 3$:

$$\int_D \omega_k(\mathbf{y}) \mathrm{d}^3\mathbf{y} = \int_D \nabla_y \cdot (y_k \boldsymbol{\omega}) \mathrm{d}^3\mathbf{y} = \int_S y_k \omega_n \mathrm{d}S = 0, \tag{7.4b}$$

because the normal component $\omega_n = 0$ should vanish on the bounding surface S of D, where $\nabla_y \cdot (y_k \boldsymbol{\omega}) = \partial_i(y_k \omega_i) = \omega_k + y_k \partial_i \omega_i$ is used together with $\partial_i \omega_i = 0$.

Introducing a vector potential $\mathbf{A}(\mathbf{x})$ defined by $\mathbf{v} = \operatorname{curl} \mathbf{A}$, the solenoidal velocity $\mathbf{v}$ is expressed as

$$\mathbf{v}(\mathbf{x}, t) = \operatorname{curl} \mathbf{A}, \quad \mathbf{A}(\mathbf{x}, t) = \frac{1}{4\pi} \int_D \frac{\boldsymbol{\omega}(\mathbf{y}, t)}{|\mathbf{x} - \mathbf{y}|} \mathrm{d}^3\mathbf{y}, \tag{7.5a}$$

where $\mathbf{y}$ denotes a position vector for the integration (with t fixed). It can be shown that $\operatorname{curl} \mathbf{v} = \boldsymbol{\omega}$ in the free space for the compact vorticity field $\boldsymbol{\omega}(\mathbf{x})$ of (7.4a) (Problem 7.1).

The velocity $\mathbf{v}$ is derived from the vector potential $\mathbf{A}$ by taking its curl:

$$\mathbf{v}(\mathbf{x}) = \operatorname{curl} \mathbf{A} = -\frac{1}{4\pi} \int \frac{(\mathbf{x} - \mathbf{y}) \times \boldsymbol{\omega}(\mathbf{y})}{|\mathbf{x} - \mathbf{y}|^3} \mathrm{d}^3\mathbf{y}, \tag{7.5b}$$

where the formula $\nabla \times (\mathbf{a}/|\mathbf{x}|) = \mathbf{a} \times \mathbf{x}/|\mathbf{x}|^3$ ($\mathbf{a}$ is a constant vector) is used. This gives the velocity field when the vorticity distribution $\boldsymbol{\omega}(\mathbf{y})$ is known, called the *Biot–Savart law*. Originally, this law was given for a magnetic field (in place of $\mathbf{v}$) produced by steady electric current (in place of $\boldsymbol{\omega}$) of volume distribution in the electromagnetic theory.

As $|\mathbf{x}| \to \infty$, the velocity (7.5b) has an asymptotic behavior:

$$\mathbf{v}(\mathbf{x}) = -\frac{1}{4\pi|\mathbf{x}|^3}\mathbf{x} \times \left(\int \boldsymbol{\omega}(\mathbf{y})\mathrm{d}^3\mathbf{y} + \mathbf{O}(|\mathbf{x}|^{-1}) \right) = O(|\mathbf{x}|^{-3}), \quad (7.5c)$$

owing to (7.4b).

7.1.3. *Invariants of motion*

When the fluid is inviscid and of constant density and the vorticity is governed by the vorticity equation (7.1), we have four invariants of motion:

$$\text{Energy}: \quad K = \frac{1}{2}\int_{V_\infty} \mathbf{v}^2 \mathrm{d}^3\mathbf{x} \tag{7.6}$$

$$= \frac{1}{2}\int_D \boldsymbol{\omega}(\mathbf{x},t) \cdot \mathbf{A}(\mathbf{x},t)\,\mathrm{d}^3\mathbf{x} \tag{7.7}$$

$$= \int_D \mathbf{v} \cdot (\mathbf{x} \times \boldsymbol{\omega})\,\mathrm{d}^3\mathbf{x}, \quad [\text{Lamb32, Sec. 153}]$$
$$\tag{7.8}$$

$$\text{Impulse}: \quad \mathbf{P} = \frac{1}{2}\int_D \mathbf{x} \times \boldsymbol{\omega}(\mathbf{x},t)\,\mathrm{d}^3\mathbf{x}, \tag{7.9}$$

$$\text{Angular impulse}: \quad \mathbf{L} = \frac{1}{3}\int_D \mathbf{x} \times (\mathbf{x} \times \boldsymbol{\omega})\,\mathrm{d}^3\mathbf{x}, \tag{7.10}$$

$$\text{Helicity}: \quad H = \int_D \mathbf{v} \cdot \boldsymbol{\omega}\,\mathrm{d}^3\mathbf{x}, \tag{7.11}$$

where div $\mathbf{v} = 0$, and V_∞ is unbounded space of inviscid fluid flow. Here, we verify invariance of the total kinetic energy K for fluid motion under the compact vorticity distribution (7.4a). Alternative expressions of $\mathbf{P}$ and $\mathbf{L}$ are given in the end.

To show $K = $ const. for an inviscid fluid under $\partial_k v_k = 0$ and (7.4a), we take scalar product of v_i and (3.14) with $\rho = $ const. and

$f_i = 0$, and obtain

$$\partial_t \left(\frac{v^2}{2} \right) = -\partial_k \left(v_k \frac{v^2}{2} \right) - \partial_k \left(v_k \frac{p}{\rho} \right), \qquad (7.5\text{d})$$

since $\partial_t(v^2/2) = v_i \partial_t v_i$ and $\partial_k(v_k v^2/2) = v_i v_k \partial_k v_i$. Integrating this over a bounded space V, the right-hand side is transformed to surface integrals. If the surface S recedes to infinity (i.e. $V \to V_\infty$) and the pressure tends to a constant value (assumed reasonably), the velocity decays as $O(|\mathbf{x}|^{-3})$ given by (7.5c), whereas the surface grows as $O(|x|^2)$. Hence both of the surface integrals vanish. Thus, we have $dK/dt = 0$, the invariance of the total kinetic energy K.

Setting $\mathbf{v} = \nabla \times \mathbf{A}$ for the second $\mathbf{v}$ of $v^2 = \mathbf{v} \cdot \mathbf{v}$, the expression (7.6) becomes

$$K = \frac{1}{2} \int_{V_\infty} \mathbf{v} \cdot (\nabla \times \mathbf{A}) \mathrm{d}^3 \mathbf{x} = \frac{1}{2} \int_{V_\infty} (\nabla \times \mathbf{v}) \cdot \mathbf{A} \mathrm{d}^3 \mathbf{x},$$

by omitting integrated terms since $\mathbf{A} = O(|\mathbf{x}|^{-2})$ obtained by the same reasoning as in the case of (7.5c). This is (7.7) since $\nabla \times \mathbf{v} = \boldsymbol{\omega}$.

In order to show (7.8), first note the following identity:

$$(\mathbf{x} \times \boldsymbol{\omega})_i = (\mathbf{x} \times (\nabla \times \mathbf{v}))_i = x_k \partial_i v_k - x_k \partial_k v_i$$
$$= 2v_i + \partial_i(x_k v_k) - \partial_k(x_k v_i), \qquad (7.5\text{e})$$

where $\partial_i x_j = \delta_{ij}$ and $\partial_j x_j = 3$ are used to obtain the last equality, and in the second equality of the first line the vector identity (A.18) is applied. Hence, the integrand of (7.8) is given by

$$\mathbf{v} \cdot (\mathbf{x} \times \boldsymbol{\omega}) = v_i(\mathbf{x} \times \boldsymbol{\omega})_i = 2v^2 + v_i \partial_i(x_k v_k) - v_i \partial_k(x_k v_i)$$

$$= \frac{1}{2}v^2 + \partial_i(x_k v_k v_i) - \partial_k \left(x_k \frac{1}{2} v^2 \right), \qquad (7.5\text{f})$$

where $\partial_i v_i = 0$ and $v_i \partial_k(x_k v_i) = (3/2)v^2 + \partial_k(x_k \frac{1}{2}v^2)$ are used to obtain the last expression. Integrating (7.5f) over V_∞, the right-hand side results in $\frac{1}{2} \int v^2 \mathrm{d}^3 \mathbf{x}$, because the last two divergence terms vanish.

The impulse $\mathbf{P}$ of (7.9) is given by integral of (7.5e) multiplied by $\frac{1}{2}$ over V_∞ which reduces to an integral of $\mathbf{v}$ over V_∞, and the other two integrals are transformed to vanishing surface integrals.

Regarding the angular impulse $\mathbf{L}$, note the following identity:

$$\frac{1}{3}\mathbf{x} \times (\mathbf{x} \times \boldsymbol{\omega}) = -\frac{1}{2}|\mathbf{x}|^2\boldsymbol{\omega} - (1/6)\partial_k(\omega_k|\mathbf{x}|^2\mathbf{x}).$$

Omitting terms of surface integrals, we obtain alternative expression of $\mathbf{L}$:

$$\mathbf{L} = -\frac{1}{2}\int_D |\mathbf{x}|^2\boldsymbol{\omega}\mathrm{d}^3\mathbf{x}. \tag{7.4g}$$

Problem 7.2 questions how to verify invariance of $\mathbf{P}$, $\mathbf{L}$ and H of (7.9)–(7.11).

7.2. Helmholtz's theorem

7.2.1. *Material line element and vortex-line*

In an inviscid incompressible fluid where $\nu = 0$ and $\operatorname{div} \mathbf{v} = 0$, the vorticity equation (7.3) reduces to

$$\frac{\mathrm{D}}{\mathrm{D}t}\boldsymbol{\omega} = (\boldsymbol{\omega} \cdot \nabla)\mathbf{v}, \tag{7.12}$$

where $\mathrm{D}\boldsymbol{\omega}/\mathrm{D}t = \partial_t\boldsymbol{\omega} + (\mathbf{v} \cdot \nabla)\boldsymbol{\omega}$. In order to see the meaning of this equation, we consider the motion of an infinitesimal line element.

We choose two fluid particles located at sufficiently close points $\mathbf{x}$ and $\mathbf{x} + \delta\mathbf{s}_a$ at an instant, $\delta\mathbf{s}_a$ denoting the *material* line element connecting the two points. Rate of change of the line element vector $\delta\mathbf{s}_a$ is given by the difference of their velocities, that is

$$\frac{\mathrm{D}}{\mathrm{D}t}\delta\mathbf{s}_a = \mathbf{v}(\mathbf{x} + \delta\mathbf{s}_a) - \mathbf{v}(\mathbf{x}) = (\delta\mathbf{s}_a \cdot \nabla)\mathbf{v} + (|\delta\mathbf{s}_a|^2).$$

Neglecting the second and higher order terms, we have

$$\frac{\mathrm{D}}{\mathrm{D}t}\delta\mathbf{s}_a = (\delta\mathbf{s}_a \cdot \nabla)\mathbf{v}. \tag{7.13}$$

The similarity between this and (7.12) is obvious. Taking the line element $\delta s_a = |\delta s_a| e$ parallel to $\omega(x) = |\omega| e$ at a point x where e is a unit vector in the direction of ω, the right-hand sides of (7.12) and (7.13) can be written as $|\omega|(e \cdot \nabla)v$ and $|\delta s_a|(e \cdot \nabla)v$, respectively. Hence, we obtain the following equation:

$$\frac{1}{|\omega|} \frac{D}{Dt}\omega = \frac{1}{|\delta s_a|} \frac{D}{Dt}\delta s_a, \qquad (7.14)$$

since both terms are equal to $(e \cdot \nabla)v$. Thus, it is found that relative rate of change of ω following the fluid particle is equal to relative rate of stretching of a material line element parallel to ω.

7.2.2. Helmholtz's vortex theorem

Regarding the total flux of vortex-lines passing through a material closed curve, Kelvin's circulation theorem of Sec. 5.3 states that it is invariant with time for the motion of an ideal fluid of homentropy (subject to a conservative external force). This is valid for flows of compressible fluids.

For the case of variable density ρ, the vorticity equation is already given by (3.33):

$$\frac{D}{Dt}\left(\frac{\omega}{\rho}\right) = \left(\frac{\omega}{\rho} \cdot \nabla\right)v, \qquad (7.15)$$

which reduces to (7.12) when $\rho = $ const. This equation states that ω/ρ behaves like a material line element parallel to ω.

For the vortex motion governed by Eq. (7.15), one can deduce the following laws of vortex motion, originally given by Helmholtz (1858):

(i)   *Fluid particles initially free of vorticity remain free of vorticity thereafter.*

(ii) *The vortex-lines move with the fluid. In other words, fluid particles on a vortex-line at any instant will be on the vortex-line at all subsequent times.*

(iii) *Strength of an infinitesimal vortex tube defined by $|\omega|\theta$ does not vary with time during the motion, where θ is the infinitesimal cross-section of the vortex tube.*

This is called the *Helmholtz's vortex theorem* derived as follows.

Multiplying a constant λ on both sides of (7.13) and subtracting it from (7.15), we obtain

$$\frac{D}{Dt}\left(\frac{\boldsymbol{\omega}}{\rho} - \lambda\delta\mathbf{s}_a\right) = \left(\frac{\boldsymbol{\omega}}{\rho} - \lambda\delta\mathbf{s}_a\right)\cdot\nabla\mathbf{v}. \tag{7.16}$$

We choose a vortex-line, which is a space curve $\mathbf{x}_\sigma = (x(\sigma), y(\sigma), z(\sigma))$ parameterized with a variable σ. Suppose that the line element $\delta\mathbf{s}_a$ coincides with a local tangent line element at s, defined by

$$\delta\mathbf{s}_a \equiv \delta\mathbf{s}_\sigma = (x'(\sigma), y'(\sigma), z'(\sigma))\,d\sigma$$

for a differential $d\sigma$ (Fig. 7.1). With a fixed σ, one may define $\mathbf{X}_\sigma$ by

$$\frac{\boldsymbol{\omega}}{\rho} - \lambda\delta\mathbf{s}_\sigma|_\sigma = (\xi(t), \eta(t), \zeta(t)) \equiv \mathbf{X}_\sigma(t).$$

Then Eq. (7.16) is regarded as the equation for $\mathbf{X}_\sigma = (\xi, \eta, \zeta)$:

$$\begin{aligned}
\frac{d\xi}{dt} &= \left(\xi\partial_x + \eta\partial_y + \zeta\partial_z\right)u\,\big|_\sigma, \\
\frac{d\eta}{dt} &= \left(\xi\partial_x + \eta\partial_y + \zeta\partial_z\right)v\,\big|_\sigma, \\
\frac{d\zeta}{dt} &= \left(\xi\partial_x + \eta\partial_y + \zeta\partial_z\right)w\,\big|_\sigma,
\end{aligned} \tag{7.17}$$

where $\mathbf{v} = (u, v, w)$. Let us investigate the evolution of the system by following the material particles on the vortex-line with a specified value of σ and $d\sigma$.

Equation (7.17) is a system of ordinary differential equations for $\xi(t)$, $\eta(t)$ and $\zeta(t)$. Suppose that we are given the initial condition,

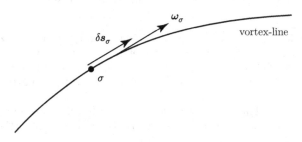

Fig. 7.1. A vortex-line $(x(\sigma), y(\sigma), z(\sigma))$ and a material line element $\delta\mathbf{s}_a$.

such that

$$\mathbf{X}_\sigma(0) = (\xi(0), \eta(0), \zeta(0)) = (0, 0, 0). \tag{7.18}$$

The above equations (7.17) give the (unique) solution,

$$\mathbf{X}_\sigma(t) = 0, \quad \text{at all } t > 0, \tag{7.19}$$

for usual smooth velocity fields $\mathbf{v}(\mathbf{x})$.[1] Hence, we obtain the relation,

$$\boldsymbol{\omega}(\mathbf{x}_\sigma, t) = \lambda \rho(\mathbf{x}_\sigma, t) \delta \mathbf{s}_\sigma(t), \tag{7.20}$$

for all subsequent times t. Therefore, the vortex-line moves with the fluid line element, and the material line element $\delta \mathbf{s}_\sigma$ forms always part of a vortex-line at $\mathbf{x}_\sigma$. This is the Helmholtz's law (ii).[2]

Thus, it is found that the length $\delta s = |\delta \mathbf{s}_\sigma|$ of the line element will vary as ω/ρ (since $\omega = \lambda \rho \delta s$), where $\omega = |\boldsymbol{\omega}|$. Using the cross-section θ of an infinitesimal vortex tube, the product $\rho \delta s \theta$ denotes the fluid mass of the line element which must be conserved during the motion. Since the product $\rho \delta s$ is proportional to ω, we obtain that the strength of the vortex defined by the product $\omega \theta$ is conserved during the motion. This verifies the Helmholtz's law (iii).

Regarding the law (i), we set $\lambda = 0$ since λ was an arbitrary constant parameter. Then, the above problem (7.18) and (7.19) states that a fluid particle initially free of vorticity remains free thereafter, verifying the Helmholtz's first law.

7.3. Two-dimensional vortex motions

Two-dimensional problems are simpler, but give us some useful information. Let us consider vortex motions of an inviscid fluid governed

---

[1] Mathematically, uniqueness of the solution is assured by the Lipschitz condition for the functions on the right-hand sides of (7.17), which should be satisfied by usual velocity fields $u(\mathbf{x})$, $v(\mathbf{x})$ and $w(\mathbf{x})$. See the footnote in Sec. 1.3.1. In [Lamb32, Sec. 146], the solution $\mathbf{X}_\sigma(t) = 0$ was given by using the Cauchy's solution.

[2] In [Lamb32, Sec. 146], it is remarked that the Helmholtz's reasoning was not quite rigorous. In fact, his equation resulted in the form (7.16), but there was no term on the right-hand side.

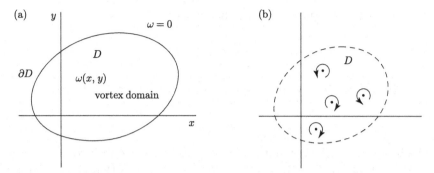

Fig. 7.2. Vorticity in a two-dimensional domain D: (a) Continuous distribution, (b) discrete distribution.

by (7.12). Suppose that the vorticity is given in an open domain D of (x, y) plane and vanishes out of D [Fig. 7.2(a)]:

$$
\begin{aligned}
\omega &= \omega(x, y), & (x, y) &\in D \\
\omega &= 0, & (x, y) &\in \partial D, \quad \text{out of } D.
\end{aligned}
\tag{7.21}
$$

It is assumed that the vorticity ω tends to zero continuously toward the boundary ∂D of D. Discrete vorticity distribution like the delta-function of a point vortex is regarded as a limit of such continuous distributions [Fig. 7.2(b)].

7.3.1. *Vorticity equation*

Assuming that the fluid motion is incompressible, the x, y components of the fluid velocity are expressed by using the stream function $\Psi(x, y, t)$ as

$$
u = \partial_y \Psi, \qquad v = -\partial_x \Psi,
\tag{7.22}
$$

(Appendix B.2). The z component of the vorticity is given by

$$
\omega = \partial_x v - \partial_y u = -\partial_x^2 \Psi - \partial_y^2 \Psi = -\nabla^2 \Psi,
\tag{7.23}
$$

where $\nabla^2 = \partial_x^2 + \partial_y^2$. The x, y components of vorticity vanish identically because $w = 0$ and $\partial_z = 0$, where w is the z component of velocity and the velocity field is independent of z.

The vorticity equation (7.12) becomes

$$\frac{D}{Dt}\boldsymbol{\omega} = 0, \tag{7.24}$$

since $(\boldsymbol{\omega} \cdot \nabla)\mathbf{v} = \omega \partial_z \mathbf{v} = 0$. This is rewritten as

$$\partial_t \omega + u\partial_x \omega + v\partial_y \omega = 0, \tag{7.25}$$

which is the vorticity equation in the two-dimensional problem.

If the function $\omega(x, y)$ is given, Eq. (7.23) is the Poisson equation $\nabla^2 \Psi = -\omega(x, y)$ of the function $\Psi(x, y)$. Its solution is expressed by the following integral form,

$$\Psi(x, y) = -\frac{1}{4\pi} \iint_D \omega(x', y') \log\big((x' - x)^2 + (y' - y)^2\big) dx' dy'. \tag{7.26}$$

u and v are obtained by differentiating $\Psi(x, y)$ from (7.22) (see the next section). Moreover, Eq. (7.25) gives the evolution of ω by

$$\partial_t \omega = -u\partial_x \omega - v\partial_y \omega. \tag{7.27}$$

7.3.2. Integral invariants

The domain D may be called a *vortex domain*. There exist four *integral invariants* for a vortex domain D:

$$\Gamma = \iint_D \omega(x, y) \, dx dy, \tag{7.28}$$

$$X = \iint_D x\omega \, dx dy, \tag{7.29}$$

$$Y = \iint_D y\omega \, dx dy, \tag{7.30}$$

$$R^2 = \iint_D (x^2 + y^2) \, \omega \, dx dy. \tag{7.31}$$

It can be shown that these four integrals are constant during the vortex motion. The first Γ is the total amount of vorticity in D, which is also equal to the circulation along any closed curve enclosing the vortex domain D. The Kelvin's circulation theorem (Sec. 5.2) assures that Γ is invariant. Below, this is shown directly by using (7.27).

From the three integrals, Γ, X and Y, we define two quantities by

$$X_c = \frac{\iint_D x\omega \, dxdy}{\iint_D \omega \, dxdy}, \quad Y_c = \frac{\iint_D y\omega \, dxdy}{\iint_D \omega \, dxdy}. \tag{7.32}$$

These are other invariants if Γ, X and Y are invariants. An interesting interpretation can be given for X_c and Y_c. If we regard ω as a hypothetical mass density, then Γ corresponds to the total mass included in D, and the pair (X_c, Y_c) corresponds to the center of mass, in other words, the *center of vorticity distribution* with a weight ω.

In order to verify that Γ, X and Y are invariant (see Problem 7.3 for the invariance of R^2), let us first derive the expressions of the velocities u and v by using the stream function (7.26):

$$u(x,y) = \partial_y \Psi = \frac{1}{2\pi} \iint_D \frac{(y' - y)\omega(x',y')}{(x' - x)^2 + (y' - y)^2} \, dx'dy', \tag{7.33}$$

$$v(x,y) = -\partial_x \Psi = -\frac{1}{2\pi} \iint_D \frac{(x' - x)\omega(x',y')}{(x' - x)^2 + (y' - y)^2} \, dx'dy'. \tag{7.34}$$

We define the function X_α by

$$X_\alpha = \iint_D x^\alpha \omega \, dxdy, \quad \alpha = 0, 1.$$

Obviously, $\Gamma = X_0$, $X = X_1$. Differentiating it with t, we obtain

$$\frac{dX_\alpha}{dt} = \iint_D x^\alpha \partial_t \omega \, dxdy = -\iint_D x^\alpha (u\partial_x \omega + v\partial_y \omega) \, dxdy$$

$$= -\iint_D \left(\partial_x(x^\alpha \omega u) + \partial_y(x^\alpha \omega v) - \alpha x^{\alpha-1} \omega u \right) dxdy, \tag{7.35}$$

where we used the continuity $\partial_x u + \partial_y v = 0$ and (7.27).

Integrating the first and second terms of (7.35) (a divergence of the vector $x^\alpha \omega \mathbf{v}$), we obtain the contour integral of $x^\alpha \omega v_n$ along the boundary ∂D by the Gauss theorem where $\omega = 0$ (v_n is the normal component). Hence, they vanish. The last term vanishes as well. It is obvious when $\alpha = 0$. For $\alpha = 1$, substituting (7.33), we obtain

$$\frac{\mathrm{d}X_1}{\mathrm{d}t} = \int_D wu\,\mathrm{d}A = \frac{1}{2\pi}\int_D\int_D (y' - y)\frac{\omega(x,y)\omega(x',y')}{(x'-x)^2 + (y'-y)^2}\,\mathrm{d}A\mathrm{d}A'$$

where $\mathrm{d}A = \mathrm{d}x\mathrm{d}y$ and $\mathrm{d}A' = \mathrm{d}x'\mathrm{d}y'$. The last integral is a four-fold integral with respect to two pairs of variables, $\mathrm{d}x\mathrm{d}y$ and $\mathrm{d}x'\mathrm{d}y'$. Interchanging the pairs, the first factor $(y' - y)$ changes its sign (anti-symmetric), whereas the second factor of the integrand is symmetric. The integral itself should be invariant with respect to the interchange of (x, y) and (x', y'). Therefore, the integral must be zero. Hence, $\mathrm{d}X/\mathrm{d}t = 0$. Similarly, one can show $\mathrm{d}Y/\mathrm{d}t = 0$. Thus, it is verified that the three integrals Γ, X and Y are invariant.

7.3.3. *Velocity field at distant points*

Let us consider the velocity field far from the vortex domain D. We choose a point (x', y') within D and another point (x, y) far from D, and assume that $r^2 = x^2 + y^2 \gg x'^2 + y'^2 = r'^2$. Using the polar coordinates (r, θ), we may write $(x, y) = r(\cos\theta, \sin\theta)$, and $(x', y') = r'(\cos\theta', \sin\theta')$. Then, we have

$$\begin{aligned}
(x' - x)^2 + (y' - y)^2 &= r^2 - 2rr'\cos(\theta - \theta') + r'^2 \\
&= r^2\left(1 - 2\frac{r'}{r}\cos(\theta - \theta') + \frac{r'^2}{r^2}\right) \\
&= r^2\left(1 + O\left(\frac{r'}{r}\right)\right).
\end{aligned}$$

The second and third terms are written simply as $O(r'/r)$, meaning the first order of the small quantity r'/r, which are negligibly small. Similarly, we may write as $y - y' = r\sin\theta\,(1 - O(r'/r))$ and $x - x' = r\cos\theta\,(1 - O(r'/r))$.

Substituting these estimates into (7.33) and (7.34), we obtain asymptotic expressions of u and v as $r/r' \to \infty$:

$$u(x, y) = -\frac{\Gamma}{2\pi}\frac{y}{r^2}\left(1 + O\left(\frac{r'}{r}\right)\right) \approx -\frac{\Gamma}{2\pi}\frac{y}{x^2 + y^2}, \qquad (7.36)$$

$$v(x,y) = \frac{\Gamma}{2\pi} \frac{x}{r^2} \left(1 + O\left(\frac{r'}{r}\right)\right) \approx \frac{\Gamma}{2\pi} \frac{x}{x^2 + y^2}, \qquad (7.37)$$

where Γ is the total vorticity included in the domain D defined by (7.28), and called the vortex strength of the domain.

7.3.4. *Point vortex*

When the vortex domain D of Fig. 7.2(a) shrinks to a point $P = (x_1, y_1)$ by keeping the vortex strength Γ to a finite value k, then we have the following relation,

$$\iint_{\lim D \to P} w(x,y)\, \mathrm{d}x\mathrm{d}y = k.$$

This implies that there is a concentrated vortex at P, and that the vorticity can be expressed by the delta function (see Appendix A.7) as

$$w(x,y) = k\delta(x - x_1)\,\delta(y - y_1), \qquad (7.38)$$

Substituting this to (7.33) and (7.34), the velocities are

$$u(x,y) = -\frac{k}{2\pi} \frac{y - y_1}{(x - x_1)^2 + (y - y_1)^2}, \qquad (7.39)$$

$$v(x,y) = \frac{k}{2\pi} \frac{x - x_1}{(x - x_1)^2 + (y - y_1)^2}, \qquad (7.40)$$

This is the same as the right-hand sides of (7.36) and (7.37) with Γ replaced by k, and also the expressions (5.47) and (5.48) (and also (5.73)) if x and y are replaced by $x - x_1$ and $y - y_1$, respectively.

From the definitions of the integral invariants (7.28)–(7.30) and (7.32), we obtain $\Gamma = k$, $X_c = x_1$ and $Y_c = y_1$. Thus, we have the following.

The strength k of the point vortex is invariant. In addition,
the position (x_1, y_1) of the vortex does not change.

Namely, the point vortex have no self-induced motion. In other words, a rectilinear vortex does not drive itself.

To represent two-dimensional flows of an inviscid fluid including point vortices, the theory of complex functions is known to be a powerful tool. A vortex at the origin $z = 0$ is represented by the following complex potential (Sec. 5.8.5),

$$F_v(z) = \frac{k}{2\pi i} \log z \quad (k : \text{real}). \tag{7.41}$$

The velocity field is given by $\mathrm{d}F_v/\mathrm{d}z = w = u - iv$, and we obtain

$$u = -\frac{k}{2\pi r} \sin \theta = -\frac{k}{2\pi} \frac{y}{r^2}, \quad v = \frac{k}{2\pi r} \cos \theta = \frac{k}{2\pi} \frac{x}{r^2}, \tag{7.42}$$

where $r^2 = x^2 + y^2$. It is seen that these are equivalent to (7.39) and (7.40).

7.3.5. *Vortex sheet*

Vortex sheet is a surface of discontinuity of tangential velocity [Fig. 4.5(a)]. Suppose that there is a surface of discontinuity at $y = 0$ of a fluid flow in the cartesian (x, y, z) plane, and that the velocity field is as follows:

$$\mathbf{v} = \left(\frac{1}{2}U, 0, 0\right) \quad \text{for } y < 0; \quad \mathbf{v} = \left(-\frac{1}{2}U, 0, 0\right) \quad \text{for } y > 0. \tag{7.43}$$

(Directions of flows are reversed from Fig. 4.5(a).) The vorticity of the flow is represented by

$$\boldsymbol{\omega} = (0, 0, \omega(y)), \quad \omega = U \, \delta(y), \tag{7.44}$$

where $\delta(y)$ is the Dirac's delta function (see A.7). This can be confirmed by using the formula (5.20) (Problem 7.4).

7.4. Motion of two point vortices

First, we consider a system of two point vortices by using the complex potential (7.41) (see Sec. 5.8.5) on the basis of the theory of complex functions (Appendix B), and derive the invariants directly from the equation of motion. After that, we learn that a system of N point vortices (of general number N) is governed by Hamiltonian's equation of motion.

Suppose that we have two point vortices which are moving under mutual interaction, and that their strengths are k_1 and k_2, and positions in (x, y) plane are expressed by the complex positions, $z_1 = x_1 + iy_1$ and $z_2 = x_2 + iy_2$, respectively. The complex potentials corresponding to each vortex are

$$F_1(z) = \frac{k_1}{2\pi i} \log(z - z_1), \quad F_2(z) = \frac{k_2}{2\pi i} \log(z - z_2). \tag{7.45}$$

The total potential is given by $F(z) = F_1(z) + F_2(z)$.

Each point vortex does not have its own proper velocity as shown in Sec. 7.3.4, and the strengths k_1 and k_2 are invariants. Hence, the vortex k_1 moves by the velocity induced by the vortex k_2, and vice versa. The complex velocity $u - iv$ is given by the derivative dF/dz. The complex velocity of vortex k_1 is the time derivative of $\bar{z}_1(t) = x_1(t) - iy_1(t)$ (complex conjugate of z_1). Thus, equating $d\bar{z}_1/dt$ with dF_2/dz at $z = z_1$, we have

$$\frac{d}{dt}\bar{z}_1 = \frac{dF_2}{dz}\bigg|_{z=z_1} = \frac{k_2}{2\pi i} \frac{1}{z_1 - z_2}. \tag{7.46}$$

Likewise, the equation of motion of vortex k_2 is

$$\frac{d}{dt}\bar{z}_2 = \frac{dF_1}{dz}\bigg|_{z=z_2} = \frac{k_1}{2\pi i} \frac{1}{z_2 - z_1}. \tag{7.47}$$

From the above two equations, we obtain immediately the following:

$$k_1 \frac{d}{dt}\bar{z}_1 + k_2 \frac{d}{dt}\bar{z}_2 = \frac{d}{dt}\left(\overline{k_1 z_1 + k_2 z_2}\right) = 0,$$

since k_1 and k_2 are constants. Therefore, we obtain

$$k_1 z_1 + k_2 z_2 = \text{const.} \tag{7.48}$$

This corresponds to the invariant $X + iY$ of the previous section. In fact, the present vorticity distribution can be written by two delta-functions,

$$\omega(x, y) = k_1 \delta(x - x_1)\,\delta(y - y_1) + k_2 \delta(x - x_2)\,\delta(y - y_2). \tag{7.49}$$

It is seen that the integral $X + iY$ defined by (7.29) and (7.30) gives the above expression $k_1 z_1 + k_2 z_2$. The center of vorticity $X_c + iY_c$ defined by (7.32) is also an invariant:

$$z_c = \frac{k_1 z_1 + k_2 z_2}{k_1 + k_2}. \tag{7.50}$$

We consider two typical examples.

Example (a). *In the system of two point vortices of the same strength k, the distance d between the two is invariant. Both vortices rotate around the fixed center $O = (X_c, Y_c)$ with the angular velocity $k/\pi d^2$ and the radius $\frac{1}{2}d$. The rotation is clockwise or counter-clockwise according as k is negative or positive* (Fig. 7.3).

In fact, denoting the positions of the two vortices by $z_1 = r_1 e^{i\theta_1}$ and $z_2 = r_2 e^{i\theta_2}$, the square of the distance d^2 between the two is given by $(z_1 - z_2)\overline{(z_1 - z_2)}$. Subtracting Eq. (7.47) from (7.46), we obtain

$$\frac{\mathrm{d}}{\mathrm{d}t}(\bar{z}_1 - \bar{z}_2) = \frac{\mathrm{d}}{\mathrm{d}t}\overline{(z_1 - z_2)} = \frac{k}{\pi i}\frac{1}{z_1 - z_2}. \tag{7.51}$$

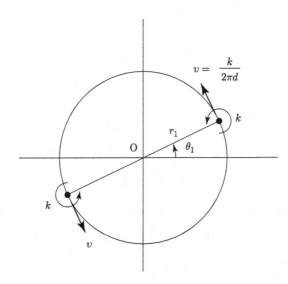

Fig. 7.3.   Rotating pair of two identical vortices: $v = r_1\,(\mathrm{d}/\mathrm{d}t)\theta_1$.

Therefore,

$$\frac{\mathrm{d}}{\mathrm{d}t}d^2 = (z_1 - z_2)\frac{\mathrm{d}}{\mathrm{d}t}\overline{(z_1 - z_2)} + \overline{(z_1 - z_2)}\frac{\mathrm{d}}{\mathrm{d}t}(z_1 - z_2) = \frac{k}{\pi i} - \frac{k}{\pi i} = 0.$$

Choosing the origin at the center O, we have $r_1 = r_2 = \frac{1}{2}d$ and $\theta_2 = \theta_1 + \pi$, leading to $z_1 = -z_2 = \frac{1}{2}de^{i\theta_1}$. Using these, Eq. (7.51) reduces to

$$\frac{\mathrm{d}}{\mathrm{d}t}\theta_1 = \frac{k}{\pi d^2}.$$

Thus, the angular velocity is $k/\pi d^2$, and the rotation is clockwise or counter-clockwise according as k is negative or positive. The rotation velocity v_θ of each vortex is given by

$$v_\theta = r_1\frac{\mathrm{d}}{\mathrm{d}t}\theta_1 = \frac{k}{2\pi d}.$$

Example (b). Vortex pair: *In the system of two point vortices of strength k and $-k$, both vortices move with the same velocity. Hence the distance d between the two is constant and a "vortex pair" is formed. The pair moves at right angles to the mutual direction with velocity $k/(2\pi d)$* (Fig. 7.4).

Writing $k_1 = k$ and $k_2 = -k$, Eqs. (7.46) and (7.47) give the same velocity,

$$\frac{\mathrm{d}}{\mathrm{d}t}\bar{z}_1 = \frac{\mathrm{d}}{\mathrm{d}t}\bar{z}_2 = \frac{k}{2\pi i}\frac{1}{z_2 - z_1} = \frac{k}{2\pi d}e^{-i(\phi + \frac{1}{2}\pi)}.$$

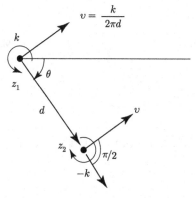

Fig. 7.4. A vortex pair $(\phi = 2\pi - \theta)$.

where $z_2 - z_1 = de^{i\phi}$. That is, $dz_1/dt = (k/2\pi d)\, e^{i(\phi + \frac{1}{2}\pi)}$. Therefore, the velocity is directed at right angles to the direction from z_1 to z_2, rotated counter-clockwise by $\frac{1}{2}\pi$. In this system, the total vorticity Γ vanishes, and the center of vorticity z_c recedes to infinity (hence of no use).

7.5.   System of N point vortices (a Hamiltonian system)

Suppose that there are N point vortices in motion under mutual interaction, and that their strengths are $k_1, k_2, \ldots, k_N$ and their complex positions are

$$z_1 = x_1 + iy_1, \quad \ldots, \quad z_N = x_N + iy_N.$$

Paying attention to the mth vortex in particular, its motion is influenced by all the vortices except itself, and the velocity of the mth vortex is given by the sum of those induced by the other vortices at the point $z_m = x_m + iy_m$. Hence, the complex velocity of the mth vortex is represented as

$$\frac{d}{dt}(x_m - iy_m) = \frac{1}{2\pi i} \sum_{n=1\,(n\neq m)}^{N} \frac{k_n}{z_m - z_n}.$$

Multiplying k_m to both sides and separating the real and imaginary parts of the resulting equation, we obtain

$$k_m \frac{dx_m}{dt} = -\frac{1}{2\pi} \sum_n{}' \frac{k_m k_n (y_m - y_n)}{r_{mn}^2}, \tag{7.52}$$

$$k_m \frac{dy_m}{dt} = \frac{1}{2\pi} \sum_n{}' \frac{k_m k_n (x_m - x_n)}{r_{mn}^2}, \tag{7.53}$$

where $\sum_n'$ denotes the summation with n from 1 to N excluding m, and

$$r_{mn}^2 = (x_m - x_n)^2 + (y_m - y_n)^2.$$

A Hamiltonian function H can be defined for the system of N vortices by

$$H = -\frac{1}{8\pi} \sum_{\substack{m=1 \ (m\neq n)}}^{N} \sum_{n=1}^{N} k_m k_n \log r_{mn}^2. \qquad (7.54)$$

Using H, the system of equations (7.52) and (7.53) can be rewritten as

$$k_m \frac{\mathrm{d}x_m}{\mathrm{d}t} = \frac{\partial H}{\partial y_m}, \quad k_m \frac{\mathrm{d}y_m}{\mathrm{d}t} = -\frac{\partial H}{\partial x_m}. \qquad (7.55)$$

It would be useful for the calculation to regard each term $(k_m k_n \log r_{mn}^2)$ in the Hamiltonian H as an entry (m, n) in a matrix, which is a symmetric matrix without diagonal elements. The system of equations (7.55) is of the form of Hamilton's equation of motion for generalized coordinate x_m and generalized momentum $k_m y_m$ (or coordinate $k_m x_m$ and momentum y_m).

The Hamiltonian H is another (fifth) invariant, because from (7.55),

$$\frac{\mathrm{d}H}{\mathrm{d}t} = \frac{\partial H}{\partial x_m}\frac{\mathrm{d}x_m}{\mathrm{d}t} + \frac{\partial H}{\partial y_m}\frac{\mathrm{d}y_m}{\mathrm{d}t} = \frac{1}{k_m}\frac{\partial H}{\partial x_m}\frac{\partial H}{\partial y_m} - \frac{1}{k_m}\frac{\partial H}{\partial y_m}\frac{\partial H}{\partial x_m} = 0.$$

7.6. Axisymmetric vortices with circular vortex-lines

We consider axisymmetric vortices in this section. An axisymmetric flow without swirl is defined by the velocity,

$$\mathbf{v}(\mathbf{x}) = (v_x(x,r), v_r(x,r), 0),$$

in the cylindrical coordinates (x, r, ϕ) (instead of (r, θ, z) of Appendix D.2). A Stokes's stream function $\Psi(x, r)$ can be defined for such axisymmetric flows of an incompressible fluid by

$$v_x = \frac{1}{r}\frac{\partial \Psi}{\partial r}, \quad v_r = -\frac{1}{r}\frac{\partial \Psi}{\partial x}. \qquad (7.56)$$

The continuity equation of an incompressible fluid, $\mathrm{div}\,\mathbf{v} = \partial_x v_x + \frac{1}{r}\partial_r(rv_r) = 0$ (see (D.9)), is satisfied identically by the expressions.

The vorticity of such a flow (defined by (D.15)) has only the azimuthal component $\omega_\phi(x, r)$:

$$\boldsymbol{\omega} = (\omega_x, \omega_r, \omega_\phi) = (0, 0, \omega_\phi(x, r)),$$

$$\omega_\phi = \partial_x v_r - \partial_r v_x = -\frac{1}{r}\left(\partial_x^2 \Psi + \partial_r^2 \Psi - \frac{1}{r}\partial_r \Psi\right). \qquad (7.57)$$

Introducing the vector potential $\mathbf{A} = (0, 0, A_\phi(x, r))$, we obtain

$$\mathbf{v} = (v_x, v_r, 0) = \operatorname{curl}\mathbf{A} = \left(r^{-1}\partial_r(r A_\phi), \; -\partial_x A_\phi, \; 0\right).$$

Therefore, we have $A_\phi = r^{-1}\Psi$. Then, the Biot–Savart's formula (7.5a) results in

$$A_\phi = \frac{1}{r}\Psi(x, r, \phi) = \frac{1}{4\pi}\int dx' \int dr' \int_0^{2\pi} r'd\phi' \; \frac{\omega_\phi(x', r')\cos(\phi' - \phi)}{|\mathbf{x} - \mathbf{y}|}, \qquad (7.58)$$

since the projection of ϕ component of $\boldsymbol{\omega}(\mathbf{y})$ to ϕ-direction at $\mathbf{x} = (x, r, \phi)$ is given by $\omega_\phi(x', r')\cos(\phi' - \phi)$, where $\mathbf{y} = (x', r', \phi')$, $d^3\mathbf{y} = r' dx' dr' d\phi'$ and

$$|\mathbf{x} - \mathbf{y}| = \left[(x - x')^2 + r^2 - 2rr'\cos(\phi - \phi') + (r')^2\right]^{1/2}. \qquad (7.59)$$

7.6.1. *Hill's spherical vortex*

Hill's spherical vortex (Hill, 1894) is an axisymmetric vortex of the vorticity field occupying a sphere of radius R with the law $\omega = Ar$ (A: a constant) and zero outside it:

$$\omega_\phi = Ar \; \left(\sqrt{x^2 + r^2} < R\right); \quad \omega_\phi = 0 \; \left(\sqrt{x^2 + r^2} > R\right). \qquad (7.60)$$

The total circulation is given by

$$\Gamma = \iint_{R_* < R} Ar \, dx dr = \frac{2}{3} AR^3,$$

where $R_* = \sqrt{x^2 + r^2}$. This vortex moves with a constant speed

$$U = \frac{2}{15} R^2 A = \frac{\Gamma}{5R}, \qquad (7.61)$$

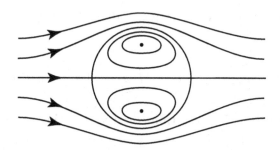

Fig. 7.5. Hill's spherical vortex.

without change of form. In the frame of reference in which the vortex is stationary, the stream function is given by

$$\Psi(x,r) = \begin{cases} \dfrac{3}{4}Ur^2\left(1 - R_*^2/R^2\right) & \text{for } R_* < R, \\[2mm] -\dfrac{1}{2}Ur^2\left(1 - R^3/R_*^3\right) & \text{for } R_* > R. \end{cases} \tag{7.62}$$

where $\Psi = -\frac{1}{2}Ur^2$ is the stream function of uniform flow of velocity $(-U,0,0)$. Obviously, the spherical surface $R_* = R$ coincides with the stream-line $\Psi = 0$ (Fig. 7.5).

The impulse of Hill's spherical vortex is obtained from (7.9) by using (7.60), and found to have only the axial component P_x:

$$\mathbf{P} = (P_x, 0, 0), \quad P_x = \int \omega_\phi r^2 \pi \mathrm{d}x\mathrm{d}r = 2\pi R^3 U.$$

7.6.2. *Circular vortex ring*

When the vortex-line is circular and concentrated on a circular core of a small cross-section, the vortex is often called a *circular vortex ring*. The stream function of a thin circular ring of radius R lying in the plane $x = 0$ is obtained from (7.58) by setting $\omega_\phi(x',r') = \gamma\delta(x')\,\delta(r'-R)$ as

$$\Psi_\gamma(x,r) = \frac{\gamma}{4\pi}rR \int_0^{2\pi} \mathrm{d}\phi \, \frac{\cos\phi}{\left[x^2 + r^2 - 2rR\cos\phi + R^2\right]^{1/2}}, \tag{7.63}$$

where $\gamma = \iint \omega_\phi \mathrm{d}x'\mathrm{d}r'$ is the strength of the vortex ring.

The impulse of the vortex ring of radius R and strength γ is given by (7.9):

$$\mathbf{P} = (P_x, 0, 0), \quad P_x = \int \omega_\phi r^2 \pi \mathrm{d}x \mathrm{d}r = \pi R^2 \gamma.$$

The stream function of a thin-cored vortex ring of radius R and strength γ, defined by (7.58), can be expressed [Lamb32, Sec. 161] as

$$\Psi_\gamma(x, r) = \frac{\gamma}{2\pi} \sqrt{rR} \left[\left(\frac{2}{k} - k \right) K(k) - \frac{2}{k} E(k) \right], \qquad (7.64)$$

where $K(k)$ and $E(k)$ are the complete elliptic integrals of the first and second kind of modulas $k^2 = 4rR/(x^2 + (r + R)^2)$, defined by

$$K(k) = \int_0^{\frac{1}{2}\pi} \frac{1}{(1 - k^2 \sin^2 \xi)^{1/2}} \, \mathrm{d}\xi, \quad E(k) = \int_0^{\frac{1}{2}\pi} (1 - k^2 \sin^2 \xi)^{1/2} \mathrm{d}\xi.$$

Considering a thin-cored vortex ring of radius R and a core radius a with circulation γ (Fig. 7.6), Kelvin (1867) gave a famous formula

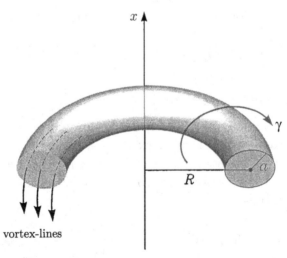

Fig. 7.6.   A vortex ring of radius R and a core radius a with circulation γ (a definition sketch).

for its speed as

$$U = \frac{\gamma}{4\pi R} \left[\log \frac{8R}{a} - \frac{1}{4} \right], \tag{7.65}$$

under the assumption $\epsilon = a/R \ll 1$. Its kinetic energy is given by

$$K = \frac{\rho R \gamma^2}{2} \left[\log \frac{8R}{a} - \frac{7}{4} \right],$$

Hicks (1885) confirmed Kelvin's result for the vortex ring with uniform vorticity in the thin-core. In addition, he calculated the speed of a ring of hollow thin-core, in which there is no vorticity within the core and pressure is constant, as

$$U_h = \frac{\gamma}{4\pi R} \left[\log \frac{8R}{a} - \frac{1}{2} \right]. \tag{7.66}$$

7.7. Curved vortex filament

In this section, we consider three-dimensional problems. The first case is a curved vortex filament $\mathcal{F}$ of very small cross-section σ with strength γ, embedded in an ideal incompressible fluid in infinite space. We assume that the vorticity at a point $\mathbf{y}$ on the filament $\mathcal{F}$ is $\boldsymbol{\omega}(\mathbf{y})$ and zero at points not on the filament.

Considering a solenoidal velocity field $\mathbf{u}(\mathbf{x})$ which satisfies $\mathrm{div}\,\mathbf{u} = 0$, the vorticity is given by $\boldsymbol{\omega} = \mathrm{curl}\,\mathbf{u}$. This vanishes by definition at all points except at $\mathbf{y}$ on $\mathcal{F}$. The vortex filament $\mathcal{F}$ of strength γ is expressed by a space curve of a small cross-section σ, assumed to move about and change its shape. We denote its volume element by $dV = \sigma\,\mathrm{dl}(\mathbf{y})$ where σ is the cross-section and $\mathrm{dl}(\mathbf{y})$ a line element of the filament at $\mathbf{y}$. Note that we have $\boldsymbol{\omega}dV = \gamma\mathrm{dl}(\mathbf{y})$, where $\gamma = |\boldsymbol{\omega}|\,\sigma$. The Biot–Savart law (7.5b) can represent velocity $\mathbf{u}$ at $\mathbf{x}$ induced by a vorticity element, $\boldsymbol{\omega}d^3\mathbf{y} = \boldsymbol{\omega}dV = \gamma\mathrm{dl} = \gamma\mathbf{t}ds$ at a point $\mathbf{y}(s)$. The velocity $\mathbf{u}(\mathbf{x})$ is expressed as

$$\mathbf{u}(\mathbf{x}) = -\frac{\gamma}{4\pi} \int_{\mathcal{F}} \frac{(\mathbf{x} - \mathbf{y}(s)) \times \mathbf{t}(s)}{|\mathbf{x} - \mathbf{y}(s)|^3} \, ds, \tag{7.67}$$

where s is an arc-length parameter along $\mathcal{F}$, $\mathrm{d}s$ is an infinitesimal arc length and $\mathbf{t}$ unit tangent vector to $\mathcal{F}$ at $\mathbf{y}(s)$ (hence $\mathrm{d}\mathbf{l} = \mathbf{t}\mathrm{d}s$).

We consider the velocity induced in the neighborhood of a point O on the filament $\mathcal{F}$. We define a local rectilinear frame K at O determined by three mutually-orthogonal vectors $(\mathbf{t}, \mathbf{n}, \mathbf{b})$, where $\mathbf{n}$ and $\mathbf{b}$ are unit vectors in the principal normal and binormal directions at O (and $\mathbf{t}$ the unit tangent to $\mathcal{F}$), as given in Appendix D.1. With the point O as the origin of K, the position vector $\mathbf{x}$ of a point in the plane normal to the filament $\mathcal{F}$ (i.e. perpendicular to $\mathbf{t}$) at O can be written as

$$\mathbf{x} = y\mathbf{n} + z\mathbf{b}$$

(Fig. 7.7). We aim to find the velocity $\mathbf{u}(\mathbf{x})$ obtained in the limit as $\mathbf{x}$ approaches the origin O, i.e. $r = (y^2 + z^2)^{1/2} \to 0$.

In this limit it is found that the Biot–Savart integral (7.67) gives

$$\mathbf{u}(\mathbf{x}) = \frac{\gamma}{2\pi}\left(\frac{y}{r^2}\mathbf{b} - \frac{z}{r^2}\mathbf{n}\right) + \frac{\gamma}{4\pi}k_0\left(\log\frac{\lambda}{r}\right)\mathbf{b} + \text{(b.t.)}, \qquad (7.68)$$

where the first two terms increase indefinitely as $r \to 0$, and the term (b.t.) denotes those that remain bounded. The first term proportional to $\gamma/2\pi$ represents the circulatory motion about the vortex filament $\mathcal{F}$, counter-clockwise in the $(\mathbf{n}, \mathbf{b})$ plane, regarded as the right motion so that this filament is said to be a *vortex*. However, there is another term, i.e. the second term proportional to the curvature k_0 which is not circulatory, but directed to $\mathbf{b}$.

The usual method to resolve the unboundedness is to use a cut-off. Namely, every vortex filament has a vortex core of finite size a, and r should be bounded below at a value of order a. If r is replaced

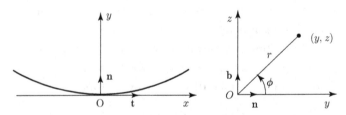

Fig. 7.7. Vortex filament and local frame of reference $(\mathbf{t}, \mathbf{n}, \mathbf{b})$.

by a, the second term is

$$\mathbf{u}_{\text{LI}} = \frac{\gamma}{4\pi} k_0 \left(\log \frac{\lambda}{a} \right) \mathbf{b}, \tag{7.69}$$

which is independent of y and z. This is interpreted such that the vortex core moves rectilinearly with velocity $\mathbf{u}_{\text{LI}}$ in the binormal direction $\mathbf{b}$.

The magnitude of velocity $\mathbf{u}_{\text{LI}}$ is proportional to the local curvature k_0 of the filament at O, and is called *local induction*. This term vanishes with a rectilinear line-vortex because k_0 becomes zero. This is consistent with the known property that a rectilinear vortex has no self-induced velocity, noted in Sec. 7.3.4.

7.8. Filament equation (an integrable equation)

When we are interested in only the motion of a filament (without seeing circulatory motion around it), the velocity would be given as $\mathbf{u}_{\text{LI}}(s)$, which can be expressed as

$$\mathbf{u}_{\text{LI}}(s) = ck(s)\,\mathbf{b}(s) = ck(s)\,\mathbf{t}(s) \times \mathbf{n}(s), \tag{7.70}$$

where $c = (\gamma/4\pi)\log(\lambda/a)$ is a constant independent of s. Rates of change of the unit vectors $(\mathbf{t}, \mathbf{n}, \mathbf{b})$ (with respect to s) along the curve are described by the Frenet–Serret equation (D.4) in terms of the curvature $k(s)$ and torsion $\tau(s)$ of the filament.[3]

A vortex ring is a vortex in the form of a circle (of radius R, say), which translates with a constant speed in the direction of $\mathbf{b}$. The binormal vector $\mathbf{b}$ of the vortex ring is independent of the position along the circle and perpendicular to the plane of circle (directed from the side where the vortex-line looks clockwise to the side where it looks counter-clockwise). The direction of $\mathbf{b}$ is the same as that of the fluid flowing inside the circle. This is consistent with the expression (7.70) since $k = 1/R = $ const.

---

[3]For a space curve $\mathbf{x}(s) = (x(s), y(s), z(s))$, we have $d\mathbf{x}/ds = \mathbf{t}(s)$, $d\mathbf{t}/ds = k(s)\,\mathbf{n}(s)$, $d\mathbf{n}/ds = -k(s)\,\mathbf{t} - \tau(s)\,\mathbf{b}$. See Appendix D.1 for more details.

It is found just above that the vortex ring in the rectilinear translational motion (with a constant speed) depicts a cylindrical surface
S_c of circular cross-section in a three-dimensional Euclidean space
$\mathbb{R}^3$. The circular vortex filament f_c coincides at every instant with
a geodesic line of the surface S_c depicted by the vortex in space $\mathbb{R}^3$.
This is true for general vortex filaments f in motion under the law
given by Eq. (7.70), because the tangent plane to the surface S_f to
be generated by vortex motion is formed by two orthogonal tangent
vectors $\mathbf{t}$ and $\mathbf{u}_{LI}dt$. Therefore, the normal $\mathbf{N}$ to the surface S_f coincides with the normal $\mathbf{n}$ to the curve f of the vortex filament. This
property is nothing but that f is a geodesic curve.[4]

Suppose that we have an *active* space curve C: $\mathbf{x}(s, t)$, which
moves with velocity $\mathbf{u}_{LI}$. Namely, the velocity $\partial_t \mathbf{x}$ at a station s
is given by the local value:

$$\partial_t \mathbf{x} = \mathbf{u}_*, \quad \mathbf{u}_* \equiv ck(s)\,\mathbf{b}(s),$$

where $\mathbf{u}_*$ is the *local induction velocity*. It can be shown that the
separation distance of two nearby particles on the curve, denoted by
Δs, is unchanged by this motion. In fact,

$$\frac{\mathrm{d}}{\mathrm{d}t}\Delta s = (\Delta s \partial_s \mathbf{u}_*) \cdot \mathbf{t}, \tag{7.71}$$

where $\partial_s \mathbf{u}_* = ck'(s)\,\mathbf{b} + ck\mathbf{b}'(s)$. From the Frenet–Serret equation
(D.4), it is readily seen that $\mathbf{b} \cdot \mathbf{t} = 0$ and $\mathbf{b}'(s) \cdot \mathbf{t} = 0$. Thus, it is
found that $(\mathrm{d}/\mathrm{d}t)\Delta s = 0$, i.e. the length element Δs of the curve is
invariant during the motion, and we can take s as the Lagrangian
parameter of material points on C.

Since $\mathbf{b} = \mathbf{t} \times \mathbf{n}$ and in addition $\partial_s \mathbf{x} = \mathbf{t}$ and $\partial_s^2 \mathbf{x} = k\mathbf{n}$
(Appendix D.1), the local relation (7.70) for the curve $\mathbf{x}(s, t)$ is
given by

$$\partial_t \mathbf{x} = \partial_s \mathbf{x} \times \partial_s^2 \mathbf{x}, \tag{7.72}$$

where the time is rescaled so that the previous ct is written as t
here. This is termed the *filament equation*. In fluid mechanics, the

---

[4]A geodesic curve on a surface S is defined by the curve connecting two given
points on S with a shortest (or an extremum) distance. See [Ka04, Sec. 2.6] for
its definition.

same equation is called the *local induction equation (approximation)*.[5] Some experimental evidences are seen in [KT71].

It is remarkable that the local induction equation can be transformed to the cubic-nonlinear Schrödinger equation. Introducing a complex function $\psi(s,t)$ by

$$\psi(s,t) = k(s) \exp\left[i \int^s \tau(s')\, ds' \right]$$

(called Hasimoto transformation [Has72]), where k and τ are the curvature and torsion of the filament. The local induction equation (7.72) is transformed to

$$\partial_t \psi = i\left(\partial_s^2 \psi + \frac{1}{2}|\psi|^2 \psi \right). \tag{7.73}$$

As is well-known, this is one of the completely integrable systems, called the *nonlinear Schrödinger equation*. This equation admits a soliton solution, which propagates with a constant speed c along an infinitely long vortex filament as $\psi(s,t) = k(s - ct) \exp[i\tau_0 s]$, where $k(s,t) = 2(\tau_0/\alpha) \operatorname{sech}(\tau_0/\alpha)(s - ct)$ with $\tau = \tau_0 = \frac{1}{2}c = $ const. and α a constant.

7.9. Burgers vortex (a viscous vortex with swirl)

There is a mechanism for spontaneous formation of a vortex in a certain flow field. In Sec. 10.4, we will see a mechanism of spontaneous enhancement of average magnitude of vorticity in turbulent flow fields.

The viscous vorticity equation without external force is written as

$$\partial_t \boldsymbol{\omega} + (\mathbf{v} \cdot \operatorname{grad})\boldsymbol{\omega} = (\boldsymbol{\omega} \cdot \operatorname{grad})\mathbf{v} + \nu \nabla^2 \boldsymbol{\omega}, \tag{7.74}$$

from (7.3). It is interesting to see that this equation has a solution which shows concentration of vorticity without external means.

---

[5]This equation was given by Da Rios (1906) and has been rediscovered several times historically.

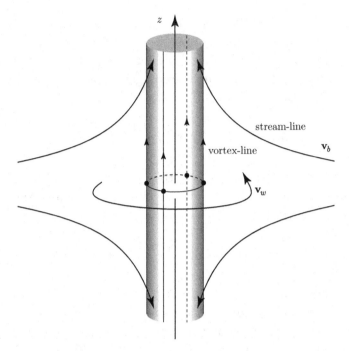

Fig. 7.8.   Burgers vortex under an external straining $\mathbf{v}_b$.

Suppose that the velocity field consists of two components (Fig. 7.8): $\mathbf{v} = \mathbf{v}_b + \mathbf{v}_\omega$. In the cylindrical frame of reference (r, θ, z), the first component $\mathbf{v}_b$ is assumed to be axisymmetric and irrotational, and represented as

$$\mathbf{v}_b = (-ar, \, 0, \, 2az), \qquad (7.75)$$

where a is a positive constant. This satisfies the solenoidal condition,

$$\text{div}\, \mathbf{v}_b = \partial_z(\mathbf{v}_b)_z + r^{-1}\partial_r(r(\mathbf{v}_b)_r) = 2a - 2a = 0.$$

The component $\mathbf{v}_b$ also has a velocity potential Φ. In fact,

$$\mathbf{v}_b = \text{grad}\,\Phi, \quad \Phi = -\frac{1}{2}\,ar^2 + az^2.$$

Hence, $\mathbf{v}_b$ is clearly irrotational and regarded as a background straining field which is acting on the second rotational component $\mathbf{v}_\omega$, assumed to be given as $(0, v_\theta(r), 0)$. Its vorticity has only the axial z component:

$$\boldsymbol{\omega} = \nabla \times \mathbf{v}_\omega = (0,\, 0,\, \omega(r,t)), \quad \omega = r^{-1}\partial_r(rv_\theta).$$

The total velocity is given as

$$\mathbf{v} = \mathbf{v}_b + \mathbf{v}_\omega = (-ar,\, v_\theta(r),\, 2az). \tag{7.76}$$

Then, the first and last terms of the vorticity equation (7.74) have only the z component. Similarly, the second term on the left and the first term on the right of (7.74) can be written respectively as

$$(\mathbf{v} \cdot \mathrm{grad})\boldsymbol{\omega} = (-ar\partial_r + v_\theta r^{-1}\partial_\theta + 2az\partial_z)\boldsymbol{\omega}(r,t)$$
$$= (0,\, 0,\, -ar\partial_r\omega(r,t)),$$
$$(\boldsymbol{\omega} \cdot \mathrm{grad})\mathbf{v} = \omega\partial_z\mathbf{v} = (0,\, 0,\, 2a\omega).$$

Hence both have only the z component as well. Thus, the vorticity equation (7.74) reduces to the following single equation for the z component:

$$\partial_t\omega - ar\partial_r\omega = 2a\omega + \nu r^{-1}\partial_r(r\partial_r\omega). \tag{7.77}$$

In order to see the significance of each term, let us consider first the equation formed only by the two terms on the left-hand side: $\partial_t\omega - ar\partial_r\omega = 0$, neglecting the terms on the right. We immediately obtain its general solution of the form $\omega = F(re^{at})$, where F is an arbitrary function. One can easily show that this satisfies the equation. At $t = 0$, the function ω takes the value $F(r_0)$ for $r = r_0$. At a later time $t > 0$, the same value is found at $r = r_0 e^{-at} < r_0$ since $a > 0$. The same can be said for all values of $r_0 \in (0, \infty)$. Therefore, the initial distribution of ω is convected inward at later times and converges to $r = 0$ as $t \to \infty$. Thus, the term $-ar\partial_r\omega$ on the left of (7.77) represents the effect of an inward convection.

Next, the influence of the first term on the right-hand side can be seen by considering another truncated equation $\partial_t\omega = 2a\omega$. This gives a solution $w(t) = w(0)e^{2at}$, which represents an amplification of the vorticity. The term $2a\omega$ from $(\boldsymbol{\omega} \cdot \mathrm{grad})\mathbf{v}$ represents vortex stretching, as seen in Sec. 7.2.1. The last term $\nu r^{-1}\partial_r(r\partial_r\omega)$ represents viscous diffusion.

The steady form of Eq. (7.77) is $ar\partial_r\omega + 2a\omega + \nu r^{-1}\partial_r(r\partial_r\omega) = 0$. Its solution is readily found as [see Problem 7.9(i)].

$$\omega_\mathrm{B}(r) = \omega_\mathrm{B}(0)\,\exp[-r^2/l_\mathrm{B}^2], \quad l_\mathrm{B} = \sqrt{2\nu/a}. \tag{7.78}$$

This steady distribution $\omega_\mathrm{B}(r)$ is called the *Burgers vortex*.

For an unsteady problem of the whole equation (7.77), a general solution can be found in the form $w(r,t) = A^2(t)\,W(\sigma,\tau)$ in Problem 7.9(iii). As time tends to infinity ($t \to \infty$ for $a > 0$), this solution tends to an asymptotic form of the Burgers vortex:

$$w(t \to \infty) \to \omega_\mathrm{B}(r) = \frac{\Gamma}{\pi l_\mathrm{B}^2}\,\exp[-r^2/l_\mathrm{B}^2], \tag{7.79}$$

where $\Gamma = \int_0^\infty \Omega_0(s)\,2\pi s\,\mathrm{d}s$ denotes the initial total vorticity [Ka84].

This asymptotic state is interpreted as follows. The vorticity is swept to the center by a converging flow $\mathbf{v}_b$. However, the effect of viscosity gives rise to an outward diffusion of ω. Thus, as a balance of the two effects, the vorticity approaches a stationary distribution $\omega_\mathrm{B}(r)$ represented by the Gaussian function of (7.79). The parameter l_B represents a length scale of the final form. The swirl velocity $v_\theta(r)$ around the Burgers vortex is given as

$$v_\theta(r) = \frac{\Gamma}{2\pi l_\mathrm{B}}\,\frac{1}{\hat{r}}\left(1 - e^{-\hat{r}^2}\right), \quad \hat{r} = r/l_\mathrm{B}. \tag{7.80}$$

As a result of detailed analyses, evidences are increasing to show that strong concentrated vortices observed in computer simulations or experiments of turbulence have this kind of Burgers-like vortex. This implies that in turbulence there exists a mechanism of spontaneous self-formation of Burgers vortices in the statistical sense.

7.10.   Problems

Problem 7.1 Vector potential A

Verify the relation curl $\mathbf{v} = \boldsymbol{\omega}$ by using the vector potential $\mathbf{A}$ defined by (7.5a). State in what cases the div $\mathbf{A} = 0$ is satisfied.

Problem 7.2 Invariants of motion

Verify the conservation of the five integrals (7.9)–(7.11) of *Impulse* **P**, *Angular impulse* **L** and *Helicity* H for the vorticity $\boldsymbol{\omega}$ evolving according to the vorticity equation (7.1).

Problem 7.3 Invariance of R^2

Verify that the integral (7.31) for R^2 is invariant during the vortex motion, according to (7.27) for 2D motions of an incompressible inviscid fluid.

Problem 7.4 Vortex sheet

Suppose that a flow is represented by (7.43) having a discontinuous surface at $y = 0$. Show that its vorticity is given by the expression (7.44).

Problem 7.5 Vortex filament

Derive the asymptotic expression (7.68) for the velocity $\mathbf{u}(\mathbf{x})$ from the Biot–Savart integral (7.67)

Problem 7.6 Helical vortex

Show that the following rotating helical vortex $\mathbf{x}_h = (x, y, z)(s, t)$ satisfies Eq. (7.72):

$$\mathbf{x}_h = a(\cos\theta,\ \sin\theta,\ hks + \lambda\omega t), \tag{7.81}$$
$$\mathbf{t}_h = ak(-\sin\theta,\ \cos\theta,\ h),\quad \theta = ks - \omega t \tag{7.82}$$

where a, k, h, ω, λ are constants and $\mathbf{t}_h$ is its tangent.

Problem 7.7 Lamb's transformation [Lamb32, Sec. 162]

Consider a thin-cored vortex ring in a fixed coordinate frame in which velocity $\mathbf{u}(\mathbf{x}, t)$ vanishes at infinity. Suppose that the fluid density is constant and the vortex is in steady motion with a constant velocity $\mathbf{U}$. Show the following relation between the kinetic energy K, impulse $\mathbf{P}$ and velocity $\mathbf{U}$:

$$K = 2\mathbf{U} \cdot \mathbf{P} + \int \mathbf{u}_* \cdot (\mathbf{x} \times \boldsymbol{\omega}) \, \mathrm{d}^3\mathbf{x}, \tag{7.83}$$

where $\mathbf{u}_* = \mathbf{u} - \mathbf{U}$ is the steady velocity field in the frame where the vortex ring is observed at rest.

Problem 7.8 Vortex ring

The kinetic energy K, impulse P and velocity U of a thin-cored vortex ring of a ring radius R and a core radius a are given by

$$K = \int \omega_\phi \Psi \pi \mathrm{d}x\mathrm{d}r$$
$$= \frac{1}{2} R\Gamma^2 \left[\log \frac{8R}{a} - \frac{7}{4} + \frac{3}{16}\epsilon^2 \log \frac{3}{8} + \cdots \right], \tag{7.84}$$

$$P = \int \omega_\phi r^2 \pi \mathrm{d}x\mathrm{d}r = \pi R^2 \Gamma \left[1 + \frac{3}{4}\epsilon^2 + \cdots \right], \tag{7.85}$$

$$U = \frac{\Gamma}{4\pi R} \left[\log \frac{8R}{a} - \frac{1}{4} + \epsilon^2 \left(-\frac{3}{8} \log \frac{8}{\epsilon} + \frac{15}{32} \right) \right.$$
$$\left. + O\left(\epsilon^4 \log \frac{8}{\epsilon} \right) \right], \tag{7.86}$$

for $\epsilon = a/R \ll 1$ (Dyson, 1893). Taking variation δR of the ring radius, show the following relation (Roberts and Donnelly, 1970):

$$U = \frac{\partial K}{\partial P}. \tag{7.87}$$

Problem 7.9 Burgers vortex

Let us try to find solutions to the vorticity equation (7.77):

$$\partial_t \omega - a r \partial_r \omega = 2 a \omega + \nu r^{-1} \partial_r (r \partial_r \omega).$$

(i) Show the steady solution to the above equation is given by

$$\omega_B(r) = \omega_B(0) \exp\left[-\frac{r^2}{l_B^2}\right], \quad l_B = \sqrt{2\nu/a}. \qquad (7.88)$$

(ii) Let us introduce the following transformation of variables:

$$\sigma = A(t) r, \quad \tau = \int_0^t A^2(t') \, dt', \quad W = \omega(r, t)/A^2(t),$$

where $A(t) = e^{\sigma t}$. Show that Eq. (7.77) reduces to the following diffusion equation for W with the new variables (σ, τ):

$$\frac{\partial}{\partial \tau} W = \nu \frac{1}{\sigma} \partial_\sigma (\sigma \partial_\sigma W), \quad \text{or} \quad \partial_\tau W = \nu \left(\partial_\xi^2 + \partial_\eta^2\right) W, \qquad (7.89)$$

where $\xi = Ax$, $\eta = Ay$ and $\sigma = \sqrt{\xi^2 + \eta^2}$.

(iii) Show that, for an arbitrary axisymmetric initial distribution $W(\tau = 0) = \Omega_0(r)$, a general solution to (7.89) is given by

$$W(\sigma, \tau) = \frac{1}{4\pi\nu\tau} \iint \Omega_0\left(\sqrt{\xi^2 + \eta^2}\right)$$

$$\times \exp\left[-\frac{(\xi - \xi_1)^2 + (\eta - \eta_1)^2}{4\nu\tau}\right] d\xi_1 d\eta_1. \qquad (7.90)$$

Chapter 8

Geophysical flows

Atmospheric motions and ocean currents are called Geophysical Flows, which are significantly influenced by the rotation of the Earth and density stratifications of the atmosphere and ocean. We consider the flows in a rotating system and the influence of density stratification on flows compactly in this chapter.[1] This subject is becoming increasingly important in the age of space science and giant computers.

8.1. Flows in a rotating frame

To study geophysical flows, we have to consider fluid motions in a rotating frame such as a frame fixed to the Earth, which is a *noninertial frame*. The equation of motion on such a noninertial frame is described by introducing a centrifugal force and Coriolis force in addition to the forces of the inertial system. It is assumed that a frame of reference $\mathcal{R}$ is rotating with angular velocity $\boldsymbol{\Omega}$ relative to an inertial frame $\mathcal{S}$ fixed to the space. The Navier–Stokes equation in the inertial frame $\mathcal{S}$ is given by (4.10), which is reproduced here:

$$\partial_t \mathbf{v} + (\mathbf{v} \cdot \nabla)\mathbf{v} = -\frac{1}{\rho}\nabla p_{\mathrm{m}} + \nu\Delta\mathbf{v}, \quad \frac{1}{\rho}\nabla p_{\mathrm{m}} = \frac{1}{\rho}\nabla p - \mathbf{g}, \qquad (8.1)$$

---

[1]For a more detailed account of the present subject (except Sec. 8.5), see for example: [Hol04; Ach90, Sec. 8.5; Hou77] or [Tri77, Secs. 15, 16].

where p_m is the *modified* pressure, defined originally by (4.11) for a constant ρ, and **g** is the acceleration due to gravity expressed by $-\nabla\chi$ (χ: the gravity potential). The formula of p_m can be extended to variable density ρ. Assuming that the atmosphere is *barotropic*, i.e. assuming $\rho = \rho(p)$, one can write as $\rho^{-1}\nabla p = \nabla\Pi$ by using Π defined below. Thus, we have

$$\frac{1}{\rho}\nabla p_m = \text{grad}\, P_m, \tag{8.2}$$

$$P_m(\mathbf{x}) := \Pi(p(\mathbf{x})) + \chi(\mathbf{x}), \quad \Pi := \int_{p_0}^{p} \frac{dp}{\rho(p)}, \tag{8.3}$$

where p_0 is a reference pressure.

Suppose that the position of a fluid particle is represented by $\mathbf{x}(t)$ in the inertial frame $\mathcal{S}$. Then, its velocity $(D\mathbf{x}_{\mathcal{S}}/Dt)_{\mathcal{S}} = \mathbf{v}_{\mathcal{S}}$ in the frame $\mathcal{S}$ is represented as

$$\mathbf{v}_{\mathcal{S}} = \left(\frac{D\mathbf{x}_{\mathcal{S}}}{Dt}\right)_{\mathcal{S}} = \left(\frac{D\mathbf{x}_{\mathcal{R}}}{Dt}\right)_{\mathcal{R}} + \mathbf{\Omega}\times\mathbf{x} = \mathbf{v}_{\mathcal{R}} + \mathbf{\Omega}\times\mathbf{x}, \tag{8.4}$$

where the subscript $\mathcal{R}$ refers to the rotating frame and $\mathbf{x}_{\mathcal{S}} = \mathbf{x}_{\mathcal{R}} = \mathbf{x}$ since the coordinate origins are common. The expression $(D\mathbf{x}_{\mathcal{R}}/Dt)_{\mathcal{R}}$ is the velocity relative to the rotating frame. The first term $\mathbf{v}_{\mathcal{R}}$ denotes the velocity as it appears to an observer in the rotating frame $\mathcal{R}$, while the second term,

$$\mathbf{v}_{\mathcal{F}} = \mathbf{\Omega}\times\mathbf{x}, \tag{8.5}$$

is a velocity due to the frame rotation which is an additional velocity to an observer in the inertial frame $\mathcal{S}$ (Fig. 8.1) [see (1.30) for the rotation velocity with an angular velocity $\frac{1}{2}\boldsymbol{\omega}$].

Analogously, the acceleration $(D\mathbf{v}_{\mathcal{S}}/Dt)_{\mathcal{S}}$ in the inertial frame $\mathcal{S}$ is given by taking time derivative of (8.4) in the frame $\mathcal{S}$:

$$\begin{aligned}
\left[\frac{D}{Dt}\mathbf{v}_{\mathcal{S}}\right]_{\mathcal{S}} &= \left[\frac{D}{Dt}\mathbf{v}_{\mathcal{R}}\right]_{\mathcal{S}} + \left[\frac{D}{Dt}(\mathbf{\Omega}\times\mathbf{x})\right]_{\mathcal{S}} \\
&= \left(\frac{D\mathbf{v}_{\mathcal{R}}}{Dt}\right)_{\mathcal{R}} + \mathbf{\Omega}\times\mathbf{v}_{\mathcal{R}} + \mathbf{\Omega}\times\left(\frac{D\mathbf{x}}{Dt}\right)_{\mathcal{S}} + \frac{d\mathbf{\Omega}}{dt}\times\mathbf{x} \\
&= (D\mathbf{v}_{\mathcal{R}}/Dt)_{\mathcal{R}} + 2\mathbf{\Omega}\times\mathbf{v}_{\mathcal{R}} + \mathbf{\Omega}\times(\mathbf{\Omega}\times\mathbf{x}) + \dot{\mathbf{\Omega}}\times\mathbf{x}. \tag{8.6}
\end{aligned}$$

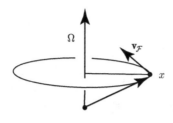

Fig. 8.1. $\mathbf{v}_{\mathcal{F}}$.

($\dot{\mathbf{\Omega}} = d\mathbf{\Omega}/dt$), where the first term on the right-hand side is represented by

$$(D\mathbf{v}_{\mathcal{R}}/Dt)_{\mathcal{R}} = \partial_t \mathbf{v}_{\mathcal{R}} + (\mathbf{v} \cdot \nabla \mathbf{v})_{\mathcal{R}}$$

[Hol04, Hou77]. The term $(D\mathbf{v}_{\mathcal{R}}/Dt)_{\mathcal{R}}$ is the acceleration relative to the rotating frame for fluid motions observed in the frame $\mathcal{R}$. Assuming $d\mathbf{\Omega}/dt = 0$,[2] and dropping the subscript $\mathcal{R}$ (therefore, all velocities will be referred to the rotating frame henceforth), and replacing the left-hand side of (8.1) by $(D\mathbf{v}_{\mathcal{S}}/Dt)_{\mathcal{S}}$ of (8.6), the equation of motion (8.1) with (8.2) is transformed to

$$\partial_t \mathbf{v} + (\mathbf{v} \cdot \nabla)\mathbf{v} = -\mathrm{grad}\, P_m - \mathbf{\Omega} \times (\mathbf{\Omega} \times \mathbf{x}) - 2\mathbf{\Omega} \times \mathbf{v} + \nu\nabla^2\mathbf{v}.$$

$$(8.7)$$

The second and third terms on the right-hand side of (8.7) are the accelerations due to *centrifugal* and *Coriolis* forces (Fig. 8.2), respectively.

Choosing a local Cartesian frame (x, y, z) with the z axis taken vertically upward and assuming that it coincides with the direction of local angular velocity $\mathbf{\Omega}$, the centrifugal force can be expressed in the following form,[3]

$$\mathbf{\Omega} \times (\mathbf{\Omega} \times \mathbf{x}) = -\mathrm{grad}\left[\frac{1}{2}\Omega^2(x^2 + y^2)\right],$$

---

[2]The rotation of the Earth is regarded as steady for the motions under consideration.

[3]This is verified by the formula (A.19): $\mathbf{\Omega} \times (\mathbf{\Omega} \times \mathbf{x}) = (\mathbf{\Omega} \cdot \mathbf{x})\mathbf{\Omega} - (\mathbf{\Omega} \cdot \mathbf{\Omega})\mathbf{x} = -(x, y, 0)\Omega^2$.

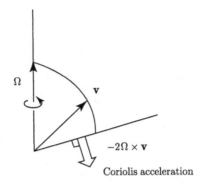

Fig. 8.2.   Coriolis force.

where $\Omega = |\mathbf{\Omega}|$ and $r \equiv (x^2 + y^2)^{1/2}$ is the distance from the rotation axis. Hence this can be combined with the modified pressure term as

$$-\operatorname{grad} P_m - \mathbf{\Omega} \times (\mathbf{\Omega} \times \mathbf{x}) = -\operatorname{grad} P, \qquad (8.8)$$

where

$$P = P_m - \frac{1}{2}\Omega^2 r^2 = \int_{p_0}^{p} \frac{\mathrm{d}p}{\rho} + \chi - \frac{1}{2}\Omega^2 r^2 = \int_{p_0}^{p} \frac{\mathrm{d}p}{\rho} + \chi_*, \qquad (8.9)$$

$$\chi_* := \chi - \frac{1}{2}\Omega^2 r^2, \quad \mathrm{d}\chi_*/\mathrm{d}z = g_*(z), \qquad (8.10)$$

where χ_* is the geo-potential (see below). Equation (8.7) is given by

$$\partial_t \mathbf{v} + (\mathbf{v} \cdot \nabla)\mathbf{v} = -\operatorname{grad} P - 2\mathbf{\Omega} \times \mathbf{v} + \nu \nabla^2 \mathbf{v}. \qquad (8.11)$$

Thus, mathematically speaking, the problem is reduced to the one that is identical with a system in which the centrifugal force is absent.

The function χ_* is the *geo-potential* (per unit mass) due to gravity and centrifugal forces, and $g_*(z)$ the corresponding acceleration. The *geopotential height* is defined by

$$z_g := \frac{\chi_*}{g_0} = \frac{1}{g_0} \int_0^z g_*(z)\,\mathrm{d}z, \qquad (8.12)$$

where $g_0 = 9.807$ m/s^{-2}, the mean value at the surface.

When there is no motion, i.e. $\mathbf{v} = 0$, the above equation becomes $\operatorname{grad} P = 0$, which reduces to the *hydrostatic equation*:

$$\frac{dp}{dz} = -\rho(z)\, g_*(z). \tag{8.13}$$

8.2. Geostrophic flows

From the previous section, it is seen that an essential difference between the dynamics in a rotating frame from a nonrotating frame is caused by the Coriolis term $-2\mathbf{\Omega} \times \mathbf{v}$. We consider flows that are dominated by the action of Coriolis force, and suppose that the effect of Coriolis force is large compared with both of the effects of convection $(\mathbf{v} \cdot \nabla)\mathbf{v}$ and viscosity terms $\nu\nabla^2\mathbf{v}$. This means

$$|\mathbf{\Omega} \times \mathbf{v}| \gg |(\mathbf{v} \cdot \nabla)\mathbf{v}| \quad \text{and} \quad |\mathbf{\Omega} \times \mathbf{v}| \gg |\nu\nabla^2\mathbf{v}|.$$

Expressing these with the order of magnitude estimation in an analogous way to that carried out in Sec. 4.3, we obtain

$$\Omega U \gg U^2/L \quad \text{and} \quad \Omega U \gg \nu U/L^2,$$

or

$$R_o := U/\Omega L \ll 1 \quad \text{and} \quad E_k := \nu/\Omega L^2 \ll 1. \tag{8.14}$$

The dimensionless number $R_o = U/\Omega L$ is known as the *Rossby number*, and $E_k = \nu/\Omega L^2$ as the *Ekman number*, respectively.

When both the Rossby number and Ekman number are small and the flow is assumed steady, the equation of motion (8.11) reduces to

$$2\mathbf{\Omega} \times \mathbf{v} = -\operatorname{grad} P, \tag{8.15}$$

where $\operatorname{grad} P$ is the resultant of $\operatorname{grad} p/\rho$ and $\operatorname{grad} \chi_* = \mathbf{g}_*$ from (8.9). The velocity field $\mathbf{v} \equiv \mathbf{v}_g(x)$ governed by this equation is called the *geostrophic flow* (Fig. 8.3). In the equipotential surface S_χ defined by $\chi_* = \text{const.}$, we have from (8.15)

$$\operatorname{grad} P \big|_{\chi_*=\text{const.}} = \frac{1}{\rho} \operatorname{grad} p = 2\,\mathbf{v}_g \times \mathbf{\Omega}. \tag{8.16}$$

Namely, $\operatorname{grad} P$ is equivalent to the pressure gradient in S_χ.

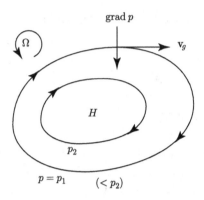

Fig. 8.3.   Geostrophic flow.

An important property of the flows on a rotating frame is now disclosed. The Coriolis force $2\mathbf{\Omega} \times \mathbf{v}$ is perpendicular to the velocity vector $\mathbf{v}$. Hence the pressure gradient in the middle of (8.16) is perpendicular to $\mathbf{v}_g$. Since the pressure gradient is perpendicular to the constant pressure surface, the pressure is constant along the stream-line. This is in marked contrast to the flows in nonrotating frame where flow velocities are often parallel to the pressure gradient (though not always so) and perpendicular to the constant pressure surface.

This feature of geostrophic flows is familiar in weather maps. The weather maps are usually compiled from data of pressure obtained at various observation stations. Iso-pressure contours are drawn on a weather map, and these are also understood as meaning such lines along which the wind is blowing, because the Earth's rotation is a dominating factor for the atmospheric wind. If it takes sufficeint time (i.e. one day or more) for the formation of a wind system such as low- or high-pressures, the atmospheric motion is strongly influenced by the Earth's rotation.[4]

In atmospheric boundary layers up to about 1 km or so from the ground, the viscosity effect cannot be neglected. In such a layer, the wind direction would not necessarily be parallel to the iso-pressure

---

[4]The wind should be observed at a sufficient height above the ground.

contours. This is known as Ekman boundary layer, which will be considered later in Sec. 8.4.

8.3. Taylor–Proudman theorem

In a coordinate frame rotating with a constant angular velocity $\boldsymbol{\Omega}$, steady geostrophic flows are governed by Eq. (8.15) for an inviscid fluid. Setting the axis of frame of rotation as z axis, one can write $\boldsymbol{\Omega} = \Omega \mathbf{k}$ where $\Omega = |\boldsymbol{\Omega}|$ and $\mathbf{k}$ is the unit vector in the direction of z-axis of the Cartesian frame (x, y, z). Denoting the velocity satisfying (8.15) as $\mathbf{v}_g = (u_g, v_g, w_g)$, the (x, y, z) components of (8.15) are written as

$$-2\Omega v_g = -\partial_x P, \quad 2\Omega u_g = -\partial_y P, \tag{8.17}$$

$$0 = -\partial_z P. \tag{8.18}$$

Hence, the modified pressure P is constant along the z-direction (parallel to $\boldsymbol{\Omega}$)[5] and may be expressed as $P = P(x, y)$, and it is found that u and v are independent of z as well from (8.17). Thus, we have

$$\partial_z u = 0, \quad \partial_z v = 0, \tag{8.19}$$

(the suffix g is omitted for brevity). In addition, the function $P/(2\Omega)$ plays the role of a stream function for the "horizontal" component (u, v), since $\mathbf{v}_h := (u, v, 0) = (1/2\Omega)(-\partial_y P, \partial_x P, 0)$ from (8.17).[6] This means that the horizontal flow is incompressible: $\partial_x u + \partial_y v = 0$. All of these (including the z-independent $\mathbf{v}_h$) are known as the *Taylor–Proudman theorem*. The equation of continuity in the form of (3.8), $D\rho/Dt + \operatorname{div} \mathbf{v} = 0$, reduces to

$$\partial_z w = -D\rho/Dt. \tag{8.20}$$

If the fluid is incompressible (i.e. $D\rho/Dt = 0$), we obtain $\partial_z w = 0$.

---

[5] grad P is the resultant of grad p/ρ and grad$\chi_* = \mathbf{g}_*$ from (8.9).
[6] If the plane $z = 0$ coincides with the equi-geopotential surface $\chi_* = $ const., then we have $\rho\, (\partial_x, \partial_y) P(x, y) = (\partial_x, \partial_y) p$, the pressure gradient in the horizontal plane.

Vectorial representation of the horizontal component of $\mathbf{v}_g$ is given by

$$\mathbf{v}_h = \frac{1}{2\Omega}\,\mathbf{k} \times \operatorname{grad} P, \qquad (8.21)$$

while the vertical z component w will be determined in the problem below (Sec. 8.4). The z component of vorticity is given by

$$\omega_g \equiv \partial_x v - \partial_y u = \frac{1}{2\Omega}\,(\partial_x^2 + \partial_y^2)P. \qquad (8.22)$$

8.4.   A model of dry cyclone (or anticyclone)

We now consider a geostrophic *swirling* flow in a rotating frame and its associated Ekman layer in which viscous effect is important. We are going to investigate a very simplified and instructive model of a cyclone or an anticyclone (without moisture) in a rotating frame. In particular, we are interested in a thin boundary layer of a geostrophic flow of a small viscosity over a flat ground at which no-slip condition applies. The boundary layer is called the *Ekman layer*. This is a coupled system of a geostrophic flow and an Ekman layer.

(a) *An axisymmetric swirling flow in a rotating frame*
Suppose that we have a steady axisymmetric swirling flow in a rotating frame, with its axis of rotation coinciding with the z axis. The flow external to the boundary layer is assumed to be a geostrophic flow over a horizontal ground at $z = 0$. Its vorticity ω_g is given by (8.22) and assumed to depend on the radial distance r only, i.e. $\omega_g = \omega(r)$ where $r = (x^2 + y^2)^{1/2}$.

A typical example would be the vorticity of Gaussian function,

$$\omega_G = \omega_0 \exp[-r^2/a^2]$$

with a as a constant. For this solution, $P = P(r)$ can be determined from (8.22).[7] If the constant ω_0 is positive, the pressure is minimum at $r = 0$ by (8.22) in each horizontal plane $z = \text{const.}$, which corresponds to a low pressure (i.e. cyclone). This may be regarded to

---

[7]By using $(\partial_x^2 + \partial_y^2)P = r^{-1}D(rDP)$, where $D \equiv d/dr$ from (D.11).

be a very simplified model of tropical cyclone (without moisture). If $\omega_0 < 0$, the pressure is maximum at $r = 0$, corresponding to a high pressure (anticyclone).

In the polar coordinate (r, θ) of the horizontal (x, y)-plane where $\theta = \arctan(y/x)$, the horizontal velocity $\mathbf{v}_h$ is given by the azimuthal component $V_\theta(r)$ with vanishing radial componet V_r:

$$V_\theta(r) = \frac{\Gamma}{2\pi} \frac{1}{r} (1 - e^{-r^2/a^2}), \quad V_r = 0, \tag{8.23}$$

(see (7.80)), where $\Gamma = \pi a^2 \omega_0$. In fact, we have $\omega = r^{-1} \partial_r (r V_\theta) = \omega_G$ by (D.15). Typical length scale of this swirling flow (8.23) is a, while typical magnitude of this flow is given by $U_s \equiv \Gamma/2\pi a$.

Cartesian components u_g and v_g of $\mathbf{v}_h$ are given by

$$u_g(x, y) = -V_\theta(r)\sin\theta, \quad v_g(x, y) = V_\theta(r)\cos\theta \tag{8.24}$$

$(x = r \cos\theta, \ y = r \sin\theta)$. Substituting these into (8.17) and assuming $P = P(r)$, we obtain the pressure gradient, $dP(r)/dr = 2\Omega V_\theta(r)$.

(b) Ekman boundary layer

The significance of the Ekman boundary layer is seen clearly and is instructive in this axisymmetric model. Motion of the main body of fluid away from the boundary is the axisymmteric flow of swirl considered in (a). In other words, the horizontal (x, y)-plane at $z = 0$ is rotating with the angular velocity Ω with respect to the fixed inertial frame, whereas the main body of fluid away from the plane $z = 0$ is the axisymmetric swirling flow of the velocity $\mathbf{v}_h = (0, V_\theta(r))$ in addition to the frame rotation. Namely, the z component of total vorticity is given by $2\Omega + \omega_G$ against the inertial frame since the uniform rotation of the frame has the vorticity 2Ω.

The fluid viscosity ν is assumed to be very small, so that the boundary layer adjacent to the wall is very thin in a relative sense. Then, the governing equation is given by

$$2\Omega \times \mathbf{v} = -\operatorname{grad} P + \nu \nabla^2 \mathbf{v}, \tag{8.25}$$

from (8.11),[8] instead of (8.15) for inviscid geostrophic flows. We assume that the Reynolds number R_e is much larger than unity, $R_e := U_s a/\nu \gg 1$.

Now we consider the boundary layer adjacent to the ground $z = 0$. Just like the boundary layer of nonrotating system investigated in Sec. 4.5, it is assumed that variations (derivatives) of velocity $\mathbf{v} = (u, v, w)$ with z are much larger than those with x or y. It is found that Eq. (8.25) reduces to

$$-2\Omega v = -\partial_x P + \nu \partial_z^2 u, \tag{8.26}$$

$$2\Omega u = -\partial_y P + \nu \partial_z^2 v, \tag{8.27}$$

$$0 = -\partial_z P + \nu \partial_z^2 w. \tag{8.28}$$

The continuity equation (3.8) is

$$\partial_x u + \partial_y v + \partial_z w + D\rho/Dt = 0. \tag{8.29}$$

From (8.29), we deduce that w is much smaller than the horizontal components, and estimate as $|w|/U_s = O(\delta/a) \ll 1$ according to the usual argument of boundary layer theory (see Problem 4.4(i) and its solution), where δ is a representative scale of the boundary layer.

By applying the same argument of Problem 4.4(i) to (8.28), we obtain

$$\frac{|\Delta P|}{U_s^2} = O(R_e^{-1}), \quad R_e = U_s a/\nu \gg 1.$$

Namely, variation ΔP across the boundary layer normalized by U_s^2 is very small and may be neglected. Therefore, the pressure in the boundary layer is imposed by that of the external flow. The flow external to the boundary layer is the swirling flow (8.23) whose P is essentially a function of x and y only (independent of z). The derivatives $\partial_x P$ and $\partial_y P$ are given by (8.17).

Eliminating $\partial_x P$ by using (8.17), Eq. (8.26) is reduced to the form $-2\Omega(v - v_g) = \nu \partial_z^2 u$. The right-hand side can be written as $\nu \partial_z^2(u - u_g)$ since u_g is independent of z. Therefore, we have $-2\Omega(v - v_g) =$

---

[8]In the axisymmetric swirling flow under consideration, the term $(\mathbf{v} \cdot \nabla)\mathbf{v} = \mathbf{V}_\theta \partial_\theta \mathbf{v} + w\partial_z \mathbf{v}$ on the left of (8.11) vanishes (for incompressible flows).

$\nu \partial_z^2(u-u_g)$. Similarly, Eq. (8.27) is reduced to the form $2\Omega(u-u_g) = \nu \partial_z^2(v - v_g)$ by eliminating $\partial_y P$. Thus, we obtain

$$-2\Omega(v - v_g) = \nu \partial_z^2(u - u_g), \tag{8.30}$$

$$2\Omega(u - u_g) = \nu \partial_z^2(v - v_g). \tag{8.31}$$

This can be integrated immediately. In fact, multiplying (8.31) by the imaginary unit $i = \sqrt{-1}$ and adding it to Eq. (8.30), we obtain

$$2\Omega i\, X = \nu \partial_z^2 X, \tag{8.32}$$

$$X := (u - u_g) + i(v - v_g). \tag{8.33}$$

Solving (8.32), we immediately obtain a general solution:

$$X = Ae^{-(1+i)\zeta} + Be^{(1+i)\zeta}, \quad \zeta \equiv z/\delta, \tag{8.34}$$
$$\delta = \sqrt{\nu/\Omega},$$

where $[\pm(1+i)]^2 = 2i$ is used. To match with the outer swirling flow (u_g, v_g), we require the boundary condition:

$$X \equiv (u - u_g) + i(v - v_g) \to 0, \quad \text{as } \zeta \to \infty.$$

So that, we must have $B = 0$. On the boundary wall at $\zeta = 0$, we impose $u = 0$ and $v = 0$. Hence we have $A = -(u_g + iv_g)$. Thus the solution is $X = -(u_g + iv_g)\, e^{-(1+i)\zeta}$. Eliminating X by (8.33), using (8.24), and finally separating the real and imaginary parts, we find the expressions of u and v.[9] From these, radial and azimuthal components are derived as follows:

$$v_r(r, z) = u \cos\theta + v \sin\theta = -V_\theta(r)\, e^{-\zeta} \sin\zeta, \tag{8.35}$$

$$v_\theta(r, z) = -u \sin\theta + v \cos\theta = V_\theta(r)\left(1 - e^{-\zeta} \cos\zeta\right), \tag{8.36}$$

where $V_\theta(r)$ is given by (8.23), with $\theta = \tan^{-1}(y/x)$ and $r = (x^2 + y^2)^{1/2}$. It is found that, as $\zeta = z/\delta$ increases from 0 across the *Ekman layer* of thickness $\delta = \sqrt{\nu/\Omega}$, the horizontal velocity (v_r, v_θ) rotates its direction, and finally asymptotes to $(0, V_\theta)$ at the outer edge of the layer as $\zeta \to \infty$.

---

[9] $u = u_g - e^{-\zeta}(u_g \cos\zeta + v_g \sin\zeta)$ and $v = v_g - e^{-\zeta}(v_g \cos\zeta - u_g \sin\zeta)$. See (8.24).

Note: If $\omega_0 > 0$, we have a low pressure as noted before and $V_\theta > 0$ from (8.23) since $\Gamma > 0$. Then we have $v_r < 0$, i.e. an inward flow within the boundary layer. On the other hand, if $\omega_0 < 0$, we have a high pressure and $V_\theta < 0$. Then we have $v_r > 0$, i.e. an outward flow within the layer.

It is remarkable that there is a small, but significant vertical velocity w_g at the outer edge of the boundary layer. To see this, note the following by assuming incompressibility div $\mathbf{v} = 0$ and using (D.9):

$$\partial_z w = -\partial_x u - \partial_y v = -r^{-1}\partial_r(rv_r) = \omega_G \, e^{-\zeta}\sin\zeta,$$

$$\omega_G(r) \equiv r^{-1}\partial_r(rV_\theta),$$

where $\omega_G(r)$ is the vertical vorticity in the outer geostrophic flow, assumed as $\omega_G = \omega_0 \exp[-r^2/a^2]$ in (a). On integrating this with respect to $\zeta = z/\delta$ from 0 to ∞,[10] we obtain the value of vertical velocity w_E at the outer edge of the Ekman layer (using $w = 0$ at $z = 0$), which is found to be proportional to ω_G:

$$w_E(x,y) = \omega_G(x,y)\int_0^\infty e^{-\zeta}(\sin\zeta)\,\delta\,\mathrm{d}\zeta = \frac{1}{2}\delta\,\omega_G(x,y)$$

$$= \sqrt{\frac{\nu}{4\Omega}}\,\omega_G(x,y). \tag{8.37}$$

It is remarkable to find that there is a nonzero vertical flow w_E at the outer edge of the Ekman layer, which is positive or negative according as the sign of ω_0. Namely, in the case of low pressure (or like a cyclone), it is an upward flow (Fig. 8.4). This is connected with the inward flow within the boundary layer, and termed as *Ekman pumping*. The total influx F_{in} at a position r within the Ekman layer across a cylindrical surface of radius r is given by the integral of influx, $\int 2\pi r v_r(z)\mathrm{d}z$, which is immediately calculated as

$$F_{in}(r) = -2\pi r v_\theta \frac{1}{2}\delta = -\frac{1}{2}\sqrt{\frac{\nu}{\Omega}}\,\pi a^2\,\omega_0(1 - \exp[-r^2/a^2]).$$

---

[10] $\int_0^\infty e^{-\zeta}(\sin\zeta)\,\mathrm{d}\zeta = \frac{1}{2}.$

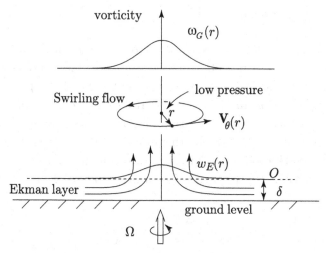

Fig. 8.4. Schematic sketch of a dry cyclone with Ekman boundary layer on a system rotating with Ω. A geostrophic swirling flow of velocity $V_\theta(r)$ with an axial vorticity $\omega_G(r)$ of Gaussian form induces upward pumping flow $w_E(r)$ at the edge of the Ekman layer.

This kind of structure of low pressure is often observed in tropical cyclones, although this model is dry and very simplified. In the case of high, w_E is downward and connected with the outward flow within the Ekman layer.

Finally, a remark must be made on *real* atmospheric conditions. Observations indicate that the wind velocity in the atmospheric boundary layer approaches its upper geostrophic value at about 1 km above the ground. If we let $5\delta = 1000$ m and $\Omega = 10^{-4}$ ($\approx 2\pi/(24 \times 60 \times 60)$) s^{-1} for the earth spin, the definition $\delta = \sqrt{\nu/\Omega}$ implies the value of ν about 4×10^4 cm^{2}s^{-1} [Hol04]. This is much larger than the molecular viscosity.

The traditional approach to this problem is to assume that turbulent eddies act in a manner analogous to molecular diffusion, so that the momentum flux is dominated by turbulent action, and the viscosity terms in (8.26)–(8.28) are replaced by $\nu_{\text{turb}}\, \partial_z^2 \mathbf{v}$, where ν_{turb} is the *turbulent* viscosity which is found as $\nu_{\text{turb}} \approx 4 \times 10^4$ cm^{2}s^{-1} in the above estimate. Later in Sec. 10.1, we will consider Reynolds stress $R_{ij} = \rho \langle u'_i u'_j \rangle$. The above is equivalent to $\partial_j R_{ij} = -\rho\nu_{\text{turb}}\, \partial_z^2 \langle u_i \rangle$. In this case, average velocity field must be considered.

8.5. Rossby waves

Rossby wave is another aspect of flows resulting from the rotation of a system, which is responsible for a certain wavy motion of the atmosphere. Waves of global scales observed on the Earth (or planets) are known as *planetary waves* (Fig. 8.5). Rossby wave is one such wave which occurs owing to the variation of Coriolis parameter (f defined below) with the latitude ϕ. For the global scales or synoptic scales L (typically 1000 km or larger in horizontal dimension which are very much larger than the vertical scale of atmosphere of about 10 km),

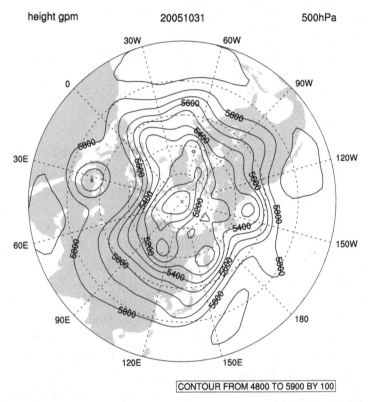

Fig. 8.5. Planetary waves, exhibited by the geopotential height z_g (m) of a constant-pressure surface (of 500 hPa corresponding to about 5000 m) in the northern hemisphere $\phi > 15°$ on 31 Oct. 2005, plotted with the NCEP/NCAR reanalysis data, by the courtesy of Dr. T. Enomoto (The Earth Simulator Center, Japan).

vertical velocities are very much smaller than horizontal velocities (see the next section), so that the vertical component of velocity **v** can be neglected in the equation of motion. Such a motion is termed as *quasi-horizontal.*

The viscous term $\nu\nabla^2\mathbf{v}$ is neglected. The equation of motion (8.11) reduces to

$$\mathrm{D}_t\mathbf{v} + 2\mathbf{\Omega} \times \mathbf{v} = -\mathrm{grad}\,P, \quad \mathrm{D}_t\mathbf{v} \equiv \partial_t\mathbf{v} + \mathbf{v} \cdot \nabla\mathbf{v}. \tag{8.38}$$

A convenient frame of reference at a point O on the earth's surface is given by the cartesian system of x directed towards the east, y towards the north and z vertically upwards, where the latitude of the origin O is denoted by ϕ (Fig. 8.6). Using $\mathbf{i}, \mathbf{j}, \mathbf{k}$ for unit vectors along respective axis and writing $\mathbf{v} = (u, v, w)$ and $\mathbf{\Omega} = \Omega(0, \cos\phi, \sin\phi)$, we have

$$2\mathbf{\Omega} \times \mathbf{v} = 2\Omega(w\cos\phi - v\sin\phi)\mathbf{i} + 2\Omega u\sin\phi\,\mathbf{j} - 2\Omega u\cos\phi\,\mathbf{k}.$$

A simplest Rossby wave solution is obtained for an atmosphere of constant density ρ under the assumption of no vertical motion $w = 0$.

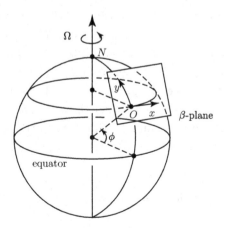

Fig. 8.6. β-plane.

The equations for a horizontal motion are, from (8.38) and (3.9),

$$D_t u - fv = -\partial_x P, \tag{8.39}$$

$$D_t v + fu = -\partial_y P, \tag{8.40}$$

$$\partial_x u + \partial_y v = 0, \tag{8.41}$$

$$f = 2\Omega \sin\phi : \text{(Coriolis parameter)}, \tag{8.42}$$

and $D_t = \partial_t + u\partial_x + v\partial_y$ by neglecting terms of $O(L/R)$ because the reference frame is a curvilinear system, where L is the scale of flow and R the radius of the earth (assumed a sphere).

Operating ∂_y on (8.39) and ∂_x on (8.40), and subtracting, we obtain[11]

$$D_t \zeta + v\partial_y f = 0, \tag{8.43}$$

(see Problem 8.2), where $\zeta = \partial_x v - \partial_y u$ is the z component of the vorticity curl $\mathbf{v}$, and the term $f(\partial_x u + \partial_y v)$ was omitted due to (8.41). Since $f (= 2\Omega \sin\phi)$ depends only on $y = a(\phi - \phi_0)$ (where $\phi(O) = \phi_0$), the above equation can be written in the form,

$$D_t [\zeta + f] = 0. \tag{8.44}$$

The quantity $\zeta + f$ is regarded as the z component of absolute vorticity, for ζ is the z component of vorticity which is twice the angular velocity of local fluid rotation and f is twice the angular velocity of frame rotation around the vertical axis. Therefore Eq. (8.44) is interpreted as the conservation of absolute vorticity $\zeta + f$.

In order to find an explicit solution of (8.44), we assume a linear variation of f with respect to y, i.e. $f = f_0 + \beta y$ where β is a constant as well as f_0. This is known as the *β-plane approximation*.[12] An unperturbed state is assumed to be a uniform zonal flow $(\bar{u}, 0)$. Writing the perturbed state as $u = \bar{u} + u'$ and $v = v'$, Eq. (8.44) is linearized to the form,

$$(\partial_t + \bar{u}\partial_x)(\partial_x v' - \partial_y u') + \beta v' = 0. \tag{8.45}$$

---

[11]$D_t u = \partial_t u + \partial_x(\mathbf{v}^2/2) - v\zeta$ and $D_t v = \partial_t v + \partial_x(\mathbf{v}^2/2) + u\zeta$ for $\zeta = \partial_x v - \partial_y u$.

[12]The horizontal plane (x, y) is called the β-plane.

Owing to the condition of nondivergent flow (8.41), one can introduce a stream function ψ (see Appendix B.2), such that

$$u' = \partial_y \psi, \quad v' = -\partial_x \psi.$$

Then we have $\partial_x v' - \partial_y u' = -\nabla^2 \psi$. Substituting these in (8.45),

$$(\partial_t + \bar{u}\partial_x)\nabla^2\psi + \beta\partial_x\psi = 0. \tag{8.46}$$

Assuming a normal mode for the perturbation, $\psi = A\,\exp[i(-\omega t + kx + ly)]$, and substituting in the above equation, we obtain

$$i(-\omega + \bar{u}k)(-k^2 - l^2)\,\psi + i\beta k\,\psi = 0,$$

since $\partial_t = -i\omega$, $\partial_x = ik$, $\partial_y = il$. This is rewritten for the phase velocity c (positive for propagation to the east) as

$$c \equiv \frac{\omega}{k} = \bar{u} - \frac{\beta}{k^2 + l^2}. \tag{8.47}$$

The velocity relative to the zonal flow is

$$c - \bar{u} = -c_{\text{Rossby}}(< 0), \quad c_{\text{Rossby}} := \frac{\beta}{k^2 + l^2}.$$

Therefore, the Rossby waves drift to the west relative to the unperturbed zonal flow $\bar{u}$. The phase speed of the wave decreases with the wavenumber, hence increases with the wavelength. Note that relative to the frame moving with $(-c_{\text{Rossby}}, 0)$ (westward), the wave motion becomes steady.

Rossby waves are observed at mid or higher latitudes and are responsible for the high-pressure and low-pressure wavy patterns observed there. A typical speed of the wave is estimated to be about a few meters per second [Hou77, Problem 8.14].

8.6. Stratified flows

Stratified fluid is named for a fluid having a vertical variation of density in a gravitational field. Flows of such a fluid are of considerable importance in geophysical flows. An example is the thermal convection in a horizontal fluid layer which is caused by vertical temperature variations under uniform vertical gravity. Salinity variations play an important role in dynamical oceanography. The density may either increase or decrease with height. In the present section, we are concerned with the latter case, that is to say the case of *stable stratification*, and in addition, frame rotation is neglected ($\Omega = 0$). The case of unstable stratification, i.e. the density increases with height, will be studied later in the chapter on instability.

Vertical motions are suppressed in the stable stratification because those flows tend to carry heavier fluid upwards and lighter fluid downwards, requiring an additional potential energy. This suppression modifies the flow pattern of laminar flows. Note that, in the previous section of Rossby waves, we studied purely horizontal wave motions.

For stratified flows, the Boussinesq approximation (see Sec. 9.4.1)[13] is usually applied to the equations of motion. In this approximation, the flow is regarded as incompressible, that is,

$$\operatorname{div} \mathbf{v} = \partial_x u + \partial_y v + \partial_z w = 0. \tag{8.48}$$

Then the equations of density (3.8) and motion (8.1) are

$$\partial_t \rho + (\mathbf{v} \cdot \nabla)\rho = 0, \tag{8.49}$$

$$\rho\big(\partial_t \mathbf{v} + (\mathbf{v} \cdot \nabla)\mathbf{v}\big) = -\operatorname{grad} p + \rho\mathbf{g} + \mu\Delta\mathbf{v}, \tag{8.50}$$

where $\mathbf{g} = (0, 0, -g_*(z))$ is the acceleration of gravity in the z-direction. Furthermore, we suppose that the flow is steady, and assume that viscous (and diffusive) effect is negligible. Then we obtain, from (8.49) and (8.50), as

$$(\mathbf{v} \cdot \nabla)\rho = 0, \tag{8.51}$$

$$\rho(\mathbf{v} \cdot \nabla)\mathbf{v} = -\operatorname{grad} p + \rho\mathbf{g}. \tag{8.52}$$

When the fluid is at rest ($\mathbf{v} = 0$), Eq. (8.51) is satisfied identically. Since the density depends on z only, Eq. (8.52) reduces to

$$0 = -\frac{\mathrm{d}}{\mathrm{d}z}p_* - \rho_s(z)g_*(z)$$

---

[13]In most analysis of thermal convection, density variation is considered for the gravity force only in the equation of motion, and is ignored in the other terms.

where $\rho_s(z)$ is the static density distribution, g_* the geo-acceleration (8.10) including centrifugal force, and p_* is the *modified* pressure given by

$$p_*(z) = \text{const.} - \int^z g_* \rho_s(z') \, dz'. \tag{8.53}$$

We now consider a flow of velocity field $\mathbf{v}(\mathbf{x}) = (u, v, w)$ with a length scale L and a magnitude of horizontal velocity U. This will modify the density field, now represented as $\rho = \rho_s(z) + \rho_1$ and $p = p_*(z) + p_1$. Then, Eq. (8.51) becomes

$$(\mathbf{v} \cdot \nabla)\rho_1 + w\rho_s'(z) = 0.$$

In an order-of-magnitude estimate, we have

$$|\rho_1| \sim \frac{WL}{U} |\rho_s'|, \tag{8.54}$$

where $\rho_s' = d\rho_s/dz$, and W denotes the scale of vertical velocity w. In the Boussinesq approximation, the density variation is taken into consideration only in the buoyancy force and the density on the left-hand side of (8.52) is given by a representative density ρ_0. Then Eq. (8.52) becomes

$$\rho_0 \, \boldsymbol{\omega} \times \mathbf{v} + \rho_0 \, \text{grad} \left(\frac{1}{2} v^2 \right) = -\text{grad} \, p_1 + \rho_1 \mathbf{g}.$$

In order to eliminate the pressure term (and the grad term on the left), we take curl of this equation and obtain

$$\rho_0 \nabla \times (\boldsymbol{\omega} \times \mathbf{v}) = \nabla \rho_1 \times \mathbf{g}, \tag{8.55}$$

since $\nabla \times \mathbf{g} = 0$. Thus, we find another estimate for the magnitude of ρ_1. Using the estimates, $|\nabla \rho_1| \sim |\rho_1|/L$, $|\mathbf{g}| = g$ and $|\boldsymbol{\omega}| \sim U/L$, we obtain the following:

$$|\rho_1| \sim \frac{\rho_0 U^2}{gL}. \tag{8.56}$$

Comparison of (8.56) with (8.54) indicates that a typical ratio of the vertical velocity W with respect to the horizontal velocity U is given as

$$\frac{W}{U} \sim \frac{\rho_0 U^2}{gL^2|\rho_s'|} = \frac{1}{R_i} = (F_r)^2, \qquad (8.57)$$

where R_i is called the *Richardson number*, and F_r the *Froud number*.[14]

When the Froud number is small (or equivalently, the Richardson number is large), the vertical motion is much weaker than the horizontal motion. That is the case for most geophysical flows in which the length scale L is of the order of earth itself, or so. Actually, the Richardson number is very large in such flows and horizontal motions are predominating. The *jet-stream* in the stratosphere is a typical example of such a horizontal flow.

In a laboratory experiment too, such a suppression of vertical motion of a stratified flow can be observed in the visualization experiment of Fig. 8.7.

8.7.   Global motions by the Earth Simulator

The Earth Simulator (ES)[15] constructed in Japan as a national project (proposed in 1997) was the largest and fastest computer in the world as of March 2003. The computer started its operation in February 2002. The project aims for computer simulations of *global* motions of atmosphere and ocean that previous computers were unable to accomplish, as well as simulations of slow plastic viscous motions of the earth's interior, and also predictions of future climate and environment of the earth. This is a gigantic computer consisting of 5,120 super-computer units. The performance of the computing speed was reported as 36Tflops, i.e. 36×10^{12} operations per second.

---

[14]The Froud number is defined by $F_r = U/\sqrt{gL}$ in hydraulics, where the density changes discontinuously at the free water surface and $L|\rho_s'|$ is replaced by ρ_0.
[15]The Earth Simulator Center is located at JAMSTEC (Japan Agency for Marine-Earth Science and Technology), Kanazawa-ku, Yokohama, Japan, 236-0001. Website: http://www.es.jamstec.go.jp/esc/eng/ESC/index.html; *Journal of the Earth Simulator* can be viewed at this site.

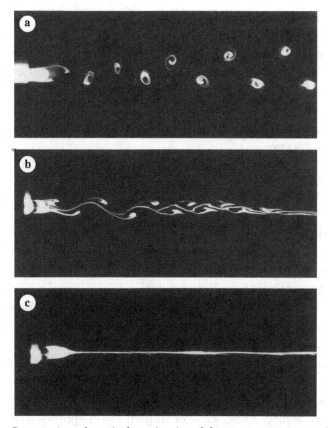

Fig. 8.7.   Suppression of vertical motion in a laboratory experiment: Flow of a water (from left to right) stratified vertically of density $\rho_s(z)$ (with a stable linear salinity component) around a horizontal circular cylinder (on the right) of diameter $d = 1.0$ cm and $U \approx 1.3$ cm/s at Re $= Ud/\nu \approx 113$. Froude number is $F_r =$ (a) ∞ (no stratification), (b) 0.2, and (c) 0.1 (with $L = 10d$). The photograph is provided through the courtesy of Prof. H. Honji (Kyushu University) [Hon88].

In numerical weather predictions carried out by previous ordinary super-computers, the horizontal mesh scale was of the order of 100 km or so, while by the Earth Simulator it is 10 km or less. This was made possible by the improvement in performance by 10^3 times of previous computers. In future, it aims for 3 km or less horizontal resolution, in order to improve predictions of local weather considerably.

In the first three years, experimental computer simulations have been carried out by new computation codes in order to test the

performance of the giant computer ES, and some innovative results
have been obtained already. The next two subsections are brief
accounts of the results of global atmospheric motions and ocean circulations, reported in the *Journal of the Earth Simulator* (see the
footnote).

8.7.1. *Simulation of global atmospheric motion by AFES code*

Experimental test simulations of global atmospheric motions were
carried out by the AFES code (Atmospheric General Circulation
Model for the Earth Simulator), which is a new computation code.
This enabled global coverage of both the large-scale general circulations and meso-scale phenomena, and their interaction [AFES04].

The first experimental computation was a long-term simulation of
twelve years, with intermediate resolutions. The initial condition was
the data set of January 1, 1979, with the boundary conditions of daily
fields of sea-surface temperature, sea-ice cover and surface topogra-
phy. Next, using the resulting data, additional short-term simulations
were performed with the finest meshes of about 10 km horizontal scale
and 20 to 500 m altitude resolution (with finer meshes in the atmospheric boundary layer), in order to find detailed local evolutions
such as (a) Winter cyclones over Japan, (b) Typhoon genesis over
the tropical western Pacific, and (c) Cyclone-genesis in the south-
ern Indian Ocean. It has been found that *the interactions among
global atmospheric motions and regional meso-scales are simulated
in a fairly realistic manner.*

8.7.2. *Simulation of global ocean circulation by OFES code*

Experimental test simulations of global ocean circulations were carried out by the OFES code, which is an optimized code of MOM3
(Oceanic General Circulation Model) developed by GFDL (USA),
adapted to the Earth Simulator. The OFES enabled a fifty-year
eddy-resolving simulation of the world ocean general circulation
within one month [OFES04].

The initial conditions for the experimental integration of 50 years are the annual mean sea temperature and salinity fields without motion of sea water. Hence, this is called a *spin-up* problem, and also called a climatological 50-year Integration. The computation domain was the area from 75°S to 75°N (except the Arctic ocean) with the horizontal grid spacing 1/10° (about 10 km) and 54 vertical levels. The computation used the monthly mean wind stresses averaged from 1950 to 1999 (hence twelve data sets) taking account of fresh water influx from both precipitation and rivers.

In the spin-up computation, five active regions have been identified (Fig. 8.8), where the sea surface height variability is significantly large: Kuroshio current in the west of northern Pacific, Gulf stream in the west of northern Atlantic, the three currents in the south western Indian Ocean, south western Pacific, south western Atlantic. In addition, the sea surface height of the Agulhas rings was clearly and realistically visualized. *These results demonstrate a promising capability of OFES, and the 50-year spin-up run represent realistic features of the world ocean both in the mean fields and eddy activities.*

Following the spin-up run, a *hindcast* run was carried out to study various ocean phenomena and compare with the past observations.

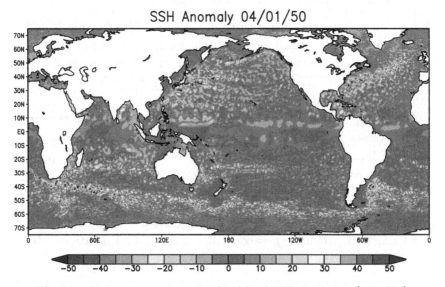

Fig. 8.8.   Five active regions visualized by OFES simulation [OFES04].

Among others, it is remarkable that the modes of El Nino and Indian Ocean Dipole have been found and are in good agreement with the observations. In view of the fact that tropical oceans play major roles in the variability of world climate, this finding is encouraging for future investigations.

These results show that hyper-resolution computations can be powerful tools for investigating not only the large-scale global circulations, but also the meso-scale phenomena such as cyclone-genesis, eddies and instability, and thirdly the interaction between global-scales and meso-scales, spatially and temporally.

Owing to the Earth Simulator, the study of *global* atmosphere and ocean has become a *science* in the sense that experimental tests can be done and quantitative comparison with observations is made possible. This is in contrast with the analyses in the previous sections: Sec. 8.2 for geostrophic flows, Sec. 8.5 for Rossby waves and Sec. 8.6 for stratified flows. They are all local analyses or approximate model analyses, not a global analysis.

8.8.   Problems

Problem 8.1 Thermal wind and jet stream

Horizontal pressure gradient arises in the atmosphere owing to temperature variation which is related to the horizontal density gradient. This is responsible to the thermal wind in a rotating system. Global distribution of average atmospheric temperature $\bar{T}(x, y, z)$ generates a thermal wind, called the jet stream (i.e. westerly).

Assuming $2\mathbf{\Omega} = f\mathbf{k}$ ($f = 2\Omega \sin \phi$) in (8.15) with $\mathbf{k}$ the unit vertical vector in the local β-plane approximation (of Sec. 8.5), we have the horizontal component $\mathbf{v}_h$ (projection to the (x, y)-plane) of the geostrophic velocity given by (8.21): $\mathbf{v}_h = (1/f)\mathbf{k} \times \operatorname{grad} P$. Answer the following questions.

(i) Suppose that $z = z_p(x, y)$ represents a surface of the altitude of constant pressure ($p = \text{const.}$). Using the hydrostatic relation (8.13), show the following equation,

$$\rho g_* \operatorname{grad} z_p(x, y) = \operatorname{grad} p(x, y; \chi_* = \text{const.}). \qquad (8.58)$$

[*Hint*: On the surface $z = z_p(x, y)$, $dp|_{y=\text{const}} = \partial_x p \, dx + \partial_z p \, dz_p = 0$.]

Thus, *the pressure gradient in the horizontal surface* $\chi_* = \text{const.}$ *is related to the inclination of the surface* $z = z_p(x, y)$. The global mean pressure in the northern hemisphere is higher toward the south (i.e. toward the equator), because the temperature is higher in the south. In the horizontal plane, we obtain $\rho \, \text{grad} \, P = \text{grad} \, p$, since $\chi_* = \text{const.}$ (see the footnotes to Sec. 8.3). Therefore, the horizontal component of geostrophic velocity $\mathbf{v}_h$ (proportional to $\mathbf{k} \times \text{grad} \, P$) is toward the east (i.e. westerly). This will be considered in terms of the temperature in the next questions.

(ii) Suppose that we consider two p-constant surfaces p_0 and $p_*(< p_0)$ and the layer in-between, by using the equation of state $p = R\rho T$ for an ideal gas ($R = R_*/\mu_m$, R_*: the gas constant; see the footnote to Sec. 1.2)). Show the following relation (8.59):

$$\text{grad} \, P|_{p=p_*} - \text{grad} \, P|_{p=p_0} = \text{grad} \, \chi_*(p_*) - \text{grad} \, \chi_*(p_0)$$

$$= RP(p_*) \, \text{grad} \, \bar{T}, \tag{8.59}$$

where
$$\chi(p_*) - \chi(p_0) = \int_{z(p_0)}^{z(p_*)} g_*(z) \, dz = -\int_{p_0}^{p_*} \frac{dp}{\rho} \tag{8.60}$$

$$= R\Pi(p_*) \, \bar{T}(x, y), \tag{8.61}$$

$$\Pi(p_*) = \int_{p_*}^{p_0} dp/p, \quad \bar{T}(x, y) = \frac{1}{z - z_0} \int_{z_0}^{z} T(x, y, z) \, dz.$$

($\bar{T}(x, y)$ is the vertical average of temperature). Thus, we obtain the equation for the thermal wind:

$$\mathbf{v}_h(p_*) - \mathbf{v}_h(p_0) = \frac{1}{f} RP(p_*)\mathbf{k} \times \text{grad} \, \bar{T}. \tag{8.62}$$

(iii) Suppose that the temperature $\bar{T}$ is higher toward the south and lower toward the north and uniform in the east-west direction. This occurs often in the mid-latitudes of the northern hemisphere and referred to as *baroclinic* [Hou77]. Which direction the geostrophic velocity $\mathbf{v}_h(p_*)$ is directed if $\mathbf{v}_h(p_0) = 0$?

Chapter 9

Instability and chaos

*Not every solution of equations of motion can actu-
ally occur in Nature, even if it is exact. Those which
do actually occur not only must obey the equations of
fluid dynamics, but must be also stable* (Landau and
Lifshitz [LL87, Sec. 26]).

In order that a specific steady state S can be observed in nature,
the state must be *stable*. In other words, when some external perturbation happens to disturb a physical system, its original state must
be recovered, i.e. the perturbation superposed on the basic state S
must decay with time. In nature there always exists a source of disturbance. One of the basic observations in physics is as follows: if a
certain macroscopic physical state repeatedly occurs in nature, then
it is highly possible that the state is characterized by a certain type
of stability.

If a small perturbation grows with time (exponentially in most
cases), then the original state S is said to be *unstable*. In such a
case, the state would have little chance to be observed in nature. If
the perturbation neither grows nor decays, but stays at the initial
perturbed level, then it is said to be *neutrally stable*.

When the initial state is unstable and the amplitude of perturbation grows, then a nonlinear mechanism which was ineffective at
small amplitudes makes its appearance in due course of time. This
nonlinearity often suppresses further exponential growth of perturbation (though not always so) and the amplitude tends to a new

finite value. The new state thus established might be a steady state, a periodic state, or irregularly fluctuating state. The last one is often said to be *turbulent*. If the initial state is a steady smooth flow, then it is said to be *laminar*.

9.1. Linear stability theory

Stability analysis of a steady state (called a *basic state*) is carried out in the following way. Basic state whose stability is to be studied is often a *laminar flow*. Various laminar flows of viscous fluids are presented in Chapter 4: boundary layer flows, parallel shear flows, rotating flows and flows around a solid body, etc.

Suppose that the basic state is described by a steady velocity field $\mathbf{v}_0(\mathbf{x})$ and an infinitesimal perturbation velocity $\mathbf{v}_1(\mathbf{x}, t)$ is superimposed on it. Then the total velocity is $\mathbf{v} = \mathbf{v}_0 + \mathbf{v}_1$. It is assumed that the velocity satisfies the divergence-free condition and the density ρ_0 is a constant. Then the equation of motion is given by

$$\partial_t \mathbf{v} + (\mathbf{v} \cdot \nabla)\mathbf{v} = -(1/\rho_0)\,\nabla p + \nu \Delta \mathbf{v} + \mathbf{f}, \qquad (9.1)$$

(see (4.8)), supplemented with $\operatorname{div} \mathbf{v} = 0$. When we consider a stability problem of an ideal fluid, the viscosity ν is set to be zero.

Let us write the velocity, pressure and force as

$$\mathbf{v} = \mathbf{v}_0 + \mathbf{v}_1, \quad p = p_0 + p_1, \quad \mathbf{f} = \mathbf{f}_0 + \mathbf{f}_1.$$

Suppose that the basic steady parts $\mathbf{v}_0, p_0, \mathbf{f}_0$ are given. The equation the basic state satisfies is

$$(\mathbf{v}_0 \cdot \nabla)\mathbf{v}_0 = -(1/\rho_0)\,\nabla p_0 + \nu \Delta \mathbf{v}_0 + \mathbf{f}_0, \qquad (9.2)$$

and $\operatorname{div} \mathbf{v}_0 = 0$. Subtracting this from (9.1), we obtain an equation for the perturbations. Since the perturbations are assumed to be infinitesimal, only linear terms with respect to perturbations $\mathbf{v}_1, p_1, \mathbf{f}_1$ are retained in the equation, and higher-order terms are omitted. Thus we obtain a *linearized* equation of motion:

$$\partial_t \mathbf{v}_1 + (\mathbf{v}_0 \cdot \nabla)\mathbf{v}_1 + (\mathbf{v}_1 \cdot \nabla)\mathbf{v}_0 = -(1/\rho_0)\,\nabla p_1 + \nu \Delta \mathbf{v}_1 + \mathbf{f}_1. \qquad (9.3)$$

One can carry out the so-called the *normal mode analysis*, since the coefficients of (9.3) to the perturbations $\mathbf{v}_1$, p_1 and $\mathbf{f}_1$ are all time-independent, so that a time factor of the fluctuating components is represented by an exponential function, say $e^{\alpha t}$. In such a case, the fluctuating components $\mathbf{v}_1$, p_1 and $\mathbf{f}_1$ can be represented by the product of $e^{\alpha t}$ and functions of the spatial variable $\mathbf{x}$. For example, the velocity $\mathbf{v}_1$ is represented as

$$\mathbf{v}_1(\mathbf{x}, t, \alpha) = e^{(\alpha_r + i\alpha_i)t}\, \mathbf{u}(\mathbf{x}, \alpha), \tag{9.4}$$

where $\alpha = \alpha_r + i\alpha_i$ is the complex frequency. We have analogous expressions for p_1 and $\mathbf{f}_1$. (This corresponds to a Fourier analysis.)

In general, the solutions satisfying the above equation (9.3) and boundary conditions (to be specified in each problem) are determined for particular *eigenvalues* of α, which may be discrete or continuous, and in addition, complex in general.[1] Correspondingly, the *eigensolutions* $\mathbf{v}_1(\mathbf{x}, t, \alpha)$, $p_1(\mathbf{x}, t, \alpha)$ and $\mathbf{f}_1(\mathbf{x}, t, \alpha)$ are complex. A general solution is given by a linear superposition of those particular solutions.

If the real part α_r of eigenvalue α is negative, then the fluctuation decays exponentially with time, whereas if it is positive, the fluctuation grows exponentially with time. Thus it is summarized as

$$\text{stable if } \alpha_r < 0; \quad \text{unstable if } \alpha_r > 0.$$

The imaginary part α_i of the eigenvalue gives the oscillation frequency of the fluctuation as seen from (9.4). Suppose that we consider a transition sequence in which the eigenvalue α_r changes from negative to positive values when $\alpha_i \neq 0$. In the beginning, the fluctuating state with $\alpha_r < 0$ decays to a steady state. Once α crosses the state $\alpha_r = 0$, a growing mode $\mathbf{v}_1$ with a frequency α_i appears. This is understood as the *Hopf bifurcation*.

The above analysis is called **linear stability theory**. In subsequent sections, we consider the instability and chaos by selecting typical example problems, i.e. the *Kelvin–Helmholtz instability*, stability of parallel shear flows and *thermal convection*.

---

[1] Very often, solutions are determined only for complex values of α.

9.2.   Kelvin–Helmholtz instability

One of the most well-known instabilities in fluid mechanics is the instability of a vortex sheet which is a surface of discontinuity of tangential velocity (Problem 7.4). The Kelvin–Helmholtz theorem states that the vortex sheet is unstable in inviscid fluids. Another well-known instability is the Rayleigh–Taylor instability, which will be considered in Problem 9.1.

Let us investigate the motion of the surface of discontinuity located at $y = 0$ in the unperturbed state [Fig. 4.5(a)]. It is assumed that the velocity of the *basic state* in (x, y)-plane is given by

$$\left(\frac{1}{2}U, 0\right) \quad \text{for } y < 0; \quad \left(-\frac{1}{2}U, 0\right) \quad \text{for } y > 0. \tag{9.5}$$

9.2.1.   *Linearization*

Suppose that due to a perturbation the surface of discontinuity is deformed and described by

$$y = \zeta(x, t). \tag{9.6}$$

Both above and below the surface, it is assumed that the flow is irrotational and the velocity potential is expressed as

$$\Phi = \begin{cases} -\dfrac{1}{2}Ux + \phi_1(x, y, t) & \text{for } y > \zeta, \\[2mm] \dfrac{1}{2}Ux + \phi_2(x, y, t) & \text{for } y < \zeta, \end{cases}$$

[Fig. 9.1(a)], where ϕ_1 and ϕ_2 are the perturbation potentials, satisfying

$$\nabla^2 \phi_1 = 0, \quad \nabla^2 \phi_2 = 0 \tag{9.7}$$

(see (5.54)). The velocity in the upper half space is expressed as

$$(u, v) = \left(-\frac{1}{2}U + \partial_x \phi_1, \ \partial_y \phi_1\right).$$

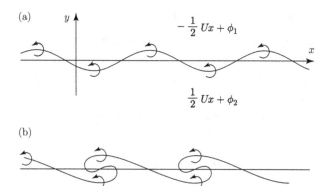

Fig. 9.1. Kelvin–Helmholtz instability: (a) Initial perturbed state, (b) nonlinear eddy formation.

The boundary condition of the velocity at the surface $y = \zeta$ is given by Eq. (6.10) of Chapter 6 with w replaced by v, which reads

$$\zeta_t + \left(-\frac{1}{2}U + \partial_x\phi_1\right)\zeta_x = \partial_y\phi_1 \quad \text{at } y = \zeta(x,t).$$

Linearizing this with respect to a small perturbation (ζ, ϕ_1), we have

$$\zeta_t - \frac{1}{2}U\zeta_x = \partial_y\phi_1\big|_{y=0}. \tag{9.8}$$

Similarly, the boundary condition of velocity of the lower half space becomes

$$\zeta_t + \frac{1}{2}U\zeta_x = \partial_y\phi_2\big|_{y=0}. \tag{9.9}$$

Furthermore, an additional condition for the pressure is necessary. The pressure formula (6.8) with z replaced by y,

$$p = C - \rho\left(\partial_t\phi + \frac{1}{2}(u^2 + v^2) + gy\right), \quad C = \text{const.}$$

holds on both sides of the surface. Denoting the pressure by p_1 and p_2 above $y = \zeta$ and below, and setting $p_1 = p_2$, we obtain the following equation by linearization,

$$\partial_t\phi_1 - \frac{1}{2}U\partial_x\phi_1 = \partial_t\phi_2 + \frac{1}{2}U\partial_x\phi_2, \quad \text{at } y = 0. \tag{9.10}$$

9.2.2. *Normal-mode analysis*

One can solve the three linear equations (9.8), (9.9) and (9.10) and determine the three functions ζ, ϕ_1 and ϕ_2, from which the stability of the system is found. All coefficients of the three equations are constants, and hence we can carry out the *normal mode analysis*[2] and express the perturbations in the following forms:

$$\zeta = A e^{\sigma t} e^{ikx}, \tag{9.11}$$

$$\phi_1 = B_1 e^{\sigma t} e^{ikx-ky}, \quad \phi_2 = B_2 e^{\sigma t} e^{ikx+ky}. \tag{9.12}$$

where A, B_1, B_2 are constants. It is readily seen that the functions ϕ_1 and ϕ_2 satisfy (9.7), in addition that $\phi_1 \to 0$ as $y \to +\infty$, whereas $\phi_2 \to 0$ as $y \to -\infty$.

Now we have arrived at an *eigenvalue* problem. For a given positive constant k, we seek an eigenvalue σ. Substituting (9.11) and (9.12) into Eqs. (9.8)–(9.10) and rearranging them, we arrive at the following matrix equation,

$$\begin{pmatrix} s_- & k & 0 \\ s_+ & 0 & -k \\ 0 & -s_- & s_+ \end{pmatrix} \begin{pmatrix} A \\ B_1 \\ B_2 \end{pmatrix} = 0, \quad s_\pm = \sigma \pm \frac{1}{2} ikU.$$

Nontrivial solution requires vanishing of the coefficient determinant:

$$\begin{vmatrix} s_- & k & 0 \\ s_+ & 0 & -k \\ 0 & -s_- & s_+ \end{vmatrix} = 0,$$

which results in

$$k(s_-^2 + s_+^2) = 2k\left(\sigma^2 - \left(\frac{1}{2}kU\right)^2\right) = 0. \tag{9.13}$$

Thus, it is found (for $k \neq 0$) that

$$\sigma = \pm \frac{1}{2} kU. \tag{9.14}$$

---

[2]This is equivalent to the Fourier mode analysis with respect to the x space. The mode e^{ikx} is orthogonal to another mode $e^{ik'x}$ for $k' \neq k$. Arbitrary perturbation can be formed by integration with k, e.g. $\int f(k,y)e^{ikx} dk = \phi(x,y)$.

The surface form (for a real A) is given by the real part of (9.11),

$$\zeta = Ae^{\sigma t} \cos kx \tag{9.15}$$

(the imaginary part is a solution as well).

The solution (9.15) with (9.14) has both modes of growing and decaying with time t corresponding to $\sigma = \frac{1}{2}kU$ and $\sigma = -\frac{1}{2}kU$, respectively. A general solution is given by a linear combination of two such modes. Therefore, it grows with t in general for any $k(> 0)$. Thus, the vortex sheet under consideration is *unstable*, and waves of sinusoidal form grow with time along the sheet. It is remarkable that the basic state (9.5) is unstable for any wavenumber $k(> 0)$. Finally, it develops into a sequential array of eddies [Fig. 9.1(b)]. This is called the **Kelvin–Helmholtz instability**.

A similar analysis can be made for the Rayleigh–Taylor problem of a heavier fluid placed over a lighter fluid in a constant gravitational field (see Problem 9.1).

9.3. Stability of parallel shear flows

One of the well-studied problems of fluid mechanics is the stability of parallel shear flows [DR81]. A typical example is the stability of a two-dimensional parallel flow of velocity in (x, y)-plane,

$$\mathbf{v}_0 = (U(y), 0), \quad -b < y < b, \tag{9.16}$$

in a channel between two parallel plane walls at $y = -b, b$. The fluid is assumed to be incompressible, either inviscid or viscous. This type of flow, i.e. a parallel shear flow, is a typical *laminar flow*. Some steady solutions $\mathbf{v}_0(y)$ of this type are presented in Sec. 4.6.

The flow is governed by the two-dimensional Navier–Stokes equations, (4.29)–(4.31). Suppose that a perturbation $\mathbf{v}_1 = (u_1, v_1)$ is superimposed on the steady flow $\mathbf{v}_0$ and the total velocity is written as $\mathbf{v} = \mathbf{v}_0 + \mathbf{v}_1$. The linearized equation for the perturbation is given by (9.3) with $\mathbf{f}_1 = 0$. In reality, perturbation must be investigated as a three-dimensional problem. However, it can be shown that a two-dimensional perturbation is most unstable (Squire's theorem,

Problem 9.4). Hence, we restrict ourselves to a two-dimensional problem here.

In the two-dimensional problem, one can introduce a stream function $\psi(x, y, t)$ for an incompressible flow (see Appendix B.2) by

$$u_1 = \partial_y \psi, \quad v_1 = -\partial_x \psi. \tag{9.17}$$

9.3.1.   *Inviscid flows ($\nu = 0$)*

Setting $\nu = 0$, and $\mathbf{v}_0 = (U(y), 0)$, $\mathbf{v}_1 = (u_1, v_1)$, $\mathbf{f}_1 = 0$ in the perturbation equation (9.3), its x and y components are reduced to

$$\partial_t u_1 + U(y)\partial_x u_1 + v_1 U'(y) = -(1/\rho_0)\partial_x p_1, \tag{9.18}$$

$$\partial_t v_1 + U(y)\partial_x v_1 = -(1/\rho_0)\partial_y p_1. \tag{9.19}$$

Eliminating p_1 between the above two equations (by applying ∂_y to (9.18) and ∂_x to (9.19), and subtracting) and using the stream function ψ of (9.17), we obtain

$$(\partial_t + U(y)\partial_x)(\partial_x^2 + \partial_y^2)\psi - U''(y)\partial_x \psi = 0. \tag{9.20}$$

The boundary condition is

$$v_1 = -\partial_x \psi = 0, \quad \text{at } y = -b, \, b. \tag{9.21}$$

In view of the property that the coefficients of Eq. (9.20) are independent of x, we can take the same normal mode e^{ikx} as before with respect to the x coordinate (with time factor specially chosen as e^{-ikct}), assuming

$$\psi(x, y, t) = \phi(y)e^{ik(x-ct)}. \tag{9.22}$$

We substitute this into (9.20) and carry out the replacements $\partial_x = ik$ and $\partial_t = -ikc$. Dropping the common exponential factor and dividing by ik, we obtain

$$(U - c)(D^2 - k^2)\phi - U''\phi = 0, \tag{9.23}$$

where $D \equiv d/dy$. This is known as the *Rayleigh's equation* (Rayleigh 1880). The boundary condition (9.21) reduces to

$$\phi = 0, \quad \text{at } y = -b, \, b. \tag{9.24}$$

Equations (9.23) and (9.24) constitute an *eigenvalue problem*. Given a real wavenumber k, this boundary-value problem would be solved with an *eigenvalue value* c_*. In general, this c_* is a complex number and written as $c_* = c_r + ic_i$. The imaginary part determines the growth rate kc_i of the perturbation, since we have from (9.22)

$$\psi(x, y, t) = \phi(y)\,e^{kc_i t}\left[\cos k(x - c_r t) + i\sin k(x - c_r t)\right].\qquad(9.25)$$

Thus the basic state of parallel flow $U(y)$ is said to be

$$\text{stable if } c_i < 0, \quad \text{unstable if } c_i > 0.$$

The perturbation is a traveling wave of phase velocity c_r.

For the purpose of stating the Rayleigh's inflexion-point theorem for inviscid parallel shear flow, let us rewrite Eq. (9.23) in the following way:

$$\phi'' - k^2\phi - \frac{U''}{U - c}\phi = 0.\qquad(9.26)$$

If $c_i \neq 0$, the factor $U - c = U - c_r - ic_i$ in the denominator of the third term never becomes zero.

Rayleigh's inflexion-point theorem: A neccesary condition for instability is that the basic velocity profile $U(y)$ should have an inflexion point. [Fig. 9.2. See Problem 9.2(i) for the proof.]

This theorem may be applied in the following way. The parabolic profile $U_P(y)$ of (4.41) has no inflexion-point [Fig. 4.3(b)]. Hence it is not unstable, i.e. neutrally stable. The perturbation solution will

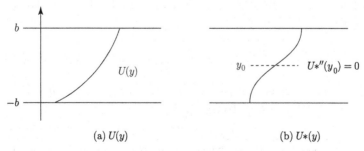

(a) $U(y)$ (b) $U_*(y)$

Fig. 9.2.   (a) No inflexion point, (b) inflexion-point at y_0.

be characterized by $c_i = 0$ and $c_r \neq 0$ within the inviscid theory according to Prob. 9.2(ii). Namely the perturbation is oscillatory with a constant amplitude. On the other hand, the profile of a jet has two inflexion-points (Fig. 4.11). This is regarded as unstable within the inviscid theory,[3] and is actually unstable in real viscous fluids if the Reynolds number is greater than a critical value R_c ($R_c \approx 4.0$ for Bickley jet $U(y) = \operatorname{sech}^2 y$ see below). As a general feature of flows of an *ideal fluid*, if there is a perturbation solution of $c_i < 0$ (stable mode), there exists always another solution of $c_i > 0$ (unstable mode) [Problem 9.2(ii)].

9.3.2. *Viscous flows*

For viscous flows, the viscosity terms $\nu \nabla^2 u_1$ and $\nu \nabla^2 v_1$ must be added on the right-hand sides of (9.18) and (9.19), respectively. Hence, the linear perturbation equation (9.20) must be replaced by

$$(\partial_t + U \partial_x) \nabla^2 \psi - U'' \partial_x \psi = \frac{1}{\mathrm{Re}} \nabla^2 \nabla^2 \psi, \qquad (9.27)$$

where $\nabla^2 = \partial_x^2 + \partial_y^2$, and

$$\partial_x v_1 - \partial_y u_1 = -\left(\partial_x^2 + \partial_y^2\right) \psi = -\nabla^2 \psi = \omega_1 \qquad (9.28)$$

is the z-component of vorticity perturbation. All the lengths and velocities are normalized by b and $U_m = \max U(y)$, respectively, with $\mathrm{Re} = U_m b / \nu$ the Reynolds number. It is seen that the above linear perturbation equation (9.27) is the equation for vorticity perturbation.

Assuming the same form of normal mode (9.22) for ψ, the above equation (9.27) becomes

$$(U - c)(D^2 - k^2)\phi - U''\phi - \frac{1}{ik\,\mathrm{Re}}(D^2 - k^2)^2 \phi = 0 \qquad (9.29)$$

($D = \mathrm{d}/\mathrm{d}y$). This is known as the *Orr–Sommerfeld equation*. See Prob. 9.4(i) for extension to 3D problem.

---

[3]Although the stability analysis here is formulated for channel flows of a finite y-width, it is not difficult to extend it to flows in unbounded y axis [DR81].

The boundary condition at the walls is the no-slip condition:

$$\phi = 0 \quad \text{and} \quad D\phi = 0, \quad \text{at } y = -1, 1. \tag{9.30}$$

Most stability analyses of parallel shear flows of viscous fluids are studied on the basis of Eq. (9.29).

In general with regard to viscous flows, there exists a certain critical value R_c for the Reynolds number Re such that steady flows $\mathbf{v}_0$ are stable when Re $< R_c$ and becomes unstable when Re exceeds R_c. This critical value R_c is called the **critical Reynolds number** which was calculated for various flows:

$R_c \approx 5772$ for the 2D Poiseuille flow $u_P(y)$ of (4.41);

$R_c \approx 520$ for the Blasius flow $u = f'(y)$ of Problem 4.4;

$R_c \approx 4.02$ for the Bickley jet $u(y) = \text{sech}^2 y$ of Problem 4.5.

Furthermore, it is known that $R_c = \infty$ (i.e. stable) for the plane Couette flow $u_C(y)$ of (4.40), and $R_c = 0$ (i.e. unstable) for the plane shear flow $u(y) = \tanh y$. It is remarked, in particular, that the Hagen–Poiseuille flow (i.e. the axisymmetric Poiseuille flow) is stable (i.e. $R_c = \infty$), namely that the imaginary part c_i of the eigenvalue is negative for all disturbance modes [DR81; SCG80].

The next stability problem is the thermal convection, which enables us to conduct detailed analysis of chaotic dynamics.

9.4. Thermal convection

9.4.1. Description of the problem

When a horizontal fluid layer is heated from below, fluid motion is driven by the buoyancy of heated fluid. However, there exist some opposing effects to control such a motion. They are viscosity and thermal conductivity. Therefore, it may be said that thermal convections are interplays between the three effects: buoyancy, viscosity and thermal conduction.

A problem which is historically well studied is the stability analysis of a horizontal fluid layer heated from below and cooled from above. Suppose that the temperature difference between the lower and upper walls is denoted by ΔT. As long as ΔT is small, the fluid layer remains static and is stable. However, once ΔT becomes larger than a certain value, the static layer becomes unstable and convective motion is set up. This kind of fluid motion is called *thermal convection*. Since the driving force is the buoyancy force, the temperature dependence of fluid density is one of the controlling factors.

In order to simplify the problem, we consider a thin horizontal fluid layer, and the flow field is assumed to be two-dimensional with the x axis taken in the horizontal direction and the y axis taken vertically upward. We apply the *Boussinesq approximation* (considered in Sec. 8.4 partly) for the equations of motion. The velocity $\mathbf{v}$ is represented as (u, v). One aspect of the Boussinesq approximation is that the velocity field is assumed to be divergence-free, div $\mathbf{v} = 0$. In the present case, this reduces to

$$\partial_x u + \partial_y v = 0. \tag{9.31}$$

Such two-dimensional divergence-free motions are described by the *stream-function* ψ (see Appendix B.2), defining the velocity components as

$$u = \partial_y \psi, \quad v = -\partial_x \psi. \tag{9.32}$$

Equation (9.31) is satisfied automatically by the above.

As the second aspect of Boussinesq approximation, the density ρ on the left of the equation of motion (8.50) is replaced by a representative density ρ_0 (constant), and only the density ρ in the buoyancy term $\rho \mathbf{g}$ is variable. The acceleration of gravity is given the form $\mathbf{g} = (0, -g)$. Thus, the x-component and y-component (vertical) are

$$\rho_0(\partial_t u + u\partial_x u + v\partial_y u) = -\partial_x p + \mu\nabla^2 u, \tag{9.33}$$

$$\rho_0(\partial_t v + u\partial_x v + v\partial_y v) = -\partial_y p + \mu\nabla^2 v - \rho g. \tag{9.34}$$

The density ρ in the buoyancy term is represented as $\rho(T)$, a function of temperature T only (its pressure-dependence being neglected). This is because the pressure variation is relatively small since the fluid layer is assumed very thin. Furthermore, the density is assumed to be a linear function of T as

$$\rho(T) = \rho_0(1 - \alpha(T - T_0)), \tag{9.35}$$

since the temperature variation is considered relatively small as well, where α is the thermal expansion coefficient of the fluid, and $\rho_0 = \rho(T_0)$ for a representative temperature T_0.

The equation governing temperature is derived from the general equation of heat transfer (4.16) of Chapter 4. Suppose that a fluid particle has gained some amount of heat, and that, as a result, its temperature is increased by δT together with the entropy increase by δs. Then the heat gained by the particle is given by a thermodynamic relation, $T\delta s = c_p\delta T$, where the specific heat c_p per unit mass at a constant pressure is used because the heat exchange proceeds in fluid usually at a constant pressure. Then the entropy term $T(Ds/Dt)$ in Eq. (4.16) is replaced by the term $c_p(DT/Dt)$, and we obtain an equation for the temperature,

$$\rho c_p(\partial_t T + (\mathbf{v} \cdot \nabla)T) = \sigma_{ik}^{(v)}\partial_k v_i + \text{div}(k\,\text{grad}\,T). \tag{9.36}$$

In the stability analysis, the fluid motion is usually very slow. The first term on the right represents the viscous dissipation of kinetic energy into heat, but it is of second order with respect to the small velocity and can be ignored because it is of higher order of smallness. The second term denotes local heat accumulation by nonuniform heat flux of thermal conduction. This term is taken as the main heat source, in which the coefficient k is assumed a constant. Thus, we obtain the equation of thermal conduction as follows:

$$\partial_t T + u\partial_x T + v\partial_y T = \lambda\nabla^2 T, \tag{9.37}$$

where $\lambda = k/(\rho_0 c_p)$ is the coefficient of the thermal diffusion.

9.4.2. *Linear stability analysis*

Suppose that a horizontal fluid layer of thickness d is at rest, and that the temperature of the lower boundary surface at $y = 0$ is maintained at T_1 and that of the upper boundary surface at $y = d$ is T_2. Hence the temperature difference is $\Delta T = T_1 - T_2 > 0$. In the steady state without fluid motion, we can set $\partial_t T = 0$, and $u = 0$, $v = 0$ in Eq. (9.37), and obtain

$$0 = \lambda \nabla^2 T = \lambda(\partial_x^2 + \partial_y^2)T. \tag{9.38}$$

The boundary conditions are given by $T(x,0) = T_1$ and $T(x,d) = T_2$. The steady temperature satisfying the above equation and the boundary conditions is given by

$$T_*(y) = T_1 + \frac{T_2 - T_1}{d}y = T_1 - \beta y, \tag{9.39}$$

$$\beta = \Delta T/d = -\partial_y T_*(y) \tag{9.40}$$

(see (2.23)), where β (> 0) is the temperature gradient within the layer. In this state, heat is being conducted steadily from the lower wall to the upper wall through the fluid at rest, and the temperature is decreasing linearly with y. For this temperature $T_*(y)$, the corresponding distribution of steady density is given by

$$\rho_*(y) = \rho_0(1 - \alpha(T_*(y) - T_0)), \quad T_*(y_0) = T_0, \tag{9.41}$$

from (9.35). The steady pressure distribution is determined by the equation, $dp_*/dy = -g\rho_*(y)$, which is obtained from (9.34) by setting the velocities zero. Thus, with defining $p_1 = p_*(0)$, we have

$$p_*(y) = p_1 - g \int_0^y \rho_*(y)dy. \tag{9.42}$$

Suppose that a motion of velocity (u, v) was excited by a certain perturbation. Let us represent the perturbed state as

$$\mathbf{v} = (u, v), \quad T = T_*(y) + T',$$

$$\rho = \rho_*(y) + \rho', \quad p = p_*(y) + p',$$

where primed variables are infinitesimal disturbances, which are functions of x, y and t. Substituting these into the governing equations (9.31), (9.33)–(9.35) and (9.37), and keeping only linear terms with respect to the perturbations, we have a set of perturbation equations:

$$\partial_x u + \partial_y v = 0, \tag{9.43}$$

$$\partial_t u = -(1/\rho_0)\partial_x p' + \nu\nabla^2 u, \tag{9.44}$$

$$\partial_t v = -(1/\rho_0)\partial_y p' + \nu\nabla^2 v - g\,(\rho'/\rho_0), \tag{9.45}$$

$$\rho'/\rho_0 = -\alpha\,T', \tag{9.46}$$

$$\partial_t T' + v\partial_y T_*(y) = \lambda\nabla^2 T'. \tag{9.47}$$

Eliminating the pressure between (9.44) and (9.45), i.e. taking curl of the equations, we obtain an equation for the z-vorticity $\omega = \partial_x v - \partial_y u$:

$$\partial_t\omega - \nu\nabla^2\omega = g\alpha\,\partial_x T', \tag{9.48}$$

where (9.46) is used. The right-hand side denotes the rate of generation of the *horizontal z* component vorticity ω by the temperature gradient $\partial_x T'$ in the horizontal x-direction, called a *baroclinic* effect (see Problem 8.2).

The velocity (u, v) satisfying (9.43) is written as $u = \partial_y\psi$ and $v = -\partial_x\psi$, and the vorticity is given by $\omega = \partial_x v - \partial_y u = -\nabla^2\psi$. From now, we will use the symbol θ instead of T' for temperature deformation from the steady distribution $T_*(y)$. Then, Eqs. (9.47) and (9.48) are rewritten as

$$\partial_t\nabla^2\psi - \nu\nabla^2\nabla^2\psi = -g\alpha\partial_x\theta, \tag{9.49}$$

$$\partial_t\theta - \lambda\nabla^2\theta = -\beta\partial_x\psi, \tag{9.50}$$

where (9.40) is used in (9.50). Thus we have obtained a pair of equations for ψ and θ, governing the *thermal convection*.

Now, we normalize the above set of equations by introducing the following dimensionless variables with a prime,

$$(x', y') = (x/d, y/d), \quad t' = t\lambda/d^2,$$

$$\psi' = \psi/\lambda, \quad \theta' = \theta/(\beta d).$$

Substituting these in (9.49) and (9.50) and rearranging the equations, and dropping the primes of dimensionless variables, we obtain the following set of dimensionless equations:

$$(\partial_t - \Delta)\theta = -\partial_x\psi, \tag{9.51}$$

$$(\sigma^{-1}\partial_t - \Delta)\Delta\psi = -R_a\partial_x\theta. \tag{9.52}$$

where $\Delta = \nabla^2$ (Laplacian), and R_a is the *Rayleigh number* and σ the *Prandtl number* defined by

$$R_a = \frac{g\alpha\beta d^4}{\nu\lambda}, \qquad \sigma = \frac{\nu}{\lambda}. \tag{9.53}$$

Furthermore, eliminating θ between (9.51) and (9.52), we finally obtain an equation for ψ only:

$$(\partial_t - \Delta)(\sigma^{-1}\partial_t - \Delta)\Delta\,\psi = R_a\partial_x^2\psi. \tag{9.54}$$

This sixth-order partial differential equation (Δ is of the second order) has a simple solution satisfying the following condition of free-boundary:

$$\psi = 0, \quad \partial_y^2\psi = 0, \quad \theta = 0: \quad \text{at } y = 0 \quad \text{and} \quad y = 1. \tag{9.55}$$

The boundary conditions ($\psi = 0$, $\partial_y^2\psi = 0$) is called *free* because the boundaries ($y = 0, 1$) are free from the viscous stress, i.e. the stress vanishes at the boundaries.[4] In addition, the boundaries coincide with a stream-line by the condition $\psi = 0$.

The boundary value problem (9.54) and (9.55) has a solution in the following form:

$$\psi = Ae^{\gamma t}\sin(\pi ax)\sin(\pi y), \tag{9.56}$$

$$\theta = Be^{\gamma t}\cos(\pi ax)\sin(\pi y), \tag{9.57}$$

where θ is given to be consistent with (9.51) and (9.56), and A, B, a, γ are constants. The parameter γ determines the growth rate of the disturbance ψ (or θ), therefore the stability of the basic static state.

---

[4]The boundary condition $\theta = 0$ at $y = 0, 1$ is equivalent to $\partial_y^4\psi = 0$ in addition to $\partial_y^2\psi = 0$ by (9.52).

In fact, it is not difficult to see that this function ψ satisfies the boundary conditions (9.55). Substituting (9.56) in (9.54), we obtain the following algebraic equation:

$$(\gamma + b^2)(\sigma^{-1}\gamma + b^2)b^2 = R_a(\pi a)^2, \qquad (9.58)$$

where $b^2 = \pi^2(a^2 + 1)$, since we have $\partial_x^2 \psi = -(\pi a)^2 \psi$ and $\Delta \psi = -b^2 \psi$. This is a quadratic equation for γ, and it is immediately shown that γ has two *real* roots. If γ is negative, the state is *stable*. If γ is positive, the state is *unstable*. The case $\gamma = 0$ corresponds to the *neutrally stable* state.

The condition of neutral stability is obtained by setting $\gamma = 0$ (because γ must be real) in (9.58) as

$$R_a(a, \gamma = 0) = \pi^4(a^2 + 1)^3/a^2. \qquad (9.59)$$

Thus, the value of $R_a(a, \gamma = 0)$ depends on the horizontal wave number πa, and has its minimum value given by

$$R_c(\text{free}) := \pi^4(27/4) \approx 657.5, \qquad (9.60)$$

which is attained at $a = a_c := 1/\sqrt{2}$. If the Rayleigh number R_a is less than R_c, we have $\gamma < 0$ for all real a, namely the state is stable. However, if $R_a > R_c$, there exists some range of a in which γ becomes positive, i.e. the corresponding mode of horizontal wavenumber πa grows *exponentially* with time, according to the framework of linear theory. The number R_c is called the *critical Rayleigh number*.

For the case of *no-slip* boundary condition on solid walls instead of the free boundary conditions (9.55), the critical Rayleigh number is known to be

$$R_c(\text{no-slip}) \approx 1708,$$

which is obtained by a numerical analysis (see, e.g. [Cha61]).

9.4.3. *Convection cell*

It is found from above that for $R_a > R_c$ there are some modes (i.e. a certain range of the wavenumber πa with positive γ) growing exponentially in time,[5] according to the linear stability analysis. As the unstable modes grow with time, a nonlinear mechanism which was ineffective during small amplitudes now becomes effective and works to suppress further growth of the modes, and a single mode of finite amplitude is selected nonlinearly. Finally, a new equilibrium steady state will be attained.[6]

In the case of the present thermal convection, the nonlinear mechanism is favorable to attain a steady equilibrium state. If the relative difference $(R_a - R_c)/R_c$ is sufficiently small, the stream function of the equilibrium state is given by

$$\psi_0(x, y) = A_0 \sin(\pi a_c x) \sin(\pi y), \qquad (9.61)$$

which corresponds to the stream function (9.56) of the critical wavenumber $a_c = 1/\sqrt{2}$ with its amplitude $Ae^{\gamma t}$ replaced by a constant A_0 which will be considered in the next section again.

The flow field described by the stream function (9.61) is composed of periodic cells of fluid *convection*, called *thermal convection cells* (Fig. 9.3). Sometimes, the convection cell is called the *Bénard* cell. Thermal convection was first studied by Bénard (1900) for a liquid state of heated wax. Later, Rayleigh (1916) solved the mathematical problem described in the previous section [Sec. 9.4.2]. However, the

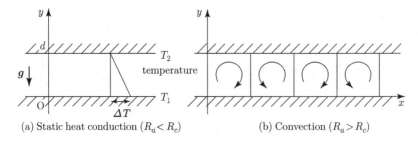

(a) Static heat conduction $(R_a < R_c)$ (b) Convection $(R_a > R_c)$

Fig. 9.3. Thermal convection cells.

---

[5] From (9.59), $R_a(a, \gamma = 0) = R_c + 9\pi^4 \xi^2 + O(\xi^3)$, where $\xi = a^2 - \frac{1}{2}$. There is a range of a^2 of positive γ since we have $d\gamma/dR_a|_{\gamma=0} = (\pi a/b^2)^2 \sigma/(1 + \sigma) > 0$, from (9.58).

[6] In some instability problems, a nonlinear mechanism is unfavorable for attaining an equilibrium and leads to explosive growth of modes.

convection observed by Bénard is now considered to be the convection driven by temperature variation of the *surface tension* of wax, while the convection solved by Rayleigh is driven by *buoyancy*. The latter is often called the Rayleigh convection, while the former is termed the Marangoni convection.

9.5. Lorenz system

It has been found so far that thermal convection is controlled by Rayleigh number R_a and Prandtl number σ within a framework of linear theory. In this section, we consider a finite-amplitude convection, which is selected by a nonlinear mechanism among unstable modes. As the value of the control parameter R_a is increased further, the state of steady convection becomes unstable and a new oscillatory mode sets in. This is termed the *Hopf bifurcation* (Sec. 9.1). When the Rayleigh number becomes sufficiently high, and hence the nonlinearity is sufficiently large with an appropriate value of Prandtl number, the time evolution of state can exhibit a chaotic behavior. One of the historically earliest examples of chaos realized on computer is the Lorenz dynamical system.

9.5.1. *Derivation of the Lorenz system*

In order to take into account the nonlinear effect appropriately, it is useful to simplify the expression of convection cells as much as possible, so that a simplest system of ordinary differential equations can be derived. The *Lorenz system* is such a kind of dynamical system. Suppose that the stream function ψ and the temperture deformation θ of convection cells are represented by

$$\psi(x, y, t) = -AX(t) \sin \frac{\pi a}{d} x \sin \frac{\pi}{d} y, \tag{9.62}$$

$$\theta(x, y, t) = BY(t) \cos \frac{\pi a}{d} x \sin \frac{\pi}{d} y - \frac{B}{\sqrt{2}} Z(t) \sin \frac{2\pi}{d} y, \tag{9.63}$$

$$A = \sqrt{2}\lambda \frac{1 + a^2}{a}, \quad B = \frac{\sqrt{2}\,\Delta T}{\pi} \frac{R_c}{R_a},$$

where x, y are dimensional variables, and R_c is given by (9.60): i.e. $R_c = \pi^4(a^2 + 1)^3/a^2$ with $a = 1/\sqrt{2}$. The time factors $e^{\gamma t}$ of (9.56) and (9.57) are replaced by $X(t)$ and $Y(t)$, respectively, and a new x-independent term with a time factor $Z(t)$ is added to the temperature deformation θ. The temperature is given by $T = T_*(y) + \theta$, where $T_*(y)$ is the steady distribution (9.39).

The two functions ψ and θ are required to satisfy the full nonlinear equations of motion. A linear equation for the vorticity $\omega = \partial_x v - \partial_y u = -\nabla^2\psi$ was given by (9.48). Nonlinear version of the vorticity equation for $-\nabla^2\psi$ can be derived by taking *curl* of the original governing equations (9.33) and (9.34), and a nonlinear version of the temperature equation for θ can be derived from (9.37). Both are written as follows:

$$\partial_t\nabla^2\psi + (\partial_y\psi\partial_x - \partial_x\psi\partial_y)\nabla^2\psi = \nu\nabla^2\nabla^2\psi - g\alpha\partial_x\theta, \quad (9.64)$$

$$\partial_t\theta + (\partial_y\psi\partial_x - \partial_x\psi\partial_y)\theta = \lambda\nabla^2\theta - \beta\partial_x\psi. \quad (9.65)$$

The Lorenz system is obtained by substituting the expressions (9.62) and (9.63). In the resulting equations, only the terms of the forms,

$$\sin\frac{\pi a}{d}x \sin\frac{\pi}{d}y, \quad \cos\frac{\pi a}{d}x \sin\frac{\pi}{d}y, \quad \sin\frac{2\pi}{d}y,$$

are retained in the equation, and the other terms are omitted. Equating the coefficients of each of the above terms on both sides, we obtain the following dynamical equations of **Lorenz system** (Problem 9.3):

$$\mathrm{d}X/\mathrm{d}\tau = -\sigma X + \sigma Y,$$
$$\mathrm{d}Y/\mathrm{d}\tau = -XZ + rX - Y, \quad (9.66)$$
$$\mathrm{d}Z/\mathrm{d}\tau = XY - bZ,$$

where

$$r = \frac{R_a}{R_c}, \quad \sigma = \frac{\nu}{\lambda}, \quad b = \frac{4}{1 + a^2}, \quad (9.67)$$

and $\tau = \lambda t(1 + a^2)\pi^2/d^2$ is the normalized time variable, which is again written as t (in order to follow the tradition). The right-hand sides of the above system of equations do not include time t. Namely, the time derivatives on the left-hand sides are determined solely by the state (X, Y, Z). Such a system is called an *autonomous* system, in general.

The trajectory $(X(t), Y(t), Z(t))$ of state is determined by integrating the above system (9.66) numerically. A set of points in the phase space (X, Y, Z) where a family of trajectories for a set of initial conditions accumulate asymptotically as $t \to \infty$ is called an *attractor*.

We can think of time derivatives $\dot{X} = dX/dt$, $\dot{Y} = dY/dt$, $\dot{Z} = dZ/dt$ as three components of a velocity vector $\mathbf{V}$ defined at the point $\mathbf{X} = (X, Y, Z)$. Then we can regard the system dynamics like a fluid motion with velocity field $\mathbf{V}(\mathbf{X})$. An important feature of such a motion of points in the phase space (X, Y, Z) is the property that the phase volume composed of points moving with $\mathbf{V} = (\dot{X}, \dot{Y}, \dot{Z})$ decreases steadily. This is verified simply as follows. Taking the divergence of velocity field $(\dot{X}, \dot{Y}, \dot{Z})$, we obtain $\text{div}\,\mathbf{V} = \partial \dot{X}/\partial X + \partial \dot{Y}/\partial Y + \partial \dot{Z}/\partial Z = -\sigma - 1 - b < 0$. Due to this property, the phase volume of an attractor where the trajectories are approaching diminishes indefinitely. This does not necessarily mean that the attractors are points, but means only that the dimension of the attractor is less than 3. This will be remarked again later. Here it is mentioned only that such a shrinking of phase volume is a common nature of *dissipative* dynamical systems, in which the kinetic energy is transformed to heat (say).

9.5.2. *Discovery stories of deterministic chaos*

There is an interesting story about the discovery of this system by Lorenz himself. He tried to solve an initial value problem of the above system (9.66) of ordinary differential equations numerically by using a calculator which was available for him at the time of 1960s, where the initial values were $(X(0), Y(0), Z(0)) = (0, 1, 0)$ with $\sigma = 10$, $r = 28$, $b = 8/3$. Needless to say, it took much time for calculations.

He wanted to obtain results at times further ahead. Without using the above initial values, he used the values obtained at an intermediate time of computation, and carried out the calculation with the

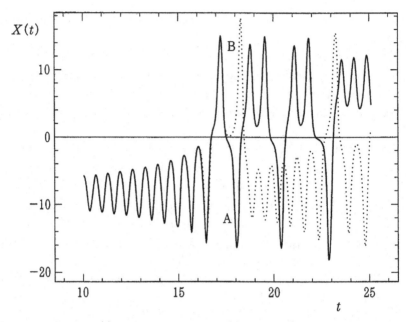

Fig. 9.4. Curve $X(t)$ obtained numerically by integrating the Lorenz system (see Sec. 9.5.2 for the details). Curves A and B are obtained from two slightly different initial conditions.

same computer program. It was expected that the computer would generate results of approximately same data for overlapping times. At this computation, he thoughtlessly used only the first three digits instead of the six digits data obtained by the calculation (e.g. curve A of Fig. 9.4). Naturally, at the beginning, the computer gave similar temporal behavior as before. However, as time went on, both results began to show differences, and finally showed a totally different time evolution from the previous computation (e.g. curve B of Fig. 9.4). Reflecting on those times in his study, he wrote that he had been shocked at this outcome. He was already convinced that the set of above equations captured a certain intrinsic chaotic nature of atmospheric phenomena, and that long-term weather prediction is doubtful.

Chaotic dynamical systems are characterized to have a property that they are *sensitively dependent on initial conditions*. The above result observed by Lorenz is just proved this fact. After detailed analysis made at later times, the temporal behaviors of the Lorenz system are found to be *stochastic* in a true sense, and thus long-term behaviors of its solution are *unpredictable* [Lor63].

Despite the fact that the time evolution of a system state is determined by a set of equations, so that its short-term evolution is *deterministic*, it is found that its long-term behaviors are *stochastic*. This phenomenon is called the *deterministic chaos*. This was found by detailed analysis of the chaos-attractor in [Lor63], now called the *Lorenz attactor*. Many nonlinear dynamical systems have such property.

In 1961, earlier than the Lorenz study, another chaos-attractor had been discovered by Y. Ueda [Ued61] during his study of analog-computer simulations of *nonautonomous* nonlinear system (corresponding to the Van der Pol equation under a *time-periodic external forcing* with a third-order term). Reminding of the long hours sitting in front of the analog-computer (made by his senior colleague M. Abe), Ueda writes "After those long exhausting vigils in front of the computer, staring at its output, chaos had become a totally natural, everyday phenomenon in my mind." This is called *Ueda attactor* [Ued61].

Solutions of the Navier–Stokes equation are believed also to have such behavior in turbulent states which we will consider in the next chapter.

9.5.3. *Stability of fixed points*

In stability analysis, steady states play an important role, because the growth or decay of perturbations to a steady state is usually investigated. A steady state corresponds to a fixed point in the dynamical system. A *fixed point* of the Lorenz system (9.66) is defined by the point where the velocity $\mathbf{V}$ vanishes, i.e. $(\dot{X}, \dot{Y}, \dot{Z}) = 0$. Setting the right-hand sides of (9.66) equal to 0, we obtain

$$X = Y, \quad -XZ + rX - Y = 0, \quad XY - bZ = 0.$$

For $r = R_a/R_c < 1$, we have only one fixed point which is the origin O: $(X, Y, Z) = (0, 0, 0)$. The point O corresponds to the static state. This is reasonable because we are considering a bifurcation problem from the static state to a convection-cell state. For $r = R_a/R_c > 1$, it can be shown without difficulty that there exist three fixed points: (a) the origin O: $(0,0,0)$, and (b) two points C and C' defined by

$$C: (q, q, r-1), \quad C': (-q, -q, r-1), \quad \text{when } r > 1,$$

where $q = \sqrt{b(r-1)}$. The fixed points C and C' are located at points of mirror symmetry with respect to the vertical plane $X + Y = 0$. Now, let us consider the stability of the three fixed points.

When $r > r_c$ ($r_c > 1$, a constant), we will encounter a strange situation in which all the three fixed points are unstable, in other words we have no stable fixed points to which the trajectory should aproach. This is a situation where the moving state has no point to head towards and trajectories become chaotic.

9.5.3.1. *Stability of the point O*

Linearizing the Lorenz system (9.66) for points in the neighborhood of O, we obtain

$$\dot{X}' = -\sigma X' + \sigma Y', \quad \dot{Y}' = rX' - Y', \quad \dot{Z}' = -bZ',$$

where X', Y', Z' are small perturbations to the fixed point $(0,0,0)$, which are represented as $X' = X_0 e^{st}$, etc. where s is the growth rate. Assuming these forms for perturbations X', Y', Z' and substituting those in the above equations, we obtain a system of linear algebraic equations, i.e. a matrix equation, for the amplitude (X_0, Y_0, Z_0):

$$\begin{pmatrix} s+\sigma & -\sigma & 0 \\ -r & s+1 & 0 \\ 0 & 0 & s+b \end{pmatrix} \begin{pmatrix} X_0 \\ Y_0 \\ Z_0 \end{pmatrix} = 0.$$

In order that there exists a *nontrivial* solution of (X_0, Y_0, Z_0), the determinant of the coefficient matrix must vanish. Thus, we obtain an eigenvalue equation for the growth rate s, which is also termed the *characteristic exponent*:

$$\begin{vmatrix} s+\sigma & -\sigma & 0 \\ -r & s+1 & 0 \\ 0 & 0 & s+b \end{vmatrix} = (s+b)\left(s^2 + (\sigma+1)s + \sigma(1-r)\right) = 0.$$

Therefore, we have

$$s_0 = -b, \quad s_\pm = -\frac{1}{2}(\sigma+1) \pm \frac{1}{2}\sqrt{(\sigma+1)^2 + 4\sigma(r-1)}.$$

If $r < 1$, the point O is stable, i.e. all $\mathrm{Re}(s_0)$ and $\mathrm{Re}(s_\pm)$ are negative. This is consistent with the linear theory because the point O corresponds to the static state which is stable for $r = R_a/R_c < 1$. However, once $r > 1$, we have $\mathrm{Re}(s_+) > 0$. Thus, it is found that the point O becomes unstable for $r > 1$, which is also consistent with the linear theory.

9.5.3.2. *Stability of the fixed points C and C'*

We can make a similar linear stability analysis near points C and C' by writing $X = \pm q + X'$, $Y = \pm q + Y'$, $Z = r - 1 + Z'$. The linearized equation with respect to X', Y', Z' leads to the following eigenvalue equation for the characteristic exponent s as a condition of nontrivial solutions:

$$F(s) = s^3 + (\sigma + b + 1)s^2 + b(\sigma + r)s + 2b\sigma(r-1) = 0.$$

This is the third-order polynomial equation for s. Obviously, we have three roots: s_1, s_2, s_3. The condition of existence of C and C' is that $r > 1$. As long as $r - 1 = \varepsilon$ is sufficiently small (and positive), it is found that all three roots are negative. Hence the two points C and C' are stable as long as ε is small. This corresponds to a stable steady convection. The two fixed points denote two possible senses of circulatory motions within a cell.

Setting $r = 1$ (when both C and C' happen to coincide with O), it can be shown immediately that the three roots are $0, -b, -(\sigma + 1)$, which are assumed to correspond to s_1, s_2, s_3, respectively in order. As r increases from 1, it can be shown from a detailed analysis (not shown here) that s_1 decreses from 0 and tends to be combined with s_2, i.e. $s_1 = s_2 < 0$ at a certain value of r. Thereafter, both together form a complex-conjugate pair, and their real part increases and reaches zero in due course. From there, the points C and C' become unstable. Namely, an instability sets in when $\text{Re}(s_1) = \text{Re}(s_2) = 0$. At that *critical* stability point, we must have $s_1 = +i\omega$, and $s_2 = -i\omega$ (where ω is real). As far as $r > 1$, s_3 is always negative. At the critical point, we have $s_1 + s_2 = 0$. On the other hand, from the cubic equation $F(s) = 0$, we obtain the sum of three roots given by $s_1 + s_2 + s_3 = s_3$ (at the critical point) $= -(\sigma + b + 1)$. Since s_3 is one of the roots, we have the following equation,

$$0 = F(-(\sigma + b + 1)) = rb(\sigma - b - 1) - b\sigma(\sigma + b + 3),$$

at the critical point. Therefore, denoting the value of r satisfying this equation by r_c, we obtain

$$r_c := \frac{\sigma(\sigma + b + 3)}{\sigma - b - 1}.$$

We have already seen that the fixed point O is unstable for $r > 1$. Now it is verified that other fixed points C and C' become unstable as well for $r > r_c$. When $r_c > 1$ for given values of σ and b, we obtain that all three fixed points are unstable for $r > r_c$, in other words, we have no stable fixed points to which the trajectory should approach. Now, we have a situation where the state point has no fixed direction. This is understood as a favorable circumstance for chaotic trajectories.

Another useful property can be verified by defining a non-negative function, $H(X, Y, Z) := X^2 + Y^2 + (Z - r - \sigma)^2$. It can be shown by using the Lorenz system (9.66) that, if $(X^2 + Y^2 + Z^2)^{1/2}$ is sufficiently large,

$$\frac{\mathrm{d}}{\mathrm{d}t}H = \frac{\mathrm{d}}{\mathrm{d}t}(X^2 + Y^2 + (Z - r - \sigma)^2) < 0.$$

Hence, the trajectory in the phase space at large distances from the origin is moving inward so as to reduce the distance between (X, Y, Z) and $(0, 0, r + \sigma)$. This sort of a positive-definite function such as $H(X, Y, Z)$ is called a *Liapounov function* and plays an important role in the *global stability* analysis as given above.

9.6. Lorenz attractor and deterministic chaos

9.6.1. *Lorenz attractor*

From the analysis of the Lorenz system of differential equations in the previous section, we have found the following properties.

(i) There is no stable fixed points for $r > r_c$.
(ii) Trajectories are directed inward at large distances from the origin.
(iii) Phase volume shrinks steadily during the orbital motion.
(iv) There is no repeller (see the footnote), no unstable limit-cycle, and no quasi-periodic orbit.

The property (iv) results from the negative value of the sum $s_1 + s_2 + s_3$. The sum $s_1 + s_2 + s_3$ must be positive for both the repeller and unstable limit-cycle.[7] If there existed a quasi-periodic orbit, it contradicts with the property (iii) because the phase volume should be conserved in the quasi-periodic motion.

Then, where should the trajectories head for? Anyway a brief time direction can be determined by the system of differential equations (9.66). The trajectory of Fig. 9.5 was computed numerically for the case $r > r_c$ with parameter values,

$$r = 28, \quad \sigma = 10, \quad b = 8/3,$$

---

[7]The repeller is a fixed point whose characteristic exponents are all positive, while the unstable limit-cycle is a periodic orbit having two positive characteristic exponents and one zero exponent corresponding to the direction of the periodic orbit.

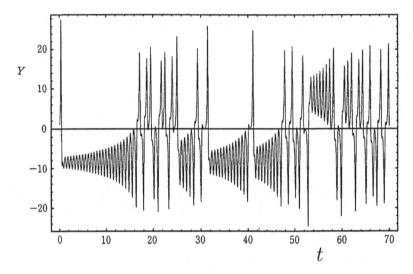

Fig. 9.5.   Curve $Y(t)$ obtained numerically by integrating the Lorenz system.

for which $r_c = 24.74$, and C and C′ are $(\pm 8.48, \pm 8.48, 27)$. The object to which the trajectories draw asymptotically is now called the **Lorenz attractor** (Fig. 9.6).

Trajectories of a dynamical system characterized by sensitive dependence on initial conditions, i.e. chaotic trajectories, accumulate in the limit as $t \to \infty$ at a manifold of a *fractal dimension*, in general. Such a set of points is called a *strange attractor*. The concept of *strange* is related not only to such a geometry, but also to an orbital dynamics. Lyapunov characteristic exponents are always defined for a dynamical system with their number equal to the dimension of the dynamical system. If the trajectory has at least one positive exponent (corresponding to one positive s), then the manifold composed of these trajectories is called a **strange attractor**.

The Lorenz attractor is a representative example of the strange attractor. Along the trajectory, the phase volume extends in the characteristic direction of positive exponents, whereas it reduces in the direction of negative exponents, and as a whole, the phase volume shrinks along the trajectory to a phase volume of dimension less than 3 in the case of the Lorenz system. As a result, the strange attractor becomes fractal.

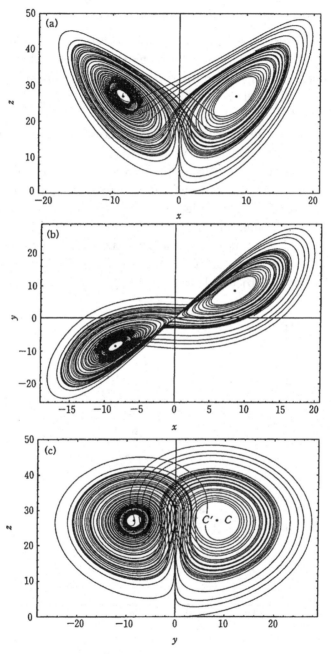

Fig. 9.6. Lorenz attractor: projection of a trajectory to (a) (X, Z)-plane, (b) (X, Y)-plane, and (c) (Y, Z)-plane (see Sec. 9.6.1).

9.6.2.   *Lorenz map and deterministic chaos*

In order to observe the time evolution of orbits, we consider the Z-coordinate of the orbit projected in the (Y, Z)-plane, which repeats up-and-down motions. Denoting the nth upper-extremum of Z-coordinate by M_n and the next upper-extremum by M_{n+1}, we create a pair (M_n, M_{n+1}) for a number of n's.

Taking M_n as the coordinate along the horizontal axis and M_{n+1} as the coordinate along the vertical axis, one can plot a point (M_n, M_{n+1}) on the (M_n, M_{n+1}) plane. Plotting those points for a number of n's, one can observe those points distribute randomly along a curve of Λ-form (Fig. 9.7), not spread over an area in the (M_n, M_{n+1})-plane. This remarkable property was discovered by Lorenz and this plot is called the **Lorenz map**. Apparently the irregular behavior of the orbit exhibits a certain regularity (lying on a one-dimensional object). However, this property itself is evidence

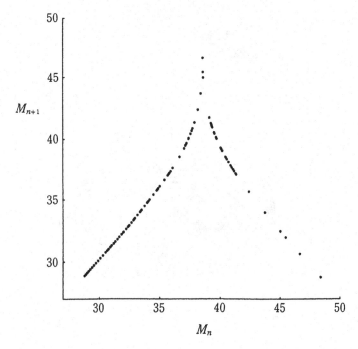

Fig. 9.7.   Lorenz map.

that the long-term evolution of the orbit is *unpredictable*, and its behavior is *chaotic*. Let us look at this property with a simplified but similar tent-map without losing the essential property of the Lorenz map.

The **tent-map** (looking like a tent, Fig. 9.8) is defined by

$$x_{n+1} = Tx_n = \begin{cases} 2x_n & \left(0 < x_n < \dfrac{1}{2}\right) \\[2mm] 2(1 - x_n) & \left(\dfrac{1}{2} < x_n < 1\right) \end{cases}$$

where x_n corresponds to $(M_n - m)/(M - m)$ with m and M being the minimum and maximum of M_n, respectively.

Let us express x_n by binary digits. If $x_1 = 0.a_1a_2a_3a_4\cdots$ by the binary, that denotes the following:

$$x_1 = a_1 2^{-1} + a_2 2^{-2} + a_3 2^{-3} + a_4 2^{-4} + \cdots,$$

where a_i takes 0 or 1. In addition, we can represent $1 - x_1$ as follows:

$$1 - x_1 = 0.1111\cdots - 0.a_1a_2a_3a_4\cdots$$
$$= 0.(Na_1)(Na_2)(Na_3)(Na_4)\cdots$$

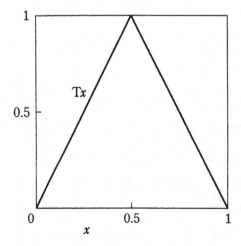

Fig. 9.8.   Tent map.

where $Na = 1 - a$. According to this, a T-operation becomes

$$Tx_1 = 0.(N^{a_1} a_2)(N^{a_1} a_3)(N^{a_1} a_4) \cdots.$$

where $N^a b = b$ (when $a = 0$), Nb (when $a = 1$). Namely, after an operation of the map T, each digit moves to the left up-shift by one place. The first digit moves up to the left of the "decimal" point and becomes 0, indicating that the number is less than 1. This is carried out as follows. If $a_1 = 0$, the point x is located on the left-half of the horizontal axis of the tent-map, and we have $N^{a_1} a_1 = a_1 = 0$ and $N^{a_1} a_k = a_k$ (for $k = 2, \ldots$). If $a_2 = 0$ ($a_1 = 0$), then the map Tx_1 is located on the left-half of the horizontal axis.

If $a_1 = 1$, the point x is located on the right-half of the horizontal axis, and we have $N^{a_1} a_1 = Na_1 = 0$ and $N^{a_1} a_k = Na_k$ (for $k = 2, \ldots$). If $a_2 = 0$ ($a_1 = 1$), then $N^{a_1} a_2 = Na_2 = 1$, and the map Tx_1 is located on the right-half of the horizontal axis. Repeating this operation n times, we obtain

$$T^n x_1 = 0.(Pa_{n+1})(Pa_{n+2})(Pa_{n+3}) \cdots$$

where $P = N^{a_1 + a_2 + \cdots + a_n}$.

Suppose that we are given two numbers x_1 and x_1' whose first six digits are identical with 0.110 001. Regarding x_1, the digits are all 0 at the seventh place and thereafter, while for the number x_1' the digits take either 0 or 1 randomly after the seventh place, such as

$$x_1 = 0.110\,001\,000 \cdots, \quad x_1' = 0.110\,001\,110 \cdots.$$

In the initial phase ($n < 6$) of the map sequence, two numbers obtained by x_1 and x_1' maps are considered to coincide practically. After T^6 map, these numbers behave in a totally different way. In particular, the behavior of $T^n x_1'$ ($n \geq 6$) will be unpredictable, because the digits are a random sequence by the presupposition.

The property that cut-off errors move up-shift to the left for each map corresponds to the exponential growth of error, i.e. *exponential instability*. Therefore in the dynamical system of the tent-map, the orbit of T^n is sensitive with respect to the accuracy of initial data, and becomes unpredictable after a certain n, say n_0 ($n_0 = 6$ in the above example). Such a kind of motion is called a **deterministic chaos**.

In the numerical experiment of the Lorenz system with $r = 28$, $\sigma = 10$, $b = 8/3$, two orbits starting from two neighboring points near the Lorenz attractor will lose their correlation before long. Such a property of sensitive dependence on initial values is a characteristic property of chaotic orbits. After the time, either of the orbits traces a path on the same attractor without mutual correlation.

The Lyapunov dimension (a kind of fractal dimension) of the Lorenz attractor for the above parameter values is estimated as

$$2.0 < D_{\mathrm{L}} < 2.401.$$

Because of this fractal dimension, *the Lorenz attractor is a strange attractor*.

9.7.   Problems

Problem 9.1 Rayleigh–Taylor instability

Suppose that a heavy fluid of constant density ρ_1 is placed above a light fluid of another constant density ρ_2, and separated by a surface S in a vertically downward gravitational field of acceleration g. In the unperturbed state, the surface S was a horizontal plane located at $y = 0$ (with the y axis taken vertically upward) and the fluid was at rest, and the density was

$$\rho_1 \quad \text{for } y > 0; \quad \rho_2(< \rho_1) \quad \text{for } y < 0.$$

Suppose that the surface S is deformed in the form,

$$y = \zeta(x, t), \tag{9.68}$$

[Fig. 9.9(a)]. Both above and below S, the flow is assumed to be irrotational and the velocity potential ϕ is expressed as

$$\phi = \phi_1(x, y, t) \quad \text{for } y > \zeta; \quad \phi_2(x, y, t) \quad \text{for } y < \zeta. \tag{9.69}$$

(i) Derive linear perturbation equations for small perturbations ζ, ϕ_1 and ϕ_2 from the boundary conditions (6.8) and (6.10).

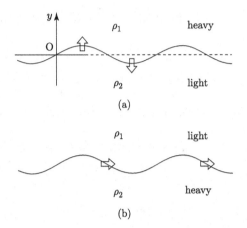

Fig. 9.9.   (a) Rayleigh–Taylor instability, (b) internal gravity wave.

(ii) Expressing the perturbations in the following forms with a growth rate σ and a wavenuumber k,

$$\zeta = Ae^{\sigma t} e^{ikx}, \quad \phi_i = B_i e^{\sigma t} e^{ikx - ky}, \quad (i = 1, 2), \qquad (9.70)$$

derive an equation to determine the growth rate σ, where A, B_1, B_2 are constants. State whether the basic state is stable or unstable.

(iii) Apply the above analysis to the case where a lighter fluid is placed above a heavy fluid, and derive a conclusion that there exists an interfacial wave (called the *internal gravity wave*, Fig. 9.9(b)). State what is the frequency.

Problem 9.2 *Rayleigh's inflexion-point theorem*

Based on the Rayleigh's equation (9.23) or (9.26), verify the followings:

(i) We have the Rayleigh's inflexion-point theorem stated in Sec. 9.3.1.

(ii) If there is a stable solution of $c_i < 0$ (stable mode), there always exists an unstable solution too, for inviscid parallel flows.

Problem 9.3 Lorenz system

Derive the Lorenz system (9.66) from the system of Eqs. (9.64) and (9.65) by using the stream function ψ and the temperture deformation θ defined by (9.62) and (9.63).

Problem 9.4 Squire's theorem

Suppose that there is a steady parallel shear flow $\mathbf{v}_0 = (U(z), 0, 0)$ of a viscous incompressible fluid of kinematic viscosity ν in a channel $-b < z < b$ in the cartesian (x, y, z) space.

(i) Derive the following normalized perturbation equations

$$\nabla^2 u - i\alpha R_e(U(z) - c)u - R_e U'(z)w = i\alpha R_e\, p, \qquad (9.71)$$
$$\nabla^2 v - i\alpha R_e(U(z) - c)v = i\beta R_e\, p, \qquad (9.72)$$
$$\nabla^2 w - i\alpha R_e(U(z) - c)w = R_e\, \mathrm{D}p, \qquad (9.73)$$
$$i\alpha u + i\beta v + \mathrm{D}w = 0, \qquad (9.74)$$

$(\mathrm{D} \equiv \mathrm{d}/\mathrm{d}z)$, for the perturbations $\mathbf{v}_1$ of the normal mode,

$$\mathbf{v}_1(\mathbf{x}, t) = (u(z), v(z), w(z)) \exp[i(\alpha x + \beta y - \alpha ct)], \qquad (9.75)$$

together with the pressure p_1/ρ_0 of the same normal mode. Here,

$$\nabla^2 = \mathrm{D}^2 - k^2, \quad k^2 = \alpha^2 + \beta^2, \quad R_e = \frac{U_m b}{\nu}, \quad U_m = \max U(z).$$

In addition, eliminating p, derive the equation for w only,

$$\nabla^2\nabla^2 w - i\alpha R_e(U(z) - c)\nabla^2 w + i\alpha R_e U''(z)\, w = 0, \qquad (9.76)$$

and the boundary conditions:

$$w = 0, \quad \text{at } z = \pm 1. \qquad (9.77)$$

[This is an extension to 3D perturbations of (9.29).]
(ii) The boundary value problem (9.76) and (9.77) determines the eigenvalue of the complex exponent $-i\alpha c = -i\alpha c_r + \alpha c_i$ for a

real α. Show the reasoning that the critical Reynolds number is given by

$$R_c = \min_{\alpha,\beta} \frac{F(\sqrt{\alpha^2 + \beta^2})}{\alpha} \qquad \text{with } c_i = 0, \qquad (9.78)$$

where $F(k)$ is a positive function of $k = \sqrt{\alpha^2 + \beta^2}$.

(iii) Show that the critical Reynolds number R_c^{2D} for two-dimensional disturbances is always less than the critical Reynolds number R_c^{3D} for three-dimensional disturbances [Squire's theorem].

Chapter 10

Turbulence

Turbulence is an irregular fluctuating flow field both spatially and temporally. It is a fundamental problem to be answered whether the fluid mechanics of continuum media can capture the turbulence.

In the past, there have been a number of discussions about how such turbulence is generated. On the one hand, turbulence is considered to be generated by spatially random initial conditions, or temporally random boundary conditions. This is a view of turbulence generated passively. In the modern view, however, turbulence is considered to be an irregularly fluctuating flow field which develops autonomously by a nonlinear mechanism of field dynamics. This change in view on turbulence is largely due to the recent development of the theory of chaos considered in the last chapter (Chapter 9).

Chaos is studied mostly for nonlinear dynamical systems of low dimensions. In a chaotic state, an initial slight difference of state grows exponentially with time. Despite the fact that the system evolves according to a governing equation, the state of the dynamical system becomes unpredictable after a certain time. This is a property termed *deterministic chaos* considered in Chapter 9.

Turbulence is a dynamical system of considerably many degrees of freedom. It is not surprising that the turbulence exhibits much more complex behaviors. In such a (turbulence) field, it would not be realistic to specify the initial condition accurately, and a statistical consideration must be made. There are some intrinsic difficulties in statistics: that is, statistical distributions of velocity or velocity difference at two points are not the Gaussian distribution and hence are not normal. This means we have to take into account all the statistical moments of variables.

Here, we take a view that turbulence is a physical system of nonlinear dynamics of a continuous field with a dissipative mechanism. In other words, turbulence exhibits every possible fluctuating motions and straining flows without breaking the fundamental laws of mechanics of a continuous medium. Under such a view, we investigate turbulence dynamics governed by the Navier–Stokes equations (4.5) and (4.6) for the velocity field u_i of an incompressible fluid of constant density ρ:

$$\partial_t u_i + \partial_j(u_i u_j) = -\rho^{-1}\partial_i p + 2\nu\partial_j e_{ij}, \tag{10.1}$$

$$D = \partial u_i / \partial x_i = 0, \tag{10.2}$$

where ν is the kinematic viscosity, p the pressure and $e_{ij} = \frac{1}{2}(\partial_i u_j + \partial_j u_i)$.[1]

When the statistical properties of turbulence are uniform in space, that is, the statistics are equivalent at every point of space, turbulence is called *homogeneous*. In addition, if the statistics of turbulence are equivalent for any direction of space, that is, the statistics such as correlation functions at n points (n: some integer) are invariant with respect to rigid-body rotation or inversion of those points, then the turbulence is called *isotropic*.

10.1. Reynolds experiment

It is generally accepted that one of the earliest studies of turbulence is the Reynolds experiment (1883). His experimental tests were carried out for flows of water through glass tubes of circular cross-sections of different diameters. The flow through a circular tube of diameter D is characterized with maximum velocity U. In Sec. 4.4, we learned that the flow is controlled by a dimensionless parameter, i.e. the Reynolds number defined by $\mathrm{Re} = UD/\nu$, where ν is the kinematic viscosity (of water). This view owes a great deal to the study of Reynolds.

---

[1]Note that $\partial_j(u_i u_j) = u_j \partial_j u_i$ and $\partial_j 2e_{ij} = \nu\nabla^2 u_i$, since $\partial_j u_j = 0$ and $\partial_j^2 = \nabla^2$.

According to his experiment, there is a critical value R_c for the Reynolds number. Below R_c, the flow is laminar, i.e. the water flows smoothly along the whole length of the tube. If the flow velocity u along the tube is represented as a function $u(r)$ of the radial coordinate r with $r = 0$ denoting the tube axis, the smooth flow takes a parabolic profile $u(r) = U(1 - r^2/a^2)$, which is called the Poiseuille flow (see (4.78) for axisymmetry;[2] (4.41) for 2D problem). Above R_c, he observed *turbulent* flows. An important point which Reynolds observed is the existence of the critical value R_c of the Reynolds number for the transition to turbulence. He also observed that the friction laws are different for flows below and above R_c. The friction was proportional to U below R_c, while it was proportional to U^2 above R_c. This suggests that the two types of flows are different in nature.

As Reynolds number becomes sufficiently high, not only pipe flows but most flows become turbulent. At each point of turbulence field, flow velocity fluctuates with time. Taking an average of the irregularly fluctuating velocity vector $\mathbf{u} = (u_i)$ at a point $\mathbf{x} = (x_i)$, we obtain an average velocity field $\langle \mathbf{u} \rangle = \langle u_i \rangle$. The average velocity $\langle \mathbf{u} \rangle (\mathbf{x})$ may vary slowly and smoothly from point to point, while the difference $\mathbf{u}' = \mathbf{u} - \langle \mathbf{u} \rangle$ may fluctuate irregularly, with its average $\langle \mathbf{u}' \rangle$ vanishing. $\mathbf{u}'$ denotes the turbulent fluctuation of velocity.

Another remarkable contribution of Reynolds in addition to the experiment mentioned above is the introduction of an average equation and recognition of "Reynolds stress" in it. Suppose that the velocity and pressure fields of an incompressible fluid are represented by superposition of an average and a fluctuation as follows:

$$u_i = \langle u_i \rangle (\mathbf{x}, t) + u_i'(\mathbf{x}, t), \quad p = \langle p \rangle (\mathbf{x}, t) + p'(\mathbf{x}, t),$$

where $\partial \langle u_i \rangle / \partial x_i = 0$. Substituting these into (10.1) and taking an average, we obtain the following equation governing the average

---

[2] The linear stability theory (Sec. 9.3.2) implies $R_c = \infty$ (stable) for the axisymmetric Poiseuille flow. Despite this, study of flows in a circular pipe is a challenging realistic problem in Fluid Mechanics.

fields:

$$\partial_t \langle u_i \rangle + \partial_j (\langle u_i \rangle \langle u_j \rangle) + \partial_j \langle u'_i u'_j \rangle = -\rho^{-1} \partial_i \langle p \rangle + \nu \partial_j 2 \langle e_{ij} \rangle, \qquad (10.3)$$

where $2 \langle e_{ij} \rangle = \partial \langle u_i \rangle / \partial x_j + \partial \langle u_j \rangle / \partial x_i$. The third term on the left-hand side $\rho \langle u'_i u'_j \rangle$ (multiplied by ρ) is called the *Reynolds stress*, because this can be combined with the viscous stress term on the right as

$$\rho \partial_j (2\nu \langle e_{ij} \rangle - \langle u'_i u'_j \rangle) = \partial_j \sigma_{ij}^{(\text{turb})},$$

where $\sigma_{ij}^{(\text{turb})} \equiv 2\mu \langle e_{ij} \rangle - \rho \langle u'_i u'_j \rangle$, where $\mu = \rho \nu$. Thus, we obtain the same form as (10.1) for the equation of average fields.

10.2.   Turbulence signals

In turbulent flows, velocity signals fluctuate irregularly with time [Fig. 10.1(a)]. In fact, there are a large number of modes from very small scales to very large scales in turbulence, which interact with each other. Their spatial distributions are also considered to be random. Statistical distribution of each (x, y, z) component of $\mathbf{u}'$ may be expected to be nearly Gaussian. However, remarkably enough, real turbulences are *not so*, and exhibit non-Gaussian properties for statistical distributions [Fig. 10.1(c)]. This implies that there may exist some nonrandom objects in turbulence field to make it deviate from a complete randomness of Gaussian statistics.

In general, a skewness factor of the distribution of longitudinal derivative[3] of a turbulent velocity-component u, denoted by $u_x = \partial u / \partial x$, is defined by

$$\text{Skewness} \quad S[u_x] = \frac{\langle (u_x)^3 \rangle}{\langle (u_x)^2 \rangle^{3/2}} \quad (\approx -0.4 \text{ in an experiment}),$$

which is always found to be negative. In the Gaussian distribution, the skewness must vanish. Furthermore, a flatness factor of u_x is

---

[3] The *longitude* denotes that the direction of derivative is the same as the velocity component u, not *lateral* to it.

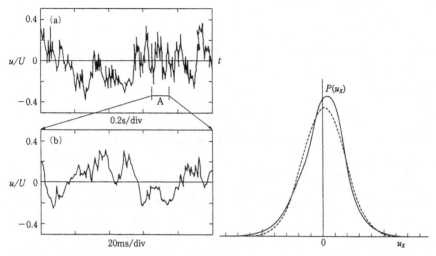

Fig. 10.1. (a) Fluctuating signal of x component u of turbulent velocity $\mathbf{u}'$, (b) Enlargement of the interval A of (a). (c) A sketch of non-Gaussian distribution function $P(u_x)$ of the *longitudinal* derivative $u_x = \partial_x u$, where the broken curve is the Gaussian distribution function [Mak91].

defined by

$$\text{Flatness} \quad F[u_x] = \frac{\langle (u_x)^4 \rangle}{\langle (u_x)^2 \rangle^2} \ (\approx 7.2 \text{ in an experiment}).$$

This is also deviating from the Gaussian value 3.0.

Nonrandom objects in the turbulence field causing these deviations from Gaussian distribution are often called *organized structures*. Recent direct numerical simulations (DNS in short) of turbulence at high Reynolds numbers present increasing evidence that homogeneous isotropic turbulence is composed of random distribution of a number of long thin vortices, often called *worms*. Their cross-sectional diameters are estimated to be of the order of l_d (defined by (10.18) below), and their axial lengths are much larger.[4]

From some experimental observation of the probability distribution of u_x, it is found that the probability of large $|u_x|$ is relatively

---

[4]Some computer simulation showed that the average separation distance between worms is of the order of Taylor micro-scale defined by a length scale of statistical correlation of two velocities at two different points.

higher than that of the Gaussian. Namely, large derivative values u_x appears intermittently in turbulence. This is one of the characteristic features of fluid turbulence and is called *intermittency*. It is not only observed by experiments, but also found by computer simulations.

10.3.   Energy spectrum and energy dissipation

10.3.1.   *Energy spectrum*

Let us consider a turbulent flow field in a three-dimensional space. It is convenient to consider turbulence in Fourier space. In order to avoid some mathematical difficulties arising from the infinity of space, the flow field is assumed to satisfy a periodic condition. Suppose that the velocity field $\mathbf{u}(x, y, z)$ is periodic with respect to the three cartesian directions (x, y, z) of periodicity length L:

$$\mathbf{u}(x + m_x L,\, y + m_y L,\, z + m_z L) = \mathbf{u}(x,\, y,\, z),$$

where m_x, m_y, $m_z \in \mathcal{Z}$ (integer). The pressure field $p(x, y, z)$ is similarly represented. Hence, the flow field is considered in a basic cubic space:

$$\mathcal{C}:\ [0 \leqq x < L,\ 0 \leqq y < L,\ 0 \leqq z < L].$$

The representation in an infinite space can be obtained by taking the limit, $L \to \infty$. Then the velocity field $\mathbf{u}(\mathbf{x}, t)$ is given by the following representation:

$$\mathbf{u}(\mathbf{x}, t) = \left(\frac{2\pi}{L}\right)^3 \sum_{\mathbf{k}} \hat{\mathbf{u}}(\mathbf{k}, t) e^{i\mathbf{k}\cdot\mathbf{x}}, \tag{10.4}$$

$$\mathbf{k} = (k_x,\, k_y,\, k_z) = (2\pi/L)(n_x, n_y, n_z). \tag{10.5}$$

where $\hat{\mathbf{u}}(\mathbf{k}, t)$ is the Fourier amplitude, $\mathbf{k}$ is the wavenumber, $n_x, n_y, n_z \in \mathcal{Z}$ (integer space). In the limit $L \to \infty$, the expression

(10.4) takes the following form[5]:

$$\mathbf{u}(\mathbf{x}, t) = \int_{\mathcal{R}^3} \hat{\mathbf{u}}(\mathbf{k}, t) \, e^{i\mathbf{k}\cdot\mathbf{x}} \, d^3\mathbf{k}. \tag{10.6}$$

Since the velocity $\mathbf{u}(\mathbf{x}, t)$ is real, the following property must be satisfied:

$$\hat{\mathbf{u}}^*(\mathbf{k}, t) = \hat{\mathbf{u}}(-\mathbf{k}, t), \tag{10.7}$$

where u^* is the complex conjugate of u. Inverse transform of (10.4) is

$$\hat{\mathbf{u}}(\mathbf{k}, t) = \frac{1}{(2\pi)^3} \int_{\mathcal{C}} \mathbf{u}(\mathbf{x}, t) \, e^{-i\mathbf{k}\cdot\mathbf{x}} \, d^3\mathbf{x}. \tag{10.8}$$

The condition of incompressibility, $\mathrm{div}\,\mathbf{u} = 0$, leads to

$$\nabla \cdot \mathbf{u} = \partial_\alpha u_\alpha = \int (i k_\alpha \hat{u}_\alpha(\mathbf{k})) \, e^{i\mathbf{k}\cdot\mathbf{x}} \, d^3\mathbf{k} = 0,$$

which results in

$$k_\alpha \hat{u}_\alpha(\mathbf{k}) = 0, \quad \text{namely } \hat{\mathbf{u}}(\mathbf{k}) \perp \mathbf{k}. \tag{10.9}$$

The wavenumber vectors $\mathbf{k}$ in (10.4) and (10.8) are discrete, whereas those in (10.6) are continuous. Corresponding to whether the wavenumber $\mathbf{k}$ is discrete or continuous, the integral $\int e^{ikx} dx$ reduces to the following:

$$\left.\begin{aligned} \text{Discrete:} \quad & \int_{\mathcal{C}} e^{ikx} dx = L\,\delta_{k,0} \equiv 2\pi\,\delta_{k,0}\,(\Delta k)^{-1}, \\ \text{Continuous:} \quad & \int_{-\infty}^{\infty} e^{ikx} dx = 2\pi\,\delta(k), \end{aligned}\right\} \tag{10.10}$$

where $\Delta k = 2\pi/L$, $\delta_{k,l}$ is the Kronecker's delta and $\delta(k)$ is the Dirac's delta function (Appendix A.7).

---

[5]By the correspondence $(2\pi/L) \Leftrightarrow \Delta k$, we have $\left(\frac{2\pi}{L}\right)^3 \sum_{\mathbf{k}} \Leftrightarrow \int_{\mathcal{R}^3} d^3\mathbf{k} = \int_{-\infty}^{\infty} dk_x \int_{-\infty}^{\infty} dk_y \int_{-\infty}^{\infty} dk_z$.

Denoting the average over the basic cubic space $\mathcal{C}$ by $\langle \cdot \rangle_{\mathcal{C}} = L^{-3} \int_{\mathcal{C}} d^3 \mathbf{x}$, we have the *average* in the limit $L \to \infty$:

$$\langle u^2 \rangle_{\mathcal{C}} = \left(\frac{2\pi}{L} \right)^6 \sum_{\mathbf{k}} |\hat{\mathbf{u}}(\mathbf{k}, t)|^2$$

$$\to \int d^3 \mathbf{k} \int d^3 \mathbf{k}' \, \hat{\mathbf{u}}(\mathbf{k}) \hat{\mathbf{u}}(\mathbf{k}') \delta(\mathbf{k} + \mathbf{k}') = \int d^3 \mathbf{k} |\hat{\mathbf{u}}(\mathbf{k})|^2. \quad (10.11)$$

Therefore, the kinetic energy per unit mass is given by

$$\left\langle \frac{u^2}{2} \right\rangle_{\mathcal{C}} = \int_{\mathcal{R}^3} \Phi(\mathbf{k}, t) d^3 \mathbf{k}, \quad \Phi(\mathbf{k}, t) = \frac{1}{2} |\hat{\mathbf{u}}(\mathbf{k}, t)|^2. \quad (10.12)$$

If we take the average with respect to a statistical ensemble of initial conditions in addition to the average over the space $\mathcal{C}$, then the average $\langle \cdot \rangle_{\mathcal{C}}$ should be replaced with the *ensemble average* $\langle \cdot \rangle$.

Provided that the turbulence is regarded as isotropic and the average $\langle \Phi(\mathbf{k}, t) \rangle$ depends on the magnitude $k = |\mathbf{k}|$ of the wavenumber $\mathbf{k}$, then we have

$$\left\langle \frac{u^2}{2} \right\rangle (t) = \int_0^\infty E(k, t) dk, \quad E(k, t) = 4\pi k^2 \Phi(|\mathbf{k}|, t), \quad (10.13)$$

where $E(k, t)$ is the **energy spectrum** at a time t.

One of the advantages of Fourier representation is that we can consider each Fourier component separately. The component of a wavenumber k corresponds to the velocity variation of wavelength $2\pi/k$. Such a component is said to have simply a *length scale $l = 1/k$*. It is often called an *eddy* of scale l. Furthermore, the Fourier representation with respect to space variables (x, y, z) enables us to reduce a *partial* differential equation to an *ordinary* differential equation with respect to time t, since ∂_x is replaced by ik_x, etc.

10.3.2. *Energy dissipation*

Let us consider an energy equation. With an analogous calculation carried out in deriving Eq. (4.23) for unbounded space in Sec. 4.3, an energy equation is written for the kinetic energy

$K = \int_{\mathcal{C}} \frac{1}{2}u^2 dV$ in the cubic space $\mathcal{C}$ as

$$\frac{\mathrm{d}K}{\mathrm{d}t} = \int_{\mathcal{C}} u_i \frac{\partial u_i}{\partial t} \mathrm{d}^3\mathbf{x} = -\nu \int_{\mathcal{C}} \left(\frac{\partial u_i}{\partial x_j}\right)^2 \mathrm{d}^3\mathbf{x}$$

$$= -2\nu \int_{\mathcal{C}} e_{ij} \frac{\partial u_i}{\partial x_j} \mathrm{d}^3\mathbf{x} = -2\nu \int_{\mathcal{C}} \|e_{ij}\|^2 \mathrm{d}^3\mathbf{x}, \qquad (10.14)$$

where e_{ij} is the rate-of-strain tensor, and the surface integral in the energy equation disappears here too because of periodic boundary conditions. Furthermore, using the vorticity vector $\boldsymbol{\omega} = \nabla \times \mathbf{u} = \varepsilon_{ijk}\partial_j u_k$, we obtain another form of energy equation (see (4.24)),

$$\frac{\mathrm{d}K}{\mathrm{d}t} = -\nu \int_{\mathcal{C}} |\boldsymbol{\omega}|^2 \mathrm{d}^3\mathbf{x}. \qquad (10.15)$$

Taking an ensemble average $\langle \cdot \rangle$ of Eqs. (10.14) and (10.15), we obtain the *average rate of energy dissipation* per unit mass ε as follows:

$$\varepsilon := -\frac{\mathrm{d}}{\mathrm{d}t}\left\langle \frac{1}{2}u^2 \right\rangle = 2\nu\langle\|e_{ij}\|^2\rangle = \nu\langle|\boldsymbol{\omega}|^2\rangle \qquad (10.16)$$

$$= 2\nu \int_0^\infty k^2 E(k)\mathrm{d}k. \quad \text{[See Problem 10.1]} \quad (10.17)$$

The expressions on the right-hand side give three different expressions of energy dissipation. The form of integrand $k^2 E(k)$ of (10.17) implies that the dissipation rate is amplified for eddies of large k (small eddies) due to the factor k^2.

10.3.3. *Inertial range and five-thirds law*

A number of experiments have been carried out so far in order to determine the energy spectrum $E(k)$ for turbulence at very high Reynolds numbers, 10^4 or larger. In fully developed turbulence, most

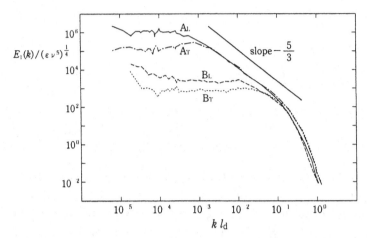

Fig. 10.2.   Log-log plot of an experimentally determined energy spectrum $E_1(k)$, showing the inertial range by a straight-line slope of $-5/3$ (from a grid turbulence with a main stream velocity 5 m/s in the x-direction [Mak91], where $E_1(k)$ is the one-dimensional spectrum which is connected to the three-dimensional energy spectrum $E(k)$ by $E(k) = k^3(d/dk)(k^{-1}dE_1(k)/dk)$ (if $E_1(k) \propto k^\alpha$, then $E(k) \propto k^\alpha$ too). [See Problem 10.3.] A_L and A_T are obtained from a *fully developed* turbulence, while B_L and B_T are from a nondeveloped turbulence. The suffices L and T denote longitudinal (x) and transverse components (y and z) of velocity.

energy spectra determined experimentally suggest the form,

$$E(k) \propto k^{-5/3}$$

for an intermediate range of wavenumbers (an inertial range). The range is defined by wavenumbers which are larger than k_Λ associated with a scale Λ (a laboratory scale) of external boundaries and less than k_d associated with a scale of viscous dissipation. This is known as the 5/3-th power law of the energy spectrum. A plot of $\log E(k)$ versus $\log k$ shows a nearly straight-line behavior of slope $-5/3$ in the corresponding range of k (Fig. 10.2). Such a range of wavenumbers is called the *inertial range*. Whether a turbulence is fully developed or not is tested with whether the energy spectrum has a sufficient inertial range. Thus, a *fully developed turbulence* means that the turbulent flow has a sufficient range of the energy spectrum of $E(k) \propto k^{-5/3}$.

10.3.4. *Scale of viscous dissipation*

Magnitude of the wavenumber associated with dominant dissipation of energy is estimated according to the following physical reasoning helped by dimensional arguments. The fully developed turbulence is regarded as controlled by the viscosity ν (kinematic viscosity) and the rate of energy dissipation ε per unit mass. It is hypothetically assumed that viscous dissipation of kinetic energy is predominant at small scales of order l_d. To compensate the energy loss at those scales, an energy flow is driven from larger to smaller scales. At scales in the inertial range, the energy flow from one scale to smaller scales is almost invariant. Arriving at scales of order l_d, the viscous loss becomes dominant. On the other hand, the energy is injected at a largest scale Λ by some external means with a laboratory scale Λ. This is an idea of *energy cascade*, envisaged by Kolmogorov and Oboukov in the 1940s. Their hypothesis results in the following dimensional arguments and estimates by scaling. It is called the *similarity hypothesis* due to Kolmogorov and Oboukov.

Denoting the dimensions of length and time by L and T, the dimensions of ν and ε are given by $[\varepsilon] = L^2 T^{-3}$ and $[\nu] = L^2 T^{-1}$, respectively. This is because, denoting the velocity by $U = L/T$, $[\varepsilon]$ is given by $U^2/T = L^2 T^{-3}$ (see (10.16)) and $[\nu]$ is given by the relation $U/T = [\nu]U/L^2$ (e.g. see (10.1)), respectively. Now, it can be immediately shown by these dimensional arguments that the rate of energy dissipation ε and the viscosity ν determine a characteristic length scale from their combination which is given by[6]

$$l_d = \left(\frac{\nu^3}{\varepsilon} \right)^{1/4}. \tag{10.18}$$

The l_d is called Kolmogorov's *dissipation scale*. The corresponding wavenumber is defined by $k_d = 1/l_d = (\varepsilon/\nu^3)^{1/4}$.

At much higher wavenumbers than k_d, the spectrum $E(k)$ decays rapidly because the viscous dissipation is much more effective at larger k's. The range $k > k_d$ is called the *dissipation range*.

---

[6]Setting $\varepsilon^\alpha \nu^\beta = L^1$, one obtains $\alpha = -1/4$ and $\beta = 3/4$.

10.3.5. *Similarity law due to Kolmogorov and Oboukov*

Based on the similarity hypothesis described above, one can derive an energy spectrum in the inertial range. Suppose that an eddy of scale l in the inertial range has a velocity of order v_l. Then, an eddy turnover time is defined by

$$\tau_l \sim l/v_l. \tag{10.19}$$

The similarity hypothesis implies a constant energy flux ε_l across each scale l within the inertial range, given by

$$\varepsilon_l \sim v_l^2/\tau_l \sim v_l^3/l. \tag{10.20}$$

Setting this to be equal to ε (constant), i.e. $\varepsilon_l = \varepsilon$, we obtain

$$v_l \sim (\varepsilon l)^{1/3}. \tag{10.21}$$

This scaling laws implies that the turbulence field is of *self-similar* structures between different eddy scales l. It is a remarkable property that the magnitude of velocity characterizing variations over a length scale l is scaled as $l^{1/3}$, showing a fractionally singular behavior as $l \to 0$.

The energy (per unit mass) of an eddy of scale l is given by $\frac{1}{2}v_l^2$. Using the equivalence $k = l^{-1}$, this is regarded as equal to $E(k)\Delta k$, which denotes the amount of energy contained in the wavenumber range Δk around k, where $\Delta k \sim l^{-1}$.[7] Thus, we obtain the relation,

$$E(k)l^{-1} \approx v_l^2 \sim (\varepsilon l)^{2/3}, \quad \therefore \ E(k) \sim \varepsilon^{2/3}l^{5/3}.$$

Repacing l with k^{-1}, we find

$$E(k) \sim \varepsilon^{2/3}k^{-5/3}. \tag{10.22}$$

This is the *Kolmogorov–Oboukov's 5/3-th law.*

---

[7] Assuming a power-law sequence of wavenumbers $k_0, k_1, \ldots$ such as $k_n = a^n k_0$, where a and k_0 are constants (say $a = 2$), we have $\Delta k = k_{n+1} - k_n = (a - 1)k_n$.

10.4. Vortex structures in turbulence

From the early times of turbulence study by GI Taylor (1938) and others, it has been recognized that there exist vortical structures in turbulence. The turbulence spectrum or energy cascade considered so far are concepts in the space of wavenumbers, while vortices are structures in real physical space. From the point of view that a vortical structure is nothing but a dissipative structure, we seek a deep relation between vortical structures and statistical laws in fully developed turbulence.

In turbulence, there exists a certain straining mechanism by which vortex lines are stretched on the average, to be described just below (Sec. 10.4.1). This is related to the *negative* value of skewness $S[u_x]$ of longitudinal derivative u_x (Sec. 10.2). Nonzero value of $S[u_x]$ implies that the statistics is *non-Gaussian*, and that there exists structures in turbulence, often called *intermittency*. As a slender vortex tube is stretched, its vorticity increases, and dissipation is enhanced around it in accordance with (10.15) and the Helmholtz law (iii) of Sec. 7.2.2.

According to the theory of energy cascade of fully developed turbulence, energy is dissipated at scales of smallest eddies of order $l_d = (\nu^3/\varepsilon)^{1/4}$, the dissipation scale. In computer simulations, the dissipative structures in turbulence are visualized as fine-scale slender objects with high level of vorticity magnitude.

Figure 10.3 shows a snapshot of a vorticity field [YIUIK02] obtained by a direct numerical simulation (DNS) of an incompressible turbulence in a periodic box with grid points 2048^3 carried out on the *Earth Simulator* (Sec. 8.5), on the basis of the Navier–Stokes equation (10.1) with an additional term of external random force **f** under the incompressible condition (10.2). Figure 10.3(b) is the enlargement of the central part of Fig. 10.3(a) by eight times. The similarity of structures between the two figures implies a self-similarity of an isotropic homogeneous turbulence in a statistical sense.

10.4.1. *Stretching of line-elements*

Although turbulence is regarded as a random field, it exhibits certain organized structures of vorticity. It is important that there

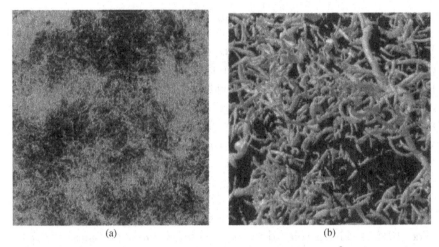

(a)   (b)

Fig. 10.3. High-vorticity isosurfaces obtained by DNS of 2048^3 grid points with $\nu = 4.4 \times 10^{-5}$, $l_d = 1.05 \times 10^{-3}$. (a) Length of a side is $2992\,l_d$, and (b) 8 times enlargement of (a), i.e. length of a side is $374\,l_d$, the area being $1/64$ of that of (a). The isosurfaces are defined by $|\boldsymbol{\omega}| = \langle\boldsymbol{\omega}\rangle + 4\sigma$ where σ is the standard deviation of the distribution of magnitude $|\boldsymbol{\omega}|$ [YIUIK02].

exits a fundamental mechanism to stretch vortex-lines in turbulence field. This is closely related to a mechanism by which a small line-element is stretched always on the average in turbulence. When a vortex-line is stretched, the vorticity is increased and dissipation is enhanced by (10.15). Let us first consider a mechanism for stretching line-elements.

Let us consider time evolution of a fluid line-element $\delta\mathbf{x}(t)$ connecting two fluid particles A and B at an infinitesimal distance:

$$\delta\mathbf{x}(t) = \mathbf{x}(B,t) - \mathbf{x}(A,t).$$

Its time evolution is described by the following equation for the ith component,

$$\frac{\mathrm{d}}{\mathrm{d}t}\delta x_i = u_i(B) - u_i(A) = \frac{\partial u_i}{\partial x_j}\delta x_j, \qquad (10.23)$$

by the linear approximation (see (7.13)). The tensor of velocity derivatives $\partial u_i/\partial x_j$ can be decomposed always into a symmetric part

S_{ij} and an anti-symmetric part Ω_{ij} (Sec. 1.4):

$$\frac{\partial u_i}{\partial x_j} = e_{ij} + \Omega_{ij}, \tag{10.24}$$

$$e_{ij} = \frac{1}{2}(\partial_j u_i + \partial_i u_j), \quad \Omega_{ij} = \frac{1}{2}(\partial_j u_i - \partial_i u_j), \tag{10.25}$$

where e_{ij} is the *rate of strain tensor*, while Ω_{ij} is the *vorticity tensor*. The latter is equivalent to the vorticity vector $\boldsymbol{\omega} = \nabla \times \mathbf{v}$ by the relation $\Omega_{ij} = -\frac{1}{2}\varepsilon_{ijk}\omega_k$ (Sec. 1.4.3 where Ω_{ij} is written as g_{ij}). The (real) tensor e_{ij} can be made *diagonal* by the symmetry $e_{ij} = e_{ji}$ at each point. In fact, fixing the point $\mathbf{x}$, one can determine three eigenvalues (E_1, E_2, E_3) of the matrix e_{ij} and eigenvectors $\mathbf{e}_1, \mathbf{e}_2, \mathbf{e}_3$. By an orthogonal transformation determined by $[\mathbf{e}_1, \mathbf{e}_2, \mathbf{e}_3]$ to the principal frame, the matrix e_{ij} is transformed to a diagonal form (Sec. 1.4.2): i.e. $(e_{ij}) = \mathrm{diag}(E_1, E_2, E_3)$.

The velocity gradients $\partial u_i/\partial x_j$ can be regarded as constant over a line-element $\delta\mathbf{x}(t)$ if it is sufficiently short. It can be verified that $\langle |\delta\mathbf{x}(t)|^2 \rangle \geq \langle |\delta\mathbf{x}(0)|^2 \rangle$. In fact, owing to the linearity of Eq. (10.23), $\delta\mathbf{x}(t)$ is related to $\delta\mathbf{x}(0)$ by a linear relation:

$$\delta x_i(t) = U_{ij}\,\delta x_j(0),$$

where U_{ij} is a random tensor (in the turbulent field) determined by the value of $\partial_j u_i$ along the trajectory of the line-element, depending on the initial position and time t, but independent of $\delta\mathbf{x}(0)$. From the above, we obtain

$$|\delta x_i(t)|^2 = W_{jk}\,\delta x_j(0)\,\delta x_k(0), \tag{A}$$

where $W_{jk} = U_{ij}U_{ik}$ is a real symmetric tensor. Let us denote the eigenvalues of W_{jk} as w_1, w_2, w_3 at a fixed time t. Because the quadratic form $W_{jk}\,\delta x_j(0)\delta x_k(0)$ is positive definite by the above definition, we should have $w_i > 0$ for all i.

Suppose that we take a small sphere of radius a with its center at a point A at an initial time with its volume $(4\pi/3)a^3$. After a small time t, it would be deformed to an ellipsoid by the action of turbulent flow,[8] and the length of the three principal axes would become $\sqrt{w_1}\,a$, $\sqrt{w_2}\,a$, $\sqrt{w_3}\,a$ with the volume given by $(4\pi/3)a^3\sqrt{w_1 w_2 w_3}$. Owing to the volume conservation, we must have $w_1 w_2 w_3 = 1$.

The initial line-element $\delta \mathbf{x}(0)$ can be chosen arbitrarily on the sphere of radius a. Since the tensor W_{jk} of the turbulent field is independent of the choice of $\delta \mathbf{x}(0)$, the values w_1, w_2, w_3 are statistically independent of $\delta \mathbf{x}(0)$. In isotropic turbulence, the eigenvectors associated with w_1, w_2, w_3 distribute isotropically. Under these conditions, one can show the property,

$$\langle |\delta \mathbf{x}(t)|^2 \rangle = \frac{w_1 + w_2 + w_3}{3}\, a^2 \geq (w_1 w_2 w_3)^{1/3}\, a^2 = a^2 = |\delta \mathbf{x}(0)|^2,$$

according to the inequality: (arithmetic mean) $\geq$ (geometric mean). Namely, any infinitesimal line-element is always stretched in turbulent fields in the statistical sense. Similarly, any infinitesimal surface area is always enlarged in a turbulent field in the statistical sense [Coc69; Ors70].

10.4.2. *Negative skewness and enstrophy enhancement*

Vortical structures in turbulent fields have dual meanings. First, those signify the structures of velocity field. Namely, the field may be irregular, but has some correlation length. Secondly, the vorticity is related with energy dissipation (see (10.15)). The latter (energy dissipation) is concerned with the dynamics and time evolution of the system, whereas the former is concerned with the statistical laws of spatial distribution. This is the subject at the moment.

---

[8]From (A) and the eigenvalues w_1, w_2, w_3 defined below it, we have

$$|\delta x_i(t)|^2 = w_1(\delta x_1(0))^2 + w_2(\delta x_2(0))^2 + w_3(\delta x_3(0))^2. \qquad (B)$$

Since $\delta \mathbf{x}(0)$ was a point on a sphere of radius a, one can write as $\delta \mathbf{x}(0) = a(\alpha, \beta, \gamma)$ where the direction cosines α, β, γ satisfy $\alpha^2 + \beta^2 + \gamma^2 = 1$. Analogously, a point $\delta \mathbf{x}(t) = (\xi, \eta, \zeta)$ on the ellipsoid satisfying the relation (B) can be expressed by $(\xi, \eta, \zeta) = a(\sqrt{w_1}\,\alpha, \sqrt{w_2}\,\beta, \sqrt{w_3}\,\gamma)$. Volume of the ellipsoid is given by $(4\pi/3)(\sqrt{w_1}\,a)(\sqrt{w_1}\,a)(\sqrt{w_1}\,a) = (4\pi/3)a^3\sqrt{w_1 w_2 w_3}$. From the isotropy requirement, we have $\langle \alpha^2 \rangle = \langle \beta^2 \rangle = \langle \gamma^2 \rangle = 1/3$.

Let us consider production of average enstrophy, defined by $\frac{1}{2}\langle\boldsymbol{\omega}^2\rangle$, in a homogeneous isotropic turbulence. The equation of vorticity $\boldsymbol{\omega}$ is written as, from (7.3),

$$\partial_t\boldsymbol{\omega} + (\mathbf{v}\cdot\mathrm{grad})\boldsymbol{\omega} = (\boldsymbol{\omega}\cdot\mathrm{grad})\mathbf{v} + \nu\nabla^2\boldsymbol{\omega}, \tag{10.26}$$

for an incompressible fluid without external force. We used this equation to obtain the solution of Burgers vortex in Sec. 7.9. Taking a scalar product of $\boldsymbol{\omega}$ with this equation, the left-hand side can be written as

$$\omega_k\partial_t\omega_k + \omega_k(\mathbf{v}\cdot\mathrm{grad})\omega_k = \partial_t\frac{1}{2}\boldsymbol{\omega}^2 + (\mathbf{v}\cdot\mathrm{grad})\frac{1}{2}\boldsymbol{\omega}^2 = \frac{\mathrm{D}}{\mathrm{D}t}\frac{1}{2}\boldsymbol{\omega}^2,$$

while the right-hand side is

$$\omega_j(\partial_j u_i)\omega_i + \nu\omega_i\partial_j\partial_j\omega_i = \frac{1}{2}\omega_j(\partial_j u_i + \partial_i u_j)\omega_i$$
$$+ \nu\partial_j(\omega_i\partial_j\omega_i) - \nu(\partial_j\omega_i)^2,$$

since $\omega_j(\partial_j u_i)\omega_i = \omega_i(\partial_i u_j)\omega_j$. Taking an ensemble average, we obtain

$$\frac{\mathrm{d}}{\mathrm{d}t}\frac{1}{2}\langle\boldsymbol{\omega}^2\rangle = \langle\omega_i e_{ij}\omega_j\rangle - \nu\left\langle\left(\frac{\partial\omega_i}{\partial x_j}\right)^2\right\rangle, \tag{10.27}$$

where e_{ij} is defined by (10.25) and the surface integral disappears when the ensemble average is taken. The right-hand side represents production or dissipation of the average enstrophy $\frac{1}{2}\langle\boldsymbol{\omega}^2\rangle$. The first term represents its production associated with the vortex-line stretching, whereas the second term is its dissipation due to viscosity. The fact that the first term expresses enstrophy production is assured by the property that the skewness of longitudinal velocity derivative is *negative* in turbulence, as observed in experiments and computer simulations.

The skewness of longitudinal velocity derivative is defined by

$$S_l = \frac{\langle(\partial u/\partial x)^3\rangle}{\langle(\partial u/\partial x)^2\rangle^{3/2}},$$

where the x axis (with u the x component velocity) may be chosen arbitrarily in the turbulence field. From the theory of homogeneous isotropic turbulence [Bat53; MY71; KD98], one can derive that

$$\langle \omega_i e_{ij} \omega_j \rangle = -\frac{35}{2} S_l \left(\frac{\varepsilon}{15\nu} \right)^{3/2}. \tag{10.28}$$

In fully developed turbulence, the skewness S_l is usually negative (see Sec. 10.2). Hence, we have

$$\langle \omega_i e_{ij} \omega_j \rangle > 0.$$

Thus, it is found that there exists a mechanism in turbulence to increase the enstrophy on the average which is associated with the mechanism of vortex-line stretching. Formation of a Burgers vortex studied in Sec. 7.9 is considered to be one of the processes in turbulence.

10.4.3.   *Identification of vortices in turbulence*

In order to identify the vortices in turbulence, a certain criterion is required, because the vortices in turbulence have no definite shape.

A simplest way is to apply the div operator to the Navier–Stokes equation (10.1) under the condition $\partial_i u_i = 0$. Using the relation $\partial_j u_i = e_{ij} + \Omega_{ij}$ of (10.24) and $\partial_i u_i = 0$, we obtain[9]

$$\nabla^2 p_* = -\partial_i u_j \partial_j u_i = -(e_{ij} + \Omega_{ij})(e_{ji} + \Omega_{ji}) \tag{10.29}$$

$$= -e_{ij} e_{ji} - \Omega_{ij} \Omega_{ji} = -\mathrm{tr}\{S^2 + \Omega^2\}$$

$$= -e_{ij} e_{ji} + \Omega_{ij} \Omega_{ij} = -\|S\|^2 + \|\Omega\|^2, \tag{10.30}$$

where $p_* = p/\rho_0$, and $e_{ji} = e_{ij}$, $\Omega_{ji} = -\Omega_{ij}$ are used. We have introduced the following notations:

$$\|e\|^2 \equiv e_{ij} e_{ji} = e_{ij} e_{ij}, \quad \|\Omega\|^2 \equiv \Omega_{ij} \Omega_{ij} = \frac{1}{2} |\omega|^2.$$

A vortex may be defined by a domain satisfying the inequality $\|\Omega\|^2 > \|S\|^2$ where the vorticity $|\omega|$ is supposed to be relatively stronger. In other words, a vortex would be defined by a set of points characterized by $\nabla^2 p_* > 0$. In general, vortex axes are curved lines, and lower pressure is distributing along them.

---

[9] $\partial_i \partial_j (u_i u_j) = \partial_i (u_j \partial_j u_i) = \partial_i u_j \partial_j u_i$, and $\Omega_{ij} e_{ji} + \Omega_{ji} e_{ij} = \Omega_{ij} e_{ij} - \Omega_{ij} e_{ij} = 0$.

(a)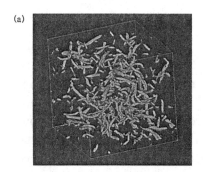

(b)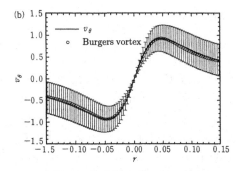

Fig. 10.4. (a) Vortex structure obtained by DNS [TMI97], (b) the circumferential velocity $v_\theta(r)$ (average by a solid curve; dispersion by vertical bars). The circles show $v_\theta(r)$ of Burgers vortex.

Figure 10.4(a) shows the vortex structures obtained by a direct numerical simulation (DNS) of Navier–Stokes equation [TMI97], visualized by the above method. The diagram (b) on the right plots the average and dispersion of the circumferential velocity $u_\theta(r)$ around such vortex structures. The average curve coincides fairly well with the solution (7.80) of Burgers vortex in Sec. 7.9.

10.4.4. *Structure functions*

It is natural to say that statistical laws of turbulence are connected with the distribution of vortex structures in turbulence. This should provide non-Gaussian statistical laws for velocity correlations at different points in the turbulent field.

In order to study a stochastic system of *non-Gaussian* property, one of the approaches is to examine structural functions of higher orders, where the pth order structure function $F_p(s)$ of longitudinal velocity difference Δv_l is defined by

$$F_p(s) = \langle (\Delta v_l(\mathbf{s}))^p \rangle, \qquad (10.31)$$

the symbol $\langle \cdot \rangle$ denoting ensemble average. Suppose that there are a number of Burgers vortices in turbulence and their axes are being distributed randomly in the field. Choosing two neighboring points $\mathbf{x}$ and $\mathbf{x} + \mathbf{s}$ in the turbulence field, the longitudinal velocity difference (with $s = |\mathbf{s}|$) is defined by

$$\Delta v_l(\mathbf{s}, \mathbf{x}) = (\mathbf{v}(\mathbf{x} + \mathbf{s}) - \mathbf{v}(\mathbf{x})) \cdot \mathbf{s}/s, \qquad (10.32)$$

In the Gaussian system of complete randomness, all the structure functions of even orders can be represented in terms of the second-order correlations, while those of odd orders vanish. However, in non-Gaussian system, one has to know all the structure functions. With this point of view, the study of statistical properties of turbulence requires all structure functions.

The velocity field of Burgers vortex given in Sec. 7.9 by the expressions (7.76) and (7.80), are reproduced here:

$$\mathbf{v}(\mathbf{x}) = (v_r, v_\theta(r), v_z) = (-ar, v_\theta(r), 2az), \qquad (10.33)$$

$$v_\theta(r) = \frac{\Gamma}{2\pi l_B} \frac{1}{\hat{r}} (1 - e^{-\hat{r}^2}), \qquad (10.34)$$

where $\hat{r} = r/l_B$ and $l_B = \sqrt{2\nu/a}$.

At the beginning (Sec. 10.2) of this chapter, we saw that the fluid turbulence is characterized by a negative value of skewness of distribution of the longitudinal velocity derivative. The reason why we consider the Burgers vortices in turbulence is that it is one of the simplest systems that induces velocity field of negative skewness around it.

This is found by examining the third-order structure function of velocity field around a single Burgers vortex. The structure function is given by using three eigenvalues σ_1, σ_2, σ_3 of the rate of strain tensor $e_{ij} = \frac{1}{2}(\partial_j v_i + \partial_i v_j)$, given in [HK97] as

$$\sigma_1 = -a + |e_{r\theta}|, \quad \sigma_2 = 2a, \quad \sigma_3 = -a - |e_{r\theta}|,$$

where $e_{r\theta} = \frac{1}{2}(v_\theta'(r) - r^{-1}v_\theta)$. When the vortex is strong enough to satisfy $|e_{r\theta}| \gg |a|$, we have the order $\sigma_1 > \sigma_2 > \sigma_3$ in magnitude.

If the distance s is small enough, the third-order structure function F_3 is given by (10.37) of the next section, which is

$$F_3 = \langle (\Delta v_l(s))^3 \rangle_{sp} = \frac{8}{35}\sigma_1\sigma_2\sigma_3 s^3 = -\frac{16}{35}a(e_{r\theta}^2 - a^2)s^3.$$

This is negative if $|e_{r\theta}| > a > 0$. The skewness S would be given by

$$S = \left\langle \left(\frac{\Delta v_l(\mathbf{s})}{s} \right)^3 \right\rangle_{sp} = -\frac{16}{35} a \left(e_{r\theta}^2 - a^2 \right) < 0.$$

This negativeness is obtained as a result of the combined action of two basic fields. In fact, if $a = 0$, there is no background field $\mathbf{v}_b = 0$ (see (7.75)), and we would have $F_3 = 0$. On the other hand, if there is no vortex with $\Gamma = 0$ and $v_\theta = 0$, then we would have $F_3 = (16/35)a^3 s^3 > 0$. Thus only the combined field of Burgers vortex can give $F_3 < 0$, as far as $|e_{r\theta}| > a > 0$.

The scaling property of structure functions $F_p(s)$ is calculated on the basis of vortex models in [HK97] and [HDC99]. These give us fairly good estimates of scaling properties of $F_p(s)$ in homogeneous isotropic turbulence.

10.4.5. *Structure functions at small s*

When the separation distance s is small, the longitudinal structure functions can be calculated by defining the average $\langle \cdot \rangle$ with the space average,

$$F_p(s) = \langle (\Delta v_l(\mathbf{s}))^p \rangle_{sp} = \frac{1}{4\pi s^2} \int_{S_s} (\Delta v_l(\mathbf{s}))^p dS, \qquad (10.35)$$

the average being taken over the spherical surface of radius s centered at $\mathbf{x}$. The first three structure functions are given by

$$F_1(s) = \langle (\Delta v_l(\mathbf{s})) \rangle_{sp} = \frac{1}{3}(\sigma_1 + \sigma_2 + \sigma_3)s,$$

$$F_2(s) = \langle (\Delta v_l(\mathbf{s}))^2 \rangle_{sp} = \frac{2}{15} \left(\sigma_1^2 + \sigma_2^2 + \sigma_3^2 \right) s^2, \qquad (10.36)$$

$$F_3(s) = \langle (\Delta v_l(\mathbf{s}))^3 \rangle_{sp} = \frac{8}{35}\sigma_1\sigma_2\sigma_3 s^3. \qquad (10.37)$$

These formulae are verified in Appendix E.

10.5. Problems

Here we learn some well-known formulae from the statistical theory of turbulence.

Problem 10.1 Spectrum function $\Psi(k)$ of dissipation

Rate of dissipation per unit mass in turbulence is given by $\nu\langle|\boldsymbol{\omega}|^2\rangle = \int \Psi(k)\mathrm{d}k$. By using the definition of vorticity $\omega_i = \varepsilon_{ijk}\partial_j u_k$ and the Fourier representation (10.6) for $u_k(\mathbf{x})$, derive the following relations:

$$(i) \qquad |\hat{\boldsymbol{\omega}}(\mathbf{k})|^2 = k^2|\hat{\mathbf{u}}(\mathbf{k})|^2, \tag{10.38}$$

$$(ii) \qquad \langle|\boldsymbol{\omega}(\mathbf{x})|^2\rangle_C = 2\int_0^\infty k^2 E(k)\mathrm{d}k \qquad \text{for isotropic field.} \tag{10.39}$$

The function $\Psi(k) = 2\nu k^2 E(k)$ thus obtained is the *dissipation spectrum*.

Problem 10.2 Homogeneous isotropic turbulence

In *homogeneous* turbulence, statistical correlation of velocity component u_i at a point $\mathbf{x}$ and u_j at $\mathbf{x}' = \mathbf{x} + \mathbf{s}$ is expressed by $B_{ij} := \langle u_i(\mathbf{x})u_j(\mathbf{x}+\mathbf{s})\rangle = B_{ij}(\mathbf{s})$ which is independent of $\mathbf{x}$ by the assumption of homogeneity. The velocity field $u_i(\mathbf{x})$ is assumed to satisfy the condition of incompressibility: $\partial_i u_i(\mathbf{x}) = 0$.

(i) Suppose that the turbulence is *isotropic* in addition to homogeneity. Show that the second-order tensor B_{ij} has the following form:

$$B_{ij}(\mathbf{s}) = G(s)\delta_{ij} + F(s)e_i e_j, \tag{10.40}$$

where $s = |\mathbf{s}|$, $e_i = s_i/s$ (unit vector in the direction of $\mathbf{s}$), and $F(s)$ and $G(s)$ are scalar functions.

(ii) For turbulence of an incompressible fluid, show that the following relation must be satisfied:

$$2F(s) + sF'(s) + sG'(s) = 0, \tag{10.41}$$

where $F'(s) = dF/ds$ and the same for $G'(s)$.

(iii) *Longitudinal* velocity correlation at $\mathbf{x}$ and $\mathbf{x} + \mathbf{s}$ is defined by

$$\langle u_l(\mathbf{x}) u_l(\mathbf{x} + \mathbf{s}) \rangle = \bar{u}^2 f(s), \tag{10.42}$$

where u_l is the velocity component parallel (i.e. *longitudinal*) to the separation vector $\mathbf{s}$, and $\bar{u}^2 := \langle (u_l)^2 \rangle$. Derive the following:

$$\bar{u}^2 f(s) = F(s) + G(s). \tag{10.43}$$

(iv) Correlation spectrum $F_{ij}(\mathbf{k})$ is defined by

$$F_{ij}(\mathbf{k}) = \frac{1}{(2\pi)^3} \int B_{ij}(\mathbf{s}) e^{-i\mathbf{k}\cdot\mathbf{s}} d^3\mathbf{s}, \tag{10.44}$$

$$B_{ij}(\mathbf{s}) = \int F_{ij}(\mathbf{k}) e^{i\mathbf{k}\cdot\mathbf{s}} d^3\mathbf{k}. \tag{10.45}$$

Show the following relations, for $B(s) = B(|\mathbf{s}|) := B_{ii}(\mathbf{s})$ and $F_{ii}(k) = F_{ii}(|\mathbf{k}|)$ (by isotropy):

$$\left\langle \frac{u^2}{2} \right\rangle = \int_0^\infty E(k,t) \, dk = \frac{1}{2} \int F_{ii}(\mathbf{k}) \, d^3\mathbf{k}, \tag{10.46}$$

$$E(k) = 2\pi k^2 F_{ii}(|\mathbf{k}|) = \frac{1}{\pi} \int_0^\infty B(s) \, ks \, \sin ks \, ds \tag{10.47}$$

$$B(s) = 2 \int_0^\infty E(k) \frac{\sin ks}{ks} \, dk, \tag{10.48}$$

$$B(s) = \bar{u}^2 \left(3f(s) + sf'(s) \right). \tag{10.49}$$

Problem 10.3 Energy spectra: $E(k)$ and $E_1(k_1)$

On the basis of the properties of homogeneous isotropic turbulence elucidated in Problem 10.2, we consider two spectra of turbulence.

The longitudinal correlation function $f(s)$ of (10.42) is dimensionless and its Fourier transform gives one-dimensional spectrum $E_1(k)$ defined by

$$E_1(k) := \frac{1}{2\pi} \int_{-\infty}^{\infty} \bar{u}^2 f(s) e^{iks} \mathrm{d}s. \tag{10.50}$$

One-dimensional spectrum $E_1(k)$ can be related to the energy specrum $E(k)$ of three dimensions in *isotropic homogeneous* turbulence. In order to derive it, verify the following relations (i)~(iv):

(i) Relation between $f(s)$ and $E(k)$:

$$\bar{u}^2 f(s) = 2 \int_0^{\infty} E(k) \frac{1}{k^2 s^2} \left(\frac{\sin ks}{ks} - \cos ks \right) \mathrm{d}k. \tag{10.51}$$

(ii)

$$\bar{u}^2 f(s) = 2 \int_0^{\infty} G(k) \frac{k \sin ks}{s} \mathrm{d}k, \quad G(k) := \int_k^{\infty} \frac{E(k)}{k^3} \mathrm{d}k. \tag{10.52}$$

(iii)

$$\frac{1}{2\pi} \int_{-\infty}^{\infty} \bar{u}^2 f(s) e^{ik_1 s} \mathrm{d}s = \frac{1}{2} \int_{k_1}^{\infty} \left(1 - \frac{k_1^2}{k^2} \right) \frac{E(k)}{k} \mathrm{d}k. \tag{10.53}$$

For the derivation, the following integral formula may be useful:

$$\int_0^{\infty} \frac{\sin ax \cos bx}{x} \mathrm{d}x = \begin{cases} \pi/2 & (\text{if } a > |b|), \\ \pi/4 & (\text{if } a = |b|), \\ 0 & (\text{if } a = |b|). \end{cases} \tag{10.54}$$

(iv)

$$E(k) = k^3 \frac{\mathrm{d}}{\mathrm{d}k} \left(\frac{1}{k} \frac{\mathrm{d}E_1(k)}{\mathrm{d}k} \right). \tag{10.55}$$

Chapter 11

Superfluid and quantized circulation

At very low temperatures close to absolute zero, *quantum* effects begin to acquire primary importance in the properties of fluids. It is well-known that helium becomes a liquid phase below the (critical) temperature $T_c = 4.22\,K$ (under atmospheric pressure), and superfluid properties appear below $T_\lambda = 2.172\,K$ (discovered by P. L. Kapitza, 1938).[1]

Recently, there has been dramatic improvement in the Bose–Einstein condensation of (magnetically) trapped alkali-atomic gases at ultra-low temperatures. Such an atomic-gas Bose–Einstein condensation differs from the liquid-helium condensate in several ways. An example of this is that condensates of alkali-atomic gases are dilute. As a result, at low temperatures, the Gross–Pitaevskii equation (11.23) below gives an extremely precise description of the atomic condensate, and its dynamics is described by potential flows of an ideal fluid with a uniform (vanishing) entropy.

In traditional fluid dynamics, the ideal fluid is a virtual idealized fluid which is characterized by vanishing transport coefficients such as viscosity and thermal conductivity. The superfluid at $T = 0°K$ is a *real* ideal fluid which supports only potential flows of zero entropy. However, it is remarkable that it can support quantized circulations

---

[1]From T_λ down to $T = 0°K$, the liquid is called *helium II*. Under saturated vapor pressure, liquid helium is an ordinary classical viscous fluid called *helium I* from T_c down to T_λ. The theory of superfluidity was developed by L. D. Landau (1941).

as well at excited states. The purpose of this chapter is to introduce such ideas of superfluid flows.[2]

11.1.   Two-fluid model

At the temperature of absolute zero, helium II is supposed to show superfluidity of a Bose liquid obeying the Bose–Einstein statistics,[3] and it flows without viscosity in narrow capillaries or along a solid surface. Besides the absense of viscosity, the superfluid flow has two other important properties. Namely, the flow is always a potential flow. Hence, the velocity $\mathbf{v}_s$ of such a superfluid flow has a velocity potential Φ defined as

$$\mathbf{v}_s = \operatorname{grad} \Phi. \tag{11.1}$$

In addition, its entropy is zero.

At temperatures other than zero, helium II behaves as if it were a mixture of two different liquids. One of them is a *superfluid*, and moves with zero viscosity. The other is a *normal fluid* with viscosity.[4] Helium II is regarded as a mixture of normal fluid and superfluid with total density ρ:

$$\rho = \rho_s + \rho_n. \tag{11.2}$$

The densities of the superfluid ρ_s and the normal fluid ρ_n are known as a function of temperature and pressure. At $T = T_\lambda$, $\rho_n = \rho$ and

---

[2]For the details, see [LL87] and [Don91].

[3]A Bose particle (called a boson) is characterized by a particle with an integer spin. The wavefunction for a system of same bosons is symmetric with respect to exchange of arbitrary two particles.This is in contrast with the fermions of semi-integer spins whose wavefunction is anti-symmetric with respect to the exchange. Any number of bosons can occupy the same state, unlike the fermions.

[4]The liquid helium is regarded as a mixture of normal fluid and superfluid parts. This is no more than a convenient description of the phenomena in a quantum fluid. Just like any description of quantum phenomena in classical terms, it falls short of adequacy. In reality, it should be said that a quantum fluid such as helium II can execute two motions at once, each of which involves its own different effective mass. One of these motions is normal, while the other is superfluid-flow.

$\rho_s = 0$. At $T = 0$, $\rho_n = 0$ and $\rho_s = \rho$. The total density ρ is nearly constant in the helium II temperature range. At temperatures lower than $1K$, $\rho \approx \rho_s$.

The flow of fluid is characterized by the normal velocity $\mathbf{v}_n$ and the superfluid velocity $\mathbf{v}_s$. The flow of the superfluid is irrotational:

$$\text{curl } \mathbf{v}_s = 0, \tag{11.3}$$

which must hold throughout the volume of the fluid. Hence, $\mathbf{v}_s$ has a velocity potential Φ defined by (11.1).

Total mass flux is defined by the sum of two components:

$$\mathbf{j} = \rho_s \mathbf{v}_s + \rho_n \mathbf{v}_n. \tag{11.4}$$

The total density ρ and mass flux $\mathbf{j}$ must satisfy the continuity equation:

$$\partial_t \rho + \text{div} \, \mathbf{j} = 0. \tag{11.5}$$

The law of conservation of momentum is given by

$$\partial_t j_i + \partial_k P_{ik} = 0, \tag{11.6}$$

where the momentum flux density tensor P_{ik} is defined by

$$P_{ik} = p\delta_{ik} + \rho_s(v_s)_i(v_s)_k + \rho_n(v_n)_i(v_n)_k. \tag{11.7}$$

See the expressions (3.21) (with $f_i = 0$) and (3.22) in the case of an ideal fluid.

The superfluid part involves no heat transfer, and therefore no entropy transfer. The entropy of superfluid is regarded as zero. The heat transfer in helium II is caused only by the normal fluid. The entropy flux density is given by the product $\rho s \mathbf{v}_n$ where s is the entropy per unit mass and ρs the entropy per unit volume. Neglecting the dissipative processes, the entropy of the fluid is conserved, which is written down as

$$\partial_t(\rho s) + \text{div}(\rho s \mathbf{v}_n) = 0. \tag{11.8}$$

The heat flux density $\mathbf{q}$ is expressed as

$$\mathbf{q} = \rho T s \mathbf{v}_n. \tag{11.9}$$

Equations (11.5)–(11.8) are supplemented by the equation for the superfluid part $\mathbf{v}_s$. The equation must be such that the superfluid flow is irrotational at all times, which is given by Euler's equation of motion for irrotational flows:

$$\partial_t \mathbf{v}_s + \mathrm{grad}\left(\frac{1}{2}v_s^2 + \mu\right) = 0, \qquad (11.10)$$

where μ is a scalar function (corresponding to the chemical potential in thermodynamics). In Sec. 5.4, we obtained Eq. (5.24) for an irrotational flow, where there were scalar functions of enthalpy h and force potential χ in place of μ. Using the expression (11.1), Eq. (11.10) is rewritten as

$$\mathrm{grad}\,\partial_t \Phi + \mathrm{grad}\left(\frac{1}{2}v_s^2 + \mu\right) = 0.$$

This results in the following first integral of motion (see (5.28)):

$$\partial_t \Phi + \frac{1}{2}(\nabla\Phi)^2 + \mu = \mathrm{const.} \qquad (11.11)$$

11.2. Quantum mechanical description of superfluid flows

11.2.1. *Bose gas*

At present there is no truly fundamental microscopic picture of superfluidity of helium II. However, there are some ideas about superfluidity based on quantum mechanics.

^{4}He atoms are bosons and hence we try to draw an analogy between helium II and an ideal Bose gas (see the footnote in Sec. 11.1 for Bose–Einstein statistics). An ideal Bose gas begins to condense into the lowest ground energy level below the critical temperature T_c. When $T = 0°K$, all particles of a number N are in the ground state. At a finite temperature $T\,(<T_c)$, the fraction of condensation to the ground state is given by the form, $N_0(T) = N\left[1 - (T/T_c)^{3/2}\right]$. The condensed phase consists of $N_0(T)$ particles which are in the zero-momentum state, while the uncondensed particles of the number $N - N_0 = N(T/T_c)^{2/3}$ are distributing over excited states. It is

tempting to associate the superfluid with particles in the ground state, while a system of excitations is regarded as a normal fluid. The normal flow (flow of the normal fluid) is actually the flow of an *excitation gas*.[5]

11.2.2. *Madelung transformation and hydrodynamic representation*

Superfluid flows can be described by Schrödinger's equation (Sec. 12.1) in quantum mechanics. Schrödinger's equation for the wave function $\psi(\mathbf{r}, t)$ can be transformed to the forms familiar in the fluid mechanics (Madelung, 1927) by expressing $\psi(\mathbf{r}, t)$ in terms of its amplitude $|\psi| = A(\mathbf{r}, t)$[6] and its phase $\arg(\psi) = \varphi(\mathbf{r}, t)$ where A and φ are real functions of $\mathbf{r}$ and t. It is remarkable that the velocity $\mathbf{u}$ can be connected with the gradient of phase φ.

Schrödinger's wave equation for a single particle of mass m in a fixed potential field $V(\mathbf{r})$ is given by

$$i\hbar \, \partial\psi/\partial t = -(\hbar^2/2m)\nabla^2\psi + V\psi, \qquad (11.12)$$

where $\hbar = h/2\pi$ and h is the Planck constant ($h = 6.62 \times 10^{-34}$ J · s). Substituting

$$\psi = A \exp[i\varphi] \qquad (11.13)$$

into (11.12), dividing it into real and imaginary parts, we obtain

$$\hbar \, \partial_t\varphi - \frac{\hbar^2}{2m}\frac{\nabla^2 A}{A} + \frac{\hbar^2}{2m}(\nabla\varphi)^2 + V = 0, \qquad (11.14)$$

$$2\partial_t A + 2\frac{\hbar}{m}\nabla A \cdot \nabla\varphi + \frac{\hbar A}{m}\nabla^2\varphi = 0. \qquad (11.15)$$

The mass probability density is defined by ρ

$$\rho = m\psi\psi^* = mA^2. \qquad (11.16)$$

---

[5]We may recall that the collective thermal motion of atoms in a quantum fluid can be regarded as a system of excitations. The excitations behave like *quasi-particles* with carrying definite momenta and energies.

[6]$A^2 = |\psi|^2$ is understood as the probability density for the particle to be observed. See standard textbooks, e.g. [LL77].

where ψ^* is the complex conjugate of ψ. The mass current is defined by

$$\mathbf{j} = (\hbar/2i)(\psi^* \nabla \psi - \psi \nabla \psi^*) = \hbar A^2 \nabla \varphi. \qquad (11.17)$$

From the last two equations, we can derive a velocity $\mathbf{u}$ by requiring that the mass flux must have the form $\mathbf{j} = \rho \mathbf{u}$. Thus, we obtain

$$\mathbf{u} = \frac{\hbar}{m} \nabla \varphi, \qquad (11.18)$$

which has the expected form (11.1) with $\Phi = (\hbar/m)\varphi$.

Multiplying (11.15) by mA and using (11.16), we find the continuity equation,

$$\partial_t \rho + \nabla \cdot (\rho \mathbf{u}) = 0, \quad \mathbf{u} = (\hbar/m) \nabla \varphi. \qquad (11.19)$$

Using $\Phi = (\hbar/m)\varphi$, Eq. (11.14) becomes

$$\partial_t \Phi + \frac{1}{2}(\nabla \Phi)^2 + B + V/m = 0, \qquad (11.20)$$

where $B = -(\hbar^2/2m^2)(\nabla^2 A/A)$. This is seen to be equivalent to the integral (11.11) if $B + V/m$ is replaced by the scalar function μ.

Thus, it is found that the Schrödinger equation (11.12) is equivalent to the system of equations (11.4), (11.5) and (11.10) for the superfluid flow $\rho = \rho_s$ with $\rho_n = 0$. By normalization of the wavefunction given by

$$\int |\psi|^2 \mathrm{d}^3 \mathbf{r} = 1, \qquad (11.21)$$

we have the relation $\int \rho \mathrm{d}^3 \mathbf{r} = m$.

11.2.3.   *Gross–Pitaevskii equation*

Instead of a single particle, we now consider an assembly of N_0 identical bosons of mass m in a potential field $W(\mathbf{r})$ in order to represent the macroscopic condensate of bosons (macroscopic Bose–Einstein condensation). If these bosons do not interact, the wavefunction $\psi(\mathbf{r}, t)$ of the system would be given by a symmetrized product of N_0 one-particle wavefunctions with the normalization condition (11.21) replaced by

$$\int |\psi|^2 \, \mathrm{d}^3 \mathbf{r} = N_0, \tag{11.22}$$

or $\int \rho \mathrm{d}^3 \mathbf{r} = \bar{\rho} \mathcal{V}$, where $\mathcal{V}$ is the volume of the system, and $\bar{\rho} \mathcal{V}$ is defined by $m N_0$.

The condensate phase has a typical interatomic potential including a strong repulsive potential, which is expressed by a short-range repulsive potential $V_r(\mathbf{x} - \mathbf{x}')$. In the Schrödinger equation, the following potential,

$$\int V_r(\mathbf{x} - \mathbf{x}') |\psi(\mathbf{x}')|^2 \mathrm{d}^3 \mathbf{x}',$$

is added to V. This increases, as the density of neighboring bosons increases. A simplest case is to express V_r by a delta function

$$V_r(\mathbf{x} - \mathbf{x}') = V_0 \delta(\mathbf{x} - \mathbf{x}').$$

The Schrödinger equation (11.12) is then replaced by

$$i\hbar \, \partial_t \psi = -(\hbar^2/2m) \nabla^2 \psi + (V_0 |\psi|^2 + V)\psi. \tag{11.23}$$

This is called the *Gross–Pitaevskii* equation (Gross 1961, 1963; Pitaevskii 1961). By applying the Madelung transformation $\psi = F \exp[i\varphi]$, this proceeds with the previous Schrödinger equation, yielding

$$\frac{\hbar}{m} \partial_t \varphi + \frac{1}{2} \mathbf{u}^2 + V/m + B + \frac{p}{\rho} = 0. \tag{11.24}$$

This equation differs from (11.20) principally by the additional term $p/\rho = V_0 |\psi|^2/m = (V_0/m^2)\rho$, corresponding to a *barotropic gas pressure* $p = (V_0/m^2)\rho^2$.

11.3. Quantized vortices

Vortices in the superfluid are considered to take the form of a (hollow) filament with a core of atomic dimensions, something like a vortex-line (Feynman 1955). *The well-known invariant called the hydrodynamic circulation is quantized; the quantum of circulation is h/m.*

In the case of cylindrical symmetry, the angular momentum per particle is $\hbar$ (Onsager 1949). With this structure, multiple connectivity in the liquid arises because the superfluid is somehow excluded from the core and circulates about the core.

11.3.1.   *Quantized circulation*

The fluid considered is a quantum fluid described by the wavefunction ψ, and one would expect some differences associated with the quantum character. Suppose that the fluid is confined to a multiply-connected region. Then, around any closed contour C (not reducible to a point by a continuous deformation), the phase φ of (11.13) can change by a multiple of 2π. This is written as

$$\oint_C \nabla\varphi \cdot \mathrm{dl} = 2\pi n \quad (n = 0, \pm 1, \pm 2, \ldots).$$

Using (11.18), this is converted to the circulation integral:

$$\Gamma = \oint_C \mathbf{u} \cdot \mathrm{dl} = n\frac{h}{m} \quad (n = 0, \pm 1, \pm 2, \ldots). \tag{11.25}$$

Thus, it has been shown that the circulation Γ around any closed contour can have only integer multiples of a unit value h/m. This coincides with the above statement of Onsager.

Suppose that a steady vortex is centered at $r = 0$. By (11.25), we have the circumferential velocity of magnitude,

$$u_\theta = \frac{\hbar}{mr}, \tag{11.26}$$

for a singly quantized vortex $(n = 1)$.[7]

We can consider a vortex-like steady solution of the system, (11.19) and (11.20), in a potential $V = -W$ with W a positive

---

[7]This velocity can be deduced also from the following argument. Suppose that the wavefunction is given by $\psi = A(r)e^{ik\theta}$, where k must be an integer for the uniqueness of the function. The associated θ-component of momentum is given by $p_\theta = -i\hbar r^{-1}\partial_\theta\psi = \hbar(k/r)\psi$. Setting $j_\theta = \mathrm{Re}(p_\theta\psi^*)$ of (11.17) equal to mu_θ, we have, $mu_\theta = \hbar(k/r)$. Thus we obtain (11.26) for $k = 1$.

constant. Since $\partial_t \Phi = 0$ and $u = |\nabla \Phi| = \hbar/mr$, Eq. (11.20) becomes

$$\frac{1}{2}\frac{\hbar^2}{m^2 r^2} - \frac{\hbar^2}{2m^2}\frac{\nabla^2 A}{A} - \frac{W}{m} = 0. \tag{11.27}$$

In the cylindrical coordinates (r, θ, z), this equation can be written as

$$r^2 \frac{d^2}{dr^2} A + r \frac{d}{dr} A - A + \frac{2mW}{\hbar^2} r^2 A = 0, \tag{11.28}$$

(see (D.10) for $\nabla^2 A$ in the cylindrical frame). Redefining a new r by $\sqrt{2mW/\hbar^2}\, r$, (11.28) becomes

$$r^2 \frac{d^2}{dr^2} A + r \frac{d}{dr} A + (r^2 - 1) A = 0. \tag{11.29}$$

This has a solution $A(r) = J_1(r)$ (Bessel function of the first order; see Problem 11.1), which satisfies $J_1(0) = 0$ and $J_1(r) \propto r$ for $r \ll 1$, and also $J_1(r_1) = 0$ for r_1 where r_1 is the first zero (say) of $J_1(r)$.

11.3.2. *A solution of a hollow vortex-line in a BEC*

According to the Gross–Pitaevskii equation for macroscopic condensate of bosons (Bose–Einstein Condensation, BEC in short) considered in Sec. 11.2.3, a hollow vortex-line exists in BEC. Like in the single particle case, we have $u = \hbar/mr$. We scale the distance with a *coherence length a*,

$$\zeta = r/a, \quad \text{where} \quad a := \frac{\hbar}{\sqrt{2mW}} = \frac{\hbar}{\sqrt{2\rho_\infty V_0}},$$

and normalize A by $\sqrt{W/V_0}$, so that

$$\mathcal{A} := \frac{A}{\sqrt{W/V_0}} \to 1, \tag{11.30}$$

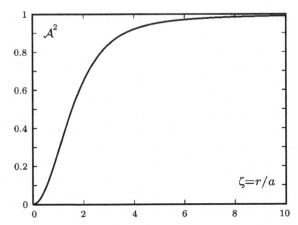

Fig. 11.1. A solution of a hollow vortex: The vertical axis is A^2, proportional to density ρ [GP58; KP66].

at large distances as $\zeta \to \infty$.[8] Then, from (11.24), we get the following equation,

$$\frac{\mathrm{d}^2}{\mathrm{d}\zeta^2}\mathcal{A} + \frac{1}{\zeta}\frac{\mathrm{d}}{\mathrm{d}\zeta}\mathcal{A} - \frac{1}{\zeta^2}\mathcal{A} + \mathcal{A} - \mathcal{A}^3 = 0. \qquad (11.31)$$

It is readily seen that there is a solution $\mathcal{A}(\zeta)$ satisfying this equation. For small distances ζ from the axis, the solution $\mathcal{A}(\zeta)$ varies linearly as ζ, and at large distances ($\zeta \gg 1$), $\mathcal{A} = 1 - \frac{1}{2}\zeta^{-2} + \cdots$ (Problem 11.2). To obtain the whole solution $\mathcal{A}(\zeta)$, Eq. (11.31) must be solved numerically (Fig. 11.1). This was carried out by [GP58] and [KP66].

The "coherence" here refers to the condensate wavefunction falling to zero inside the quantum liquid, which is the result of interparticle interactions. In this model, the vortex is characterized by a line-node (zero-line) in the wavefunction, around which circulation can take a value of an integer multiple of h/m.

---

[8]There are N_0 condensate particles in volume $\mathcal{V}$ (see Sec. 11.2.3), and the density is given by $\rho_\infty = mN_0/\mathcal{V} = mW/V_0$ by the normalization (11.30).

11.4. Bose–Einstein Condensation (BEC)

11.4.1. *BEC in dilute alkali-atomic gases*

Recently, there has been a dramatic improvement in Bose–Einstein condensation of magnetically trapped alkali-atomic gases at ultra-low temperatures [PS02]. Such atomic gas Bose–Einstein condensation differs from liquid helium condensates in several ways. First, the condensates of alkali-atomic gases are dilute, having the mean particle density n with $na^3 \ll 1$ for interaction length a. As a result, at low temperatures, the GP equation (11.23) gives an extremely precise description of atomic condensate and their dynamics. In superfluid ^{4}He, relatively high density and strong repulsive interactions make the analysis more complicated. Furthermore, because of the relatively strong interactions, the condensate fraction in bulk superfluid ^{4}He is only about 10% of the total particles, even at zero temperature. In contrast, almost all atoms participate in the condensate in an atomic-gas BEC. In 1995, BEC was first realized for a cluster of dilute Rb and Na atoms. Since then, with many alkali-atomic gases, BEC was realized experimentally. After 1998, BEC was confirmed also for H and He.

BEC is different from the liquefied gas in the normal phase change at ordinary temperatures. The liquefaction requires interaction between atoms, while no interaction is required with BEC. Moreover, all the atoms composing the BEC quantum state are identical and cannot be distinguished, and represented by a single wavefunction of a macroscopic scale (of 1 mm, say) for a macroscopic number of atoms.

However, the number density must be sufficiently low, i.e. dilute. Usually, atomic gases are liquefied when the temperature is lowered. Atoms in a condensate state as in liquid or solid are bounded by a certain interatomic attractive potential such as the van der Waals potential. In other words, atoms must release the bounding energy to some agent. This is carried out by a three-body collision (or higher order) in a gas away from solid walls. This is not the case in the two-body collision in which energy and momentum are conserved by two particles only and any bound state cannot be formed. When the number density n is low enough, two-body collision becomes dominant since probability of m-body collision is proportional to n^m.

Realization of BEC of dilute alkali-atomic gases is the most remarkable experimental achievement in the last decades at the end of 20th century. For experimental realization of BEC, three techniques must be combined: (a) *laser cooling*, (b) *magnetic trap* of gas particles in a bounded space, and (c) *evaporation cooling*.

11.4.2. *Vortex dynamics in rotating BEC condensates*

Madison *et al.* [MCBD01] reported the observation of nonlinear dynamics of quantum vortices such as vortex nucleation and lattice formation in a rotating condensate trapped by a potential, together

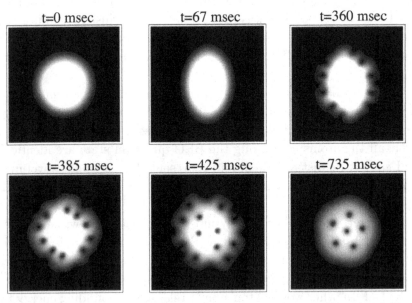

Fig. 11.2. Time evolution of the density of a rotating two-dimensional Bose condensate from the initial circular state, governed by the GP equation in a rotating frame with angular velocity Ω: $(i - \gamma)\hbar\, \partial_t \psi = -(\hbar^2/2m)\nabla^2\psi + (V_0|\psi|^2 + V - \Omega L_z - \mu)\psi$, where γ is a damping constant, μ a chemical potential and L_z the z-angular momentum [Tsu03].

with detailed observation of time evolution of condensate deformation. Such a direct observation of vortex phenomena has been made possible only in the BEC of alkali-atomic gases. Computer simulation [TKA03] of corresponding vortex dynamics was carried out successfully by numerically solving an extended form of the GP equation (Fig. 11.2). The physics of quantized vortices has a novel feature. Since the density is dilute, the relatively large coherence length a makes it possible to directly visualize the quantized vortices by using optical means. Secondly, because the order of coherence length a is close to the size of the condensate, the vortex dynamics is closely connected with the collective motion of the condensate density.

This opens a new area of vortex dynamics of an ideal fluid.

11.5. Problems

Problem 11.1 Bessel's differential equation

Bessel's differential equation of the nth order is given by

$$\frac{d^2}{dr^2}A + \frac{1}{r}\frac{d}{dr}A + \left(1 - \frac{n^2}{r^2}\right)A = 0, \qquad (11.32)$$

where $n = 0, 1, 2, \ldots$ (zero or positive integers). Show that this has a following power series solution $A(r) = J_n(r)$:

$$
\begin{aligned}
J_n(r) &= \left(\frac{r}{2}\right)^n \sum_{k=0}^{\infty} (-1)^k \frac{1}{k!\,(n+k)!} \left(\frac{r}{2}\right)^{2k} \\
&= \left(\frac{r}{2}\right)^n \left[\frac{1}{n!} - \frac{1}{1!(n+1)!}\left(\frac{r}{2}\right)^2 + \frac{1}{2!(n+2)!}\left(\frac{r}{2}\right)^4 - \cdots\right],
\end{aligned}
$$

$$(11.33)$$

where $0! := 1$, $1! := 1$, and $k! := 1 \cdot 2 \cdot \cdots \cdot k$ for an integer $k > 0$. [Note: Equation (11.29) corresponds to $n = 1$.]

Problem 11.2 Axisymmetric Gross–Pitaevskii equation

Axisymmetric Gross–Pitaevskii equation is given by (11.31):

$$\frac{\mathrm{d}^2}{\mathrm{d}\zeta^2}\mathcal{A} + \frac{1}{\zeta}\frac{\mathrm{d}}{\mathrm{d}\zeta}\mathcal{A} - \frac{1}{\zeta^2}\mathcal{A} + \mathcal{A} - \mathcal{A}^3 = 0. \qquad (11.31)$$

Show that, (i) for small ζ, $\mathcal{A}(\zeta) = c\zeta + O(\zeta^3)$ (c: a constant); (ii) for large ζ ($\gg 1$), $\mathcal{A} = 1 - \frac{1}{2}\zeta^{-2} + \cdots$.

Chapter 12

Gauge theory of ideal fluid flows

Fluid mechanics is a field theory in Newtonian mechanics, i.e. a *field theory of mass flows* subject to Galilean transformation. In the gauge theory of particle physics and the theory of relativity, a guiding principle is that laws of physics should be expressed in a form that is independent of any particular coordinate system. It is well known that there are various similarities between fluid mechanics and electromagnetism. Gauge theory provides us a basis to reflect on this property.

There are obvious differences between the field of fluid flows and other fields of electromagnetism, quantum physics, or particle physics. Firstly, the field of fluid flow is nonquantum. However this causes no problem since the gauge principle is independent of the quantization principle. In addition, the fluid flow is subject to the Galilean transformation instead of the Lorentz transformation in the field theory. This is not an obstacle because the former (Galilean) is a limiting transformation of the latter (Lorentz) as the ratio of a flow velocity to the light speed tends to an infinitesimally small value. Thirdly, the gauge principle in the particle physics requires the system to have a *symmetry*, i.e. the gauge invariance, which is an invariance with respect to a certain group of transformations. The symmetry groups (i.e. gauge groups) are different between different physical systems. This will be considered below in detail.

In the formulation of fluid flows, we seek a scenario which has a formal equivalence with the gauge theory in physics. An essential building block of the gauge theory is the *covariant derivative.* The convective derivative $\mathrm{D}/\mathrm{D}t$, i.e. the Lagrange derivative, in fluid mechanics can be identified as the covariant derivative in the framework of gauge theory. Based on this, we define appropriate Lagrangian functions for motions of an ideal fluid below. Euler's equation of motion can be derived from the variational principle. In addition, the continuity equation and entropy equation are derived simultaneously.

12.1. Backgrounds of the theory

We review the background of gauge theory first and illustrate the scenario of the gauge principle in a physical system.

12.1.1. *Gauge invariances*

In the theory of electromagnetism, it is well known that there is an invariance under a *gauge transformation* of electromagnetic potentials consisting of a scalar potential ϕ and a vector potential $\mathbf{A}$. The electric field $\mathbf{E}$ and magnetic field $\mathbf{M}$ are represented as

$$\mathbf{E} = -\nabla\phi - \frac{1}{c}\frac{\partial\mathbf{A}}{\partial t}, \quad \mathbf{B} = \nabla \times \mathbf{A}, \tag{12.1}$$

where c is the light velocity. The idea is that the fields $\mathbf{E}$ and $\mathbf{M}$ are unchanged by the following transformation: $(\phi, \mathbf{A}) \to (\phi', \mathbf{A}')$, where $\phi' = \phi - c^{-1}\partial_t f$, $\mathbf{A}' = \mathbf{A} + \nabla f$ with $f(\mathbf{x}, t)$ being an arbitrary differentiable scalar function of position $\mathbf{x}$ and time t. It is readily seen that

$$\mathbf{B}' \equiv \nabla \times \mathbf{A}' = \nabla \times \mathbf{A} = \mathbf{B} \quad (\text{since } \nabla \times \nabla f = 0),$$
$$\mathbf{E}' \equiv -c^{-1}\partial_t\mathbf{A}' - \nabla\phi' = -c^{-1}\partial_t\mathbf{A} - \nabla\phi = \mathbf{E}.$$

In fluid mechanics, it is interesting to find that there is an analogous invariance under (gauge) transformation of a velocity potential Φ of irrotational flows of an ideal fluid, where the velocity is represented as $\mathbf{v} = \nabla\Phi$. This was noted briefly as a gauge invariance in the footnote in Sec. 5.4. There, a potential flow of a homentropic fluid had an integral of motion (5.28) expressed as

$$\partial_t \Phi + \frac{1}{2} v^2 + H = \text{const.}, \tag{12.2}$$

where H denotes the summation of the enthalpy h and a force potential χ ($H = h + \chi$). It is evident that all of the velocity $\mathbf{v}$, the integral (12.2) and the equation of motion (5.24) are unchanged by the transformations: $\Phi \to \Phi + f(t)$ and $H \to H - \partial_t f(t)$ with an arbitrary scalar function $f(t)$. This case is simpler because only a scalar potential Φ and a scalar function (of t only) are concerned.

In short, there exists arbitrariness in the expressions of physical fields in terms of potentials in both systems.

12.1.2. Review of the invariance in quantum mechanics

First we highlight a basic property in quantum mechanics. Namely, for a charged particle of mass m in electromagnetic fields, Schrödinger's equation and the electromagnetic fields are invariant with respect to a gauge transformation. This is as follows.

In the *absence* of electromagnetic fields, Schrödinger's equation for a wave function ψ of a free particle m is written as[1]

$$S[\psi] \equiv i\hbar \partial_t \psi - \frac{1}{2m} p_k^2 \psi = 0,$$

where $p_k = -i\hbar \partial_k$ is the momentum operator ($\partial_k = \partial/\partial x^k$ for $k = 1, 2, 3$).

In the *presence* of electromagnetic fields, Schrödinger's equation for a particle with an electric charge e is obtained by the following transformation from the above equation:

$$\partial_t \to \partial_t + (e/i\hbar)A_0, \quad \partial_k \to \partial_k + (e/i\hbar c)A_k, \quad (k = 1, 2, 3), \tag{12.3}$$

---

[1]This is an equation for a *free* particle since the potential V is set to zero in (11.12).

where $(A_\mu) = (A_0, A_1, A_2, A_3) = (A_0, \mathbf{A}) = (-\phi, \mathbf{A})$ is the four-vector potential.[2] Thus, we obtain the equation with electromagnetic fields:

$$S_A[\psi] \equiv i\hbar\partial_t\psi - e\phi\psi - \frac{1}{2m}\left[(-i\hbar)\left(\partial_k + \frac{e}{i\hbar c}A_k\right)\right]^2\psi = 0. \quad (12.4)$$

From the four-vector potential $A_\mu = (-\phi, \mathbf{A})$, we obtain the electric field $\mathbf{E}$ and magnetic field $\mathbf{B}$ represented by (12.1).

Suppose that a wave function $\psi(x^\mu)$ satisfies the equation $S_A[\psi] = 0$ of (12.4). Consider the following set of transformations of $\psi(x^\mu)$ and A_μ:

$$\psi'(x_\mu) = \exp[i\alpha(x^\mu)]\psi(x^\mu), \quad (12.5)$$

$$A_k \rightarrow A'_k = A_k + \partial_k\beta, \quad \phi \rightarrow \phi' = \phi - c^{-1}\partial_t\beta, \quad (12.6)$$

where $\alpha = (e/\hbar c)\beta$. Then, it can be readily shown that the transformed function $\psi'(x^\mu)$ satisfies the Schrödinger equation $S_{A'}[\psi] = 0$, obtained by replacing the potential A_μ with the transformed one A'_μ. In addition, if

$$\Box\alpha \equiv (\nabla^2 - c^{-2}\partial_t^2)\alpha = 0$$

is satisfied, then the electromagnetic fields $\mathbf{E}$ and $\mathbf{B}$ are *invariant*, and the probability density is also invariant: $|\psi'(x^\mu)|^2 = |\psi(x^\mu)|^2$. This is the *gauge invariance* of the system of an electric charge in electromagnetic fields.

In summary, the Schrödinger equation (12.4) and the electromagnetic fields $\mathbf{E}$ and $\mathbf{B}$ are *invariant* with respect to the *gauge transformations*, (12.5) and (12.6). This system is said to have a *gauge symmetry*. Namely, the system has a certain kind of freedom. This freedom (or symmetry) allows us to formulate the following gauge principle.

---

[2]A point in the space-time frame of reference is expressed by a four-component vector, $(x^\mu) = (x^0, x^1, x^2, x^3)$ with $x^0 = ct$ (upper indices). Corresponding form of the electromagnetic potential is represented by a four-covector (lower indices) $A_\mu = (-\phi, A_k)$ with ϕ the electric potential and $\mathbf{A} = (A_k)$ the magnetic three-vector potential ($k = 1, 2, 3$). Transformation between a vector A^ν and a covector A_μ is given by the rule $A_\mu = g_{\mu\nu}A^\nu$ with the metric tensor $g_{\mu\nu}$ defined in Appendix F.

12.1.3. *Brief scenario of gauge principle*

In the gauge theory of particle physics,[3] a free-particle Lagrangian $\mathcal{L}_{\text{free}}$ is first defined for a particle with an electric charge. Next, a gauge principle is applied to the Lagrangian, requiring it to have a gauge invariance.

Gauge Transformation: Suppose that a Lagrangian $\mathcal{L}_{\text{free}}[\psi]$ is defined for the wave function ψ of a free charged particle. Let us consider the following (gauge) transformation:

$$\psi \mapsto e^{i\alpha}\psi.$$

If $\mathcal{L}_{\text{free}}$ is invariant with this transformation when α is a constant, it is said that $\mathcal{L}_{\text{free}}$ has a *global* gauge invariance. In spite of this, $\mathcal{L}_{\text{free}}$ may *not* be invariant for a function $\alpha = \alpha(x)$. Then, it is said that $\mathcal{L}_{\text{free}}$ is not gauge-invariant *locally*.

In order to acquire local gauge invariance, let us introduce a new field. Owing to this new field, if local gauge invariance is recovered, then the new field is called a *gauge field*.

In the previous section, it was found that *local* gauge invariance was acquired by replacing ∂_μ with

$$\nabla_\mu = \partial_\mu + \mathcal{A}_\mu \qquad (12.7)$$

(see (12.3)), where $\mathcal{A}_\mu = (e/i\hbar c)A_\mu$, and $A_\mu(x)$ is the electromagnetic potential, termed as a *connection form* in mathematics. The operator ∇_μ is called the *covariant derivative*.

Thus, when the original Lagrangian is not locally gauge invariant, the *principle of local gauge invariance* requires a new gauge field to be introduced in order to acquire local gauge invariance, and the Lagrangian is to be altered by replacing the partial derivative with the *covariant* derivative including a gauge field. This is the *Weyl's gauge principle*. Electromagnetic field is a typical example of such a kind of gauge field.

In mathematical terms, suppose that we have a group $\mathcal{G}$ of transformations and an element $g(x) \in \mathcal{G}$, and that the wave function is transformed as $\psi' = g(x)\psi$. In the previous example, $g(x) = e^{i\alpha(x)}$

---

[3][Wnb95] or [Fr97].

and the group is $\mathcal{G} = U(1)$.[4] An important point of introducing the gauge field A is to define a covariant derivative $\nabla = \partial + A$ as a generalization of the derivative ∂ that transforms as $g\nabla = g(\partial + A) = (\partial' + A')g$, so that we obtain

$$\nabla'\psi' = (\partial' + A')g(x)\psi = g(\partial + A)\psi = g\nabla\psi. \qquad (12.8)$$

Thus, $\nabla\psi$ transforms in the same way as ψ itself.

In dynamical systems which evolve with time t, such as fluid flows (in an Euclidean space), the replacement is to be made only for the t derivative: $\partial_t \to \nabla_t = \partial_t + A(x)$. This is considered in the next section.

12.2. Mechanical system

To start with, we review the variational formulation in Newtonian mechanics of a system of point masses, aiming at presenting a new formulation later. Then we consider an invariance property with respect to certain group of transformations, i.e. a symmetry of the mechanical system.

12.2.1. *System of n point masses*

Suppose that we have a system of n point masses m_j $(j = 1, \ldots, n)$ whose positions $\mathbf{x}_j$ are denoted by $q = (q^i)$, where

$$\mathbf{x}_1 = (q^1, q^2, q^3), \ \ldots, \ \mathbf{x}_n = (q^{3n-2}, q^{3n-1}, q^{3n}).$$

Their velocities $\mathbf{v}_j = (v_j^1, v_j^2, v_j^3)$ are written as $(q_t^{3j-2}, q_t^{3j-1}, q_t^{3j})$. Usually, a Lagrangian function L is defined by (*Kinetic energy*) − (*Potential energy*) for such a mechanical system. However, we start here with the following form of *Lagrangian*:

$$L = L[q, q_t], \qquad (12.9)$$

which depends on the coordinates $q = q(t)$ and the velocities $q_t = \partial_t q(t) = (q_t^i)$ for $i = 1, 2, \ldots, 3n$ $(\partial_t = d/dt)$. The Lagrangian L describes a dynamical system of $3n$ degrees of freedom.

---

[4]Unitary group $U(1)$ is the group of complex numbers $z = e^{i\theta}$ of absolute value 1.

For derivation of the equation of motion, we rely on the variational principle as follows. Given two end times t_1 and t_2, the action functional I is defined by $I = \int_{t_1}^{t_2} L[q, q_t] dt$. Consider a reference trajectory $q = q(t)$ and velocity $q_t = \dot{q}(t)$ (where $\dot{q} = \partial_t q = dq/dt$). Their variations q' and q_t' are defined by $q'(t, \varepsilon) = q(t) + \varepsilon \xi(t)$ and $q_t' = \dot{q}(t) + \varepsilon \dot{\xi}(t)$ for a virtual displacement $\xi(t)$ satisfying the condition that $\xi(t)$ vanishes at the end points: i.e. $\xi(t_1) = \xi(t_2) = 0$.

The principle of least action, i.e. the Hamilton's principle, is given by

$$\delta I = \int_{t_1}^{t_2} \delta L[q, q_t] dt = 0, \tag{12.10}$$

for the variations $\delta q = \varepsilon \xi$ and $\delta q_t = \partial_t(\delta q)$, where the variation of Lagrangian L is

$$\delta L = \frac{\partial L}{\partial q} \delta q + \frac{\partial L}{\partial q_t} \delta q_t = \frac{\partial L}{\partial q^i} \delta q^i + \frac{\partial L}{\partial q_t^i} \frac{d}{dt}(\delta q^i)$$

$$= \left(\frac{\partial L}{\partial q^i} - \frac{d}{dt}\left(\frac{\partial L}{\partial q_t^i} \right) \right) \delta q^i + \frac{d}{dt}\left(\frac{\partial L}{\partial q_t^i} \delta q^i \right). \tag{12.11}$$

Substituting the expression (12.11) into (12.10), the last term can be integrated, giving the difference of values at t_1 and t_2, both of which vanish by the imposed conditions: $\xi(t_1) = \xi(t_2) = 0$. Thus, we obtain

$$\delta I = \int_{t_1}^{t_2} \left(\frac{\partial L}{\partial q^i} - \frac{d}{dt}\left(\frac{\partial L}{\partial q_t^i} \right) \right) \delta q^i dt = 0.$$

The variational principle requires that this must be valid against any variation δq^i satisfying $\delta q^i(t_1) = \delta q^i(t_2) = 0$. This results in the following *Euler–Lagrange equation*:

$$\frac{d}{dt}\left(\frac{\partial L}{\partial q_t^i} \right) - \frac{\partial L}{\partial q^i} = 0. \tag{12.12}$$

If the Lagrangian is of the following form for n masses m_j ($j = 1, \ldots, n$),

$$L_f = \frac{1}{2} \sum_{j=1}^{n} m_j \langle \mathbf{v}_j, \mathbf{v}_j \rangle, \tag{12.13}$$

then L_f is called a *free* Lagrangian, where $\langle \mathbf{v}_j, \mathbf{v}_j \rangle = \sum_{k=1}^{3} v_j^k v_j^k$ is the inner product, because the above equation (12.12) results in the equation of *free motion* (using v_j^k instead of q_t^i):

$$\frac{d}{dt}\left(\frac{\partial L_f}{\partial v_j^k}\right) = \partial_t p_j^k = 0, \quad \text{where } \frac{\partial L_f}{\partial q_t^i} \Rightarrow \frac{\partial L_f}{\partial v_j^k} = m_j v_j^k \equiv p_j^k,$$

since $\partial L/\partial q^i = 0$, i.e. the momentum vector $\mathbf{p}_j = m_j \mathbf{v}_j$ is constant. In general, $\partial L/\partial v_j^k$ represents the kth component of *momentum* of the jth particle.

12.2.2. *Global invariance and conservation laws*

Let us consider a translational transformation of the system (a kind of *gauge transformation*), which is given by a displacement of the coordinate origin with the axes being fixed parallel. Namely, the position coordinates of every particle in the system is shifted (*translated*) by the same amount $\boldsymbol{\xi}$, i.e. $\mathbf{x}_j$ is replaced by $\mathbf{x}_j + \boldsymbol{\xi}$, where $\boldsymbol{\xi} = (\xi^1, \xi^2, \xi^3)$ is an arbitrary constant infinitesimal vector.[5] Resulting variation of the Lagrangian L is given by (12.11), where $\delta x_j^k = \delta q^{3j-3+k} = \xi^k$ for $j = 1, \ldots, n$ and $k = 1, 2, 3$, and $\delta q_t = d/dt(\delta q) = 0$ because ξ^k are constants. The first term of (12.11) vanishes owing to the Euler–Lagrange equation (12.12). Thus, we obtain

$$\delta L = \frac{d}{dt}\left(\frac{\partial L}{\partial q_t^i}\delta q^i\right). \tag{12.14}$$

When the displacement $\boldsymbol{\xi}$ is a constant vector for all $\mathbf{x}_j$ $(j = 1, \ldots, n)$ like in the present case, the transformation is called *global*. Requiring that the Lagrangian is invariant under this transformation, i.e. $\delta L = 0$, we obtain

$$0 = \frac{d}{dt}\left(\frac{\partial L}{\partial q_t^i}\delta q^i\right) = \xi^k \sum_{k=1}^{3} \frac{d}{dt}\left(\sum_{j=1}^{n} \frac{\partial L}{\partial v_j^k}\right).$$

---

[5]This is equivalent to a shift of the coordinate origin by $-\boldsymbol{\xi}$. This *global* gauge transformation is different from the variational principle considered in the previous section.

Since ξ^k ($k = 1, 2, 3$) are arbitrary independent constants, the following must hold:

$$\sum_{j=1}^{n} \frac{\partial L}{\partial v_j^1} = \text{const.}, \quad \sum_{j=1}^{n} \frac{\partial L}{\partial v_j^2} = \text{const.}, \quad \sum_{j=1}^{n} \frac{\partial L}{\partial v_j^3} = \text{const.}$$

Thus the three components of the total momentum are conserved. This is the *Noether's theorem* for the global invariance of the Lagrangian L of (12.9). It is well known that the Newton's equation of motion is invariant with respect to Galilean transformation, i.e. a transformation between two inertial frames of reference in which one frame is moving with a constant velocity $\mathbf{U}$ relative to the other. The Galilean transformation is a sequence of *global* translational gauge transformations with respect to the time parameter t. Global invariance of Lagrangian with respect to translational transformations is associated with the homogeneity of space.

Lagrangian of (12.9) has usually another invariance associated with the isotropy of space. Namely, the mechanical property of the system is unchanged when it is rotated as a whole in space. This is a rotational transformation, i.e. a position vector of every particle in the system is rotated by the same angle with respect to the space. The global symmetry with respect to the rotational transformations results in the conservation of total angular momentum ([LL76], Secs. 7 and 9).

It is to be noted that the gauge symmetry of the Schrödinger system considered in Sec. 12.1.2 was local. The next section investigates extension of the above global symmetry to local symmetry for a *continuous* mechanical system.

12.3. Fluid as a continuous field of mass

From now on, we consider fluid flows and try to formulate the flow field on the basis of the gauge principle. First, we investigate how the Lagrangian of the form (12.9), or (12.13), of *discrete* systems must be modified for a system of fluid flows characterized by *continuous* distribution of mass. According to the principle of gauge invariance, we consider gauge transformations, which are both global and local.

Concept of local transformation is a generalization of the global transformation. When we consider local gauge transformation, the physical system under consideration must be modified so as to allow us to consider a continuous field by extending the original discrete mechanical system. We replace the discrete variables q^i by continuous parameters $\mathbf{a} = (a^1, a^2, a^3)$ to represent continuous distribution of particles in a subspace M of three-dimensional Euclidean space E^3. Spatial position $\mathbf{x} = (x^1, x^2, x^3)$ of each massive particle of the name tag $\mathbf{a}$ (Lagrange parameter) is denoted by $\mathbf{x} = \mathbf{x}_a(t) \equiv \mathbf{X}(\mathbf{a}, t)$, a function of $\mathbf{a}$ as well as the time t. Conversely, the particle occupying the point $\mathbf{x}$ at a time t is denoted by $\mathbf{a}(\mathbf{x}, t)$.

12.3.1.   *Global invariance extended to a fluid*

Now, we consider a continuous distribution of mass (i.e. *fluid*) and its motion. The Lagrangian (12.9) must be modified to the following integral form (instead of summation):

$$L = \int \Lambda(q, q_t) \mathrm{d}^3 \mathbf{x}, \tag{12.15}$$

where Λ is a Lagrangian density. Suppose that an infinitesimal transformation is expressed by

$$\begin{aligned} q = \mathbf{x} \rightarrow q' = \mathbf{x} + \delta\mathbf{x}, \quad & \delta\mathbf{x} = \boldsymbol{\xi}(\mathbf{x}, t), \\ q_t = \mathbf{v} \rightarrow q_t' = \mathbf{v} + \delta\mathbf{v}, \quad & \delta\mathbf{v} = \partial_t(\delta\mathbf{x}), \end{aligned} \tag{12.16}$$

where $\mathbf{v} = \partial_t \mathbf{x}_a(t)$, and the vector $\boldsymbol{\xi}(\mathbf{x}, t)$ is an arbitrary differentiable variation field. In (12.16) and henceforth, ∂_t denotes the *partial derivative* $\partial/\partial t$. Resulting variation of the Lagrangian density $\Lambda(\mathbf{x}, \mathbf{v})$ is

$$\delta\Lambda = \left(\frac{\partial\Lambda}{\partial\mathbf{x}} - \partial_t \left(\frac{\partial\Lambda}{\partial\mathbf{v}} \right) \right) \cdot \delta\mathbf{x} + \partial_t \left(\frac{\partial\Lambda}{\partial\mathbf{v}} \cdot \delta\mathbf{x} \right). \tag{12.17}$$

This does not vanish in general owing to the arbitrary function $\delta\mathbf{x} = \boldsymbol{\xi}(\mathbf{x}, t)$ depending on time t. In fact, assuming the Euler–Lagrange equation $\partial\Lambda/\partial\mathbf{x} - \partial_t(\partial\Lambda/\partial\mathbf{v}) = 0$ (an extended form of (12.12)), we obtain

$$\delta \Lambda = \partial_t \left(\frac{\partial \Lambda}{\partial \mathbf{v}} \cdot \delta \mathbf{x} \right) = \partial_t \left(\frac{\partial \Lambda}{\partial \mathbf{v}} \right) \cdot \boldsymbol{\xi} + \frac{\partial \Lambda}{\partial \mathbf{v}} \cdot \partial_t \boldsymbol{\xi}. \tag{12.18}$$

In the global transformation of $\boldsymbol{\xi} = $ const., we have $\partial_t \boldsymbol{\xi} = 0$. Then, global invariance of the Lagrangian ($\delta L = 0$) for arbitrary constant $\boldsymbol{\xi}$ requires

$$\partial_t \int \frac{\partial \Lambda}{\partial \mathbf{v}} \mathrm{d}^3 \mathbf{x} = 0.$$

This states the conservation of total momentum defined by $\int (\partial \Lambda / \partial \mathbf{v}) \mathrm{d}^3 \mathbf{x}$. The same result for the global transformation ($\boldsymbol{\xi} = $ const. and $\delta \mathbf{v} = \partial_t \boldsymbol{\xi} = 0$) can be obtained directly from (12.15) since

$$\delta L = \int \boldsymbol{\xi} \cdot \frac{\partial \Lambda}{\partial \mathbf{x}} \, \mathrm{d}^3 \mathbf{x} = \boldsymbol{\xi} \cdot \int \partial_t \left(\frac{\partial \Lambda}{\partial \mathbf{v}} \right) \mathrm{d}^3 \mathbf{x} = 0, \tag{12.19}$$

by the above Euler–Lagrange equation. In the local transformation, however, the variation field $\boldsymbol{\xi}$ depends on the time t and space point $\mathbf{x}$, and the variation $\delta L = \int \delta \Lambda \mathrm{d}^3 \mathbf{x}$ does not vanish in general.

12.3.2. *Covariant derivative*

According to the gauge principle (Sec. 12.1.3), nonvanishing of δL is understood as meaning that a *new field* G must be taken into account in order to achieve *local gauge invariance* of the Lagrangian L. To that end, we try to replace the partial time derivative ∂_t by a *covariant derivative* D_t, where the derivative D_t is defined by

$$\mathrm{D}_t = \partial_t + G, \tag{12.20}$$

with G being a gauge field (an operator).

In dynamical systems like the present case, the *time derivative* is the primary object to be considered in the analysis of local gauge transformation. (The invariance property noted in Sec. 12.1.1 is an example.) Thus, the time derivatives $\partial_t \boldsymbol{\xi}$ and $\partial_t q$ are replaced by

$$\mathrm{D}_t q = \partial_t q + G q, \quad \mathrm{D}_t \boldsymbol{\xi} = \partial_t \boldsymbol{\xi} + G \boldsymbol{\xi}. \tag{12.21}$$

Correspondingly, the Lagrangian L_f of (12.13) is replaced by

$$L_f = \int \Lambda_f(\mathbf{v}) d^3\mathbf{x} := \frac{1}{2} \int \langle \mathbf{v}, \mathbf{v} \rangle d^3\mathbf{a}, \qquad (12.22)$$

$$\mathbf{v} = D_t\mathbf{x}_a, \qquad (12.23)$$

where $d^3\mathbf{a} = \rho d^3\mathbf{x}$ denotes the mass (in place of m_j) in a volume element $d^3\mathbf{x}$ of the $\mathbf{x}$-space with ρ the mass-density.[6] The action is defined by

$$I = \int_{t_1}^{t_2} L_f dt = \int_{t_1}^{t_2} dt \int_M d^3\mathbf{x} \Lambda_f(\rho, D_t\mathbf{x}_a), \qquad (12.24)$$

where M is a bounded space of E^3, and $\Lambda_f = \frac{1}{2}\rho\langle D_t\mathbf{x}_a, D_t\mathbf{x}_a \rangle$.

12.4. Symmetry of flow fields I: Translation symmetry

The symmetries of fluid flows we are going to investigate are the *translation symmetry* and *rotation symmetry*, seen in the discrete system.

The Lagrangian (12.13) has a global symmetry with respect to the translational transformations (and possibly with respect to rotational transformations). A family of translational transformations is a *group* of transformations,[7] i.e. a translation group. Lagrangian defined by (12.22) for a continuous field has the same properties globally, inheriting from the discrete system of point masses. It is a primary concern here to investigate whether the system of fluid flows satisfies local invariance. First, we consider parallel translations (without local rotation), where the coordinate q^i is regarded as

---

[6]Here the Lagrangian coordinates $\mathbf{a} = (a,b,c)$ are defined so as to represent the *mass coordinate*. Using the Jacobian of the map $\mathbf{x} \mapsto \mathbf{a}$ defined by $J = \partial(\mathbf{a})/\partial(\mathbf{x})$, we have $d^3\mathbf{a} = J d^3\mathbf{x}$, where J is ρ.

[7]A family of transformations is called a *group* G, provided that, (i) with two elements $g, h \in G$, their product $g \cdot h$ is another element of G (in the case of translations, $g \cdot h \equiv g + h$), (ii) there is an identity $e \in G$ such that $g \cdot e = e \cdot g = g$ for any $g \in G$ ($e = 0$ in the case of translations), and (iii) for every $g \in G$, there is an inverse element $g^{-1} \in G$ such that $g \cdot g^{-1} = g^{-1} \cdot g = e$ ($g^{-1} = -g$ in the case of translations). If $g \cdot h = h \cdot g$ for any $g, h \in G$, the group is called a commutative group, or an *Abelian* group.

the cartesian space coordinate $x^k = x_a^k(t)$ (kth component), and q_t^i is a velocity component $u^k = \partial_t x_a^k(t)$.

12.4.1. *Translational transformations*

Suppose that we have a differentiable function $f(x)$. Its variation by an infinitesimal translation $x \to x + \xi$ is given by $\delta f = \xi \partial_x f$ where $\partial_x \equiv \partial/\partial x$ is regarded as a translation operator (ξ a parameter).

The operator of parallel translation in three-dimensional cartesian space is denoted by $T_k = \partial/\partial x_k = \partial_k$, ($k = 1, 2, 3$). An arbitrary translation is represented by $\xi^k T_k$ ($\equiv \sum_{k=1}^{3} \xi^k T_k$) with ξ^k infinitesimal parameters. For example, a variation of x^i is given by

$$\delta x^i = (\xi^k T_k)x^i = \xi^k \delta_k^i = \xi^i,$$

since $T_k x^i = \partial_k x^i = \delta_k^i$. If the Lagrangian density Λ of (12.15) is independent of the coordinate q (such as Λ_{f} of (12.22)), then Λ is invariant with respect to the translational transformation, i.e. Λ has a *translational symmetry*.[8] If ξ^k are constants, the symmetry is global, whereas if ξ^k are functions of $\mathbf{x}$ and t, the symmetry is local. This implies that the covariant derivative may be of the form:

$$\mathrm{d}_t = \partial_t + g^k T_k, \tag{12.25}$$

for the translational symmetry, where g^k ($k = 1, 2, 3$) are scalars. This is investigated just below.

12.4.2. *Galilean transformation (global)*

Translational transformation from one frame F to another F' moving with a relative velocity $\mathbf{U}$ is called the Galilean transformation in Newtonian mechanics. The transformation law is expressed as follows:

$$x \equiv (t, \mathbf{x}) \Rightarrow x' \equiv (t', \mathbf{x}') = (t, \mathbf{x} - \mathbf{U}t). \tag{12.26}$$

---

[8]The operators T_k ($k = 1, 2, 3$) are called *generators* of the group of translational transformations. The operators are commutative: $[T_k, T_l] = \partial_k \partial_l - \partial_l \partial_k = 0$.

This is a sequence of global translational transformations with a parameter t. Corresponding transformation of velocity is

$$u \equiv (1, \mathbf{u}) \Rightarrow u' \equiv (1, \mathbf{u}') = (1, \mathbf{u} - \mathbf{U}). \qquad (12.27)$$

From (12.26), differential operators $\partial_t = \partial/\partial t$, $\partial_k = \partial/\partial x_k$ and $\nabla = (\partial_1, \partial_2, \partial_3)$ are transformed according to

$$\partial_t = \partial_{t'} - \mathbf{U} \cdot \nabla', \quad \nabla = \nabla'. \qquad (12.28)$$

It is remarkable to find that we have a transformation invariance of the convective derivative $D_c \equiv \partial_t + (\mathbf{u} \cdot \nabla)$ from (12.27) and (12.28):

$$\partial_t + (\mathbf{u} \cdot \nabla) = \partial_{t'} + (\mathbf{u}' \cdot \nabla'). \qquad (12.29)$$

12.4.3. *Local Galilean transformation*

Suppose that a velocity field $\mathbf{u}(\mathbf{x}, t)$ is defined by the velocity $(d/dt)\mathbf{x}_a(t)$ of fluid particles, i.e. $\mathbf{u}(\mathbf{x}_a, t) = (d/dt)\mathbf{x}_a(t)$. Consider the following infinitesimal transformation:

$$\mathbf{x}'(\mathbf{x}, t) = \mathbf{x} + \boldsymbol{\xi}(\mathbf{x}, t), \quad t' = t. \qquad (12.30)$$

This is regarded as a local gauge transformation between noninertial frames. In fact, the transformations (12.30) is understood to mean that the coordinate $\mathbf{x}$ of a fluid particle at $\mathbf{x} = \mathbf{x}_a(t)$ in the frame F is transformed to the new coordinate $\mathbf{x}'$ of F', which is given by $\mathbf{x}' = \mathbf{x}'_a(\mathbf{x}_a, t) = \mathbf{x}_a(t) + \boldsymbol{\xi}(\mathbf{x}_a, t)$. Therefore, its velocity $\mathbf{u} = (d/dt)\mathbf{x}_a(t)$ is transformed to the following representation:

$$\mathbf{u}'(\mathbf{x}') \equiv \frac{d}{dt}\mathbf{x}'_a = \frac{d}{dt}\left(\mathbf{x}_a(t) + \boldsymbol{\xi}(\mathbf{x}_a, t)\right) = \mathbf{u}(\mathbf{x}) + d\boldsymbol{\xi}/dt, \qquad (12.31)$$

$$d\boldsymbol{\xi}/dt \equiv \frac{d}{dt}\boldsymbol{\xi}(\mathbf{x}_a(t), t) = \partial_t\boldsymbol{\xi} + (\mathbf{u} \cdot \nabla)\boldsymbol{\xi}. \qquad (12.32)$$

This is interpreted as follows: the local coordinate origin is displaced by $-\boldsymbol{\xi}$ where the axes of the local frame are moving (without rotation) with the velocity $-d\boldsymbol{\xi}/dt$ in accelerating motion (noninertial frame).[9] This implies that the velocity $\mathbf{u}(\mathbf{x})$ at $\mathbf{x}$ is transformed

---

[9]In the previous Sec. 12.4.2 where $\boldsymbol{\xi} = -\mathbf{U}t$, the transformation was global.

locally as $\mathbf{u}'(\mathbf{x}') = \mathbf{u}(\mathbf{x}) + d\boldsymbol{\xi}/dt$. It is to be noted in this transformation that the points $\mathbf{x}$ and $\mathbf{x}'$ are the same point with respect to the space F.

In view of the transformations defined by $t' = t$ and $\mathbf{x}' = \mathbf{x} + \boldsymbol{\xi}(\mathbf{x}, t)$, the time derivative and spatial derivatives are transformed as

$$\partial_t = \partial_{t'} + (\partial_t \boldsymbol{\xi}) \cdot \boldsymbol{\nabla}', \quad \boldsymbol{\nabla}' = (\partial_k'), \tag{12.33}$$

$$\partial_k = \partial_k' + \partial_k \xi^l \partial_l', \quad \partial_k' = \partial/\partial x'^k. \tag{12.34}$$

It is remarkable and important to find that we have a transformation invariance of the convective derivative $D_c \equiv \partial_t + (\mathbf{u} \cdot \boldsymbol{\nabla})$ in this local transformation too:

$$D_c = \partial_t + (\mathbf{u} \cdot \boldsymbol{\nabla}) = \partial_{t'} + (\mathbf{u}' \cdot \boldsymbol{\nabla}'). \tag{12.35}$$

In fact, from (12.34) and (12.31), we obtain

$$\mathbf{u} \cdot \boldsymbol{\nabla} = \mathbf{u} \cdot \boldsymbol{\nabla}' + (\mathbf{u} \cdot \boldsymbol{\nabla}\boldsymbol{\xi}) \cdot \boldsymbol{\nabla}' = \mathbf{u}'(\mathbf{x}') \cdot \boldsymbol{\nabla}' + \left(-(d\boldsymbol{\xi}/dt) + \mathbf{u} \cdot \boldsymbol{\nabla}\boldsymbol{\xi}\right) \cdot \boldsymbol{\nabla}',$$

where $\mathbf{u} = \mathbf{u}' - d\boldsymbol{\xi}/dt$. The last term is $-\partial_t \boldsymbol{\xi} \cdot \boldsymbol{\nabla}'$ by (12.32). Thus, we find (12.35) by (12.33).

This invariance (12.35), i.e. the transformation symmetry, implies that the covariant derivative d_t is given by the convective derivative D_c:

$$d_t \equiv \partial_t + g_k T_k = \partial_t + \mathbf{u} \cdot \boldsymbol{\nabla}. \tag{12.36}$$

12.4.4. Gauge transformation (translation symmetry)

The transformation in the previous section is regarded as a gauge transformation, defined by $\mathbf{x} \to \mathbf{x}' = \mathbf{x} + \boldsymbol{\xi}$ and $\mathbf{u}(\mathbf{x}) \to \mathbf{u}'(\mathbf{x}') = \mathbf{u}(\mathbf{x}) + (d\boldsymbol{\xi}/dt)$ from (12.31). Namely, in the transformed space F', the variations of coordinate $q^i \equiv x^i$ and velocity u^i are given by

$$\delta q^i = \xi^k T_k \, q^i = \xi^i, \quad (T_k = \partial/\partial x^k) \tag{12.37}$$

$$\delta u^i = d\xi^i/dt = \partial_t \xi^i + u^k T_k \xi^i. \tag{12.38}$$

Correspondingly, variation of $\partial_t q_i$ is given by

$$\delta(\partial_t q^i) = \partial_t(\delta q^i) + (\delta \partial_t)q^i = \partial_t \xi^i - (\partial_t \xi^k T_k)q^i = 0, \tag{12.39}$$

where $\delta\partial_t = \partial_{t'} - \partial_t = -\partial_t\xi^k T_k$ from (12.33) (to the first order of ξ), and $T_k q^i = \delta_k^i$.

These infinitesimal transformations allow us to define the covariant derivative $d_t q$ by

$$d_t q := \partial_t q + u^k T_k\, q\,. \tag{12.40}$$

In fact, using (12.37)–(12.39) and (12.34) (giving $\delta T_k = T_k' - T_k = -(T_k\xi^l)T_l$), its transformation is given by

$$\begin{aligned} \delta(d_t q^i) &= \delta(\partial_t q^i) + (\delta u^k)T_k\, q^i + u^k(\delta T_k)q^i + u^k T_k(\delta q^i) \\ &= (d\xi^k/dt)T_k q^i - u^k(T_k\xi^l)T_l q^i + u^k T_k\xi^i \\ &= (d\xi^i/dt), \end{aligned} \tag{12.41}$$

where $T_k q^i = \delta_k^i$ is used. Thus, it is found that, in this gauge transformation, the covariant derivative $d_t q$ transforms just like the velocity field u whose transform is given by (12.38) (or (12.31)). This implies that $d_t q$ represents in fact the particle velocity. Namely,

$$d_t \mathbf{x} = (\partial_t + u^k\partial_k)\mathbf{x} = \mathbf{u}(\mathbf{x}, t) \tag{12.42}$$

where $\mathbf{u}(\mathbf{x}_a, t) = (d/dt)\mathbf{x}_a(t)$ (see the beginning of Sec. 12.4.3).

12.4.5. *Galilean invariant Lagrangian*

When a representative flow velocity v becomes higher to such a degree that it is not negligible relative to the light velocity c, the Galilean transformation must be replaced by the Lorentz transformation (Appendix F, or [LL75]). Theory of quantum fields is formulated under the framework of the Lorentz transformation (F.1), whereas fluid mechanics is constructed under the above Galilean transformation (12.26). The difference is not an essential obstacle to the formulation of gauge theory, because the Galilean transformation is considered to be a limiting transformation of the Lorentz transformation of space-time as $\beta = v/c \to 0$.

In Appendix F.1, the Lagrangian Λ_G of fluid motion in the Galilean system is derived from the Lorentz-invariant Lagrangian $\Lambda_L^{(0)}$ by taking the limit as $v/c \to 0$. Thus, the Lagrangian in the

Galilean system is given by the expression (F.6), which is reproduced here:

$$\Lambda_{\mathrm{G}}\mathrm{d}t = \mathrm{d}t \int_M \mathrm{d}^3 x \rho(x) \left(\frac{1}{2} \langle \mathbf{u}(x), \mathbf{u}(x) \rangle - \epsilon(\rho, s) \right), \qquad (12.43)$$

where ρ is the fluid density, and M is a bounded space under consideration with $\mathbf{x} \in M \subset \mathbb{R}^3$. Namely, the Lagrangian L_ϵ defined by

$$L_\epsilon = - \int_M \epsilon(\rho, s) \rho \mathrm{d}^3 \mathbf{x}, \qquad (12.44)$$

must be introduced for the invariance, where $\epsilon(\rho, s)$ is the internal energy and s the entropy. It is understood that the background continuous material is characterized by the internal energy ϵ of the fluid, given by a function $\epsilon(\rho, s)$ where ϵ and s are defined per unit mass.[10] For the fields of density $\rho(\mathbf{x})$ and entropy $s(\mathbf{x})$, the Lagrangian L_ϵ is invariant with respect to the gauge transformation (12.30), since the transformation is a matter of the coordinate origin under the invariance of mass: $\rho(\mathbf{x})\mathrm{d}^3\mathbf{x} = \rho'(\mathbf{x}')\mathrm{d}^3\mathbf{x}'$ and the coordinate $\mathbf{x}$ does not appear explicitly.

According to the scenario of the gauge principle, an additional Lagrangian (called a *kinetic* term, [Fr97]) is to be defined in connection with the background field (the material field *in motion* in the present context), in order to get nontrivial field equations (for ρ and s). *Possible* type of two Lagrangians are proposed as

$$L_\phi = - \int_M \mathrm{d}_t \phi \, \rho \mathrm{d}^3 \mathbf{x}, \qquad L_\psi = - \int_M \mathrm{d}_t \psi \rho s \mathrm{d}^3 \mathbf{x} \qquad (12.45)$$

where $\phi(\mathbf{x}, t)$ and $\psi(\mathbf{x}, t)$ are scalar gauge potentials associated with the density ρ and entropy s respectively.[11] These Lagrangians are invariant by the same reasoning as that of L_ϵ above. In particular, d_t is invariant by local Galilean transformation. It will be found later

---

[10]In thermodynamics, a physical material of a single phase is characterized by two thermodynamic variables such as ρ, s etc., which are regarded as gauge fields in the present formulation.

[11]The minus signs in L's are a matter of convenience, as become clear later.

that the *equations* of mass conservation and entropy conservation are deduced as the results of variational principle. Thus the total Lagrangian is defined by $L_T := L_f + L_\epsilon + L_\phi + L_\psi$, as far as the translation symmetry is concerned.

12.5. Symmetry of flow fields II: Rotation symmetry

12.5.1. *Rotational transformations*

Rotation (with a point $\mathbf{x}_0$ fixed) is described by a 3×3 matrix defined as follows. Rotational transformation of a vector $v = (v_j)$ is represented by $v' = Rv$, i.e. $v'_i = R_{ij}v_j$ in component representation, where $R = (R_{ij})$ is a matrix of rotational transformation. *Rotational transformation* requires that the magnitude $|v|$ is invariant, that is *isometric*. Therefore, we have the invariance of the inner product: $\langle v', v' \rangle = \langle Rv, Rv \rangle = \langle v, v \rangle$. This is not more than the definition of the *orthogonal transformation*. In matrix notation,

$$\langle v', v' \rangle = (v')_i(v')_i = R_{ij}v_j R_{ik}v_k = v_k v_k = \langle v, v \rangle.$$

Therefore, the orthogonal transformation is defined by

$$R_{ij}R_{ik} = (R^T)_{ji}R_{ik} = (R^T R)_{jk} = \delta_{jk}, \tag{12.46}$$

where R^T is the transpose of R: $(R^T)_{ji} = R_{ij}$. Using the unit matrix $I = (\delta_{jk})$, this is rewritten as

$$R^T R = R R^T = I. \tag{12.47}$$

Hence, R^T is equal to the inverse R^{-1} of R.

There is a group of special significance, which is $SO(3)$.[12] When we mention a rotational symmetry of fluid flows, we consider

---

[12] $SO(3)$: Special Orthogonal group. An element $g \in SO(3)$ is characterized by $\det(g) = 1$. This is a subgroup of a larger orthogonal group satisfying (12.47), which leads to $\det(g) = \pm 1$. If the inner products such as $\langle A, A \rangle$, $\langle B, B \rangle$, etc. are invariant, then invariance of the cross inner product $\langle A, B \rangle$ can be verified from the invariance of $\langle A + B, A + B \rangle = \langle A, A \rangle + 2\langle A, B \rangle + \langle B, B \rangle$.

$SO(3)$ always because it is connected with the identity transformation $R = I$.

It is almost obvious now that the Lagrangian density Λ_f of (12.22) is *invariant* under the rotational transformation of $SO(3)$, $v(x) \mapsto v'(x) = R(x)v(x)$ depending on any point x, where $R(x) \in SO(3)$ at $^\forall x \in M$:

$$\int \langle R(x)v(x), R(x)v(x) \rangle R(x) \mathrm{d}^3\mathbf{a} = \int \langle v(x), v(x) \rangle \mathrm{d}^3\mathbf{a}.$$

This implies that the Lagrangian density Λ_f satisfies local rotational gauge invariance. Thus, the total kinetic energy $\mathcal{K} \equiv \mathcal{L}_f$ is invariant, where the mass in a volume element d^3x is invariant by the rotational transformation, $R[\rho(x_0)\mathrm{d}^3x] = \rho(x_0)\mathrm{d}^3x = \mathrm{d}^3\mathbf{a}$.

12.5.2. *Infinitesimal rotational transformation*

In rotational transformations, an arbitrary vector v_0 is sent to $v = Rv_0$ with an element $R \in SO(3)$. Now we consider an *infinitesimal* transformation. Suppose that a varied orthogonal transformation R' is written as $R' = R + \delta R$ for an finitesimal variation δR. We then have $\delta v = \delta R v_0 = (\delta R)R^{-1}v$, (since $v_0 = R^{-1}v$) so that we obtain the infinitely near vector $v + \delta v$ by the action of a general R by

$$v \rightarrow v + \delta v = \left(I + (\delta R)R^{-1}\right)v,$$

where $(\delta R)R^{-1}$ is skew-symmetric for othogonal matrices R and $R + \delta R$.[13] Writing $I + (\delta R)R^{-1}$ as $\exp[\theta]$ for $R \in SO(3)$, we have

$$I + (\delta R)R^{-1} = \exp[\theta] = I + \theta + O(|\theta|^2), \qquad (12.48)$$

where $|\theta| \ll 1$, and θ is a skew-symmetric 3×3 matrix since $(\delta R)R^{-1}$ is skew-symmetric, neglecting $O(|\delta R|^2)$ as in the footnote. Associated with $R \in SO(3)$, the linear term θ is represented as $\theta = \theta^k e_k$ by using

---

[13]Since $RR^{-1} = RR^T = I$ and $(R + \delta R)(R + \delta R)^T = I$ from (12.47), we have $(\delta R)R^{-1} + (R^{-1})^T(\delta R)^T = 0$, neglecting $\delta R(\delta R)^T$. This implies that $(\delta R)R^{-1}$ is skew-symmetric, i.e. $[(\delta R)R^{-1}]_{ij} = -[(\delta R)R^{-1}]_{ji}$.

the three bases e_1, e_2 and e_3 of skew-symmetric 3×3 matrices:

$$e_1 = \begin{bmatrix} 0 & 0 & 0 \\ 0 & 0 & -1 \\ 0 & 1 & 0 \end{bmatrix}, \quad e_2 = \begin{bmatrix} 0 & 0 & 1 \\ 0 & 0 & 0 \\ -1 & 0 & 0 \end{bmatrix}, \quad e_3 = \begin{bmatrix} 0 & -1 & 0 \\ 1 & 0 & 0 \\ 0 & 0 & 0 \end{bmatrix}.$$

$$(12.49)$$

In fact, we have $\theta_{ij} = -\theta_{ji}$, where

$$\theta = (\theta_{ij}) \equiv \theta^k e_k = \begin{bmatrix} 0 & -\theta^3 & \theta^2 \\ \theta^3 & 0 & -\theta^1 \\ -\theta^2 & \theta^1 & 0 \end{bmatrix}$$

represents a general skew-symmetric matrix with three infinitesimal parameters θ^1, θ^2 and θ^3. The commutation between bases are given by the following noncommutative relations:

$$[e_j, e_k] = \varepsilon_{jkl} e_l, \tag{12.50}$$

where ε_{jkl} is the completely skew-symmetric third-order tensor (see (A.1)). Corresponding to the group $SO(3)$ (which is a Lie group), the space of infinitesimal rotational transformations (consisting of skew-symmetric 3×3 matrices) is termed as Lie algebra $\mathbf{so}(3)$, and (e_1, e_2, e_3) are bases of $\mathbf{so}(3)$.[14]

An operator of infinitesimal rotation is defined by $\theta = (\theta_{ij}) = \theta^k e_k$. Let us define an axial vector $\hat{\theta}$ by $\hat{\theta} = (\theta^1, \theta^2, \theta^3)$, an infinitesimal angle vector. Then, an infinitesimal rotation of the displacement vector $\mathbf{s} = (s^1, s^2, s^3)$ is given by $(\theta \mathbf{s})^i = \theta_{ij} s^j = (\theta_k e_k \mathbf{s})^i$:

$$(\theta \mathbf{s})^j = \theta^k e_k s^j = (\theta^2 s^3 - \theta^3 s^2, \theta^3 s^1 - \theta^1 s^3, \theta^1 s^2 - \theta^2 s^1).$$

The right-hand side is nothing but the components of vector product of two vectors $\hat{\theta}$ and $\mathbf{s}$. Thus, we have two equivalent expressions: $\theta \mathbf{s}$ (tensor multiplication) $\Leftrightarrow \hat{\theta} \times \mathbf{s}$ (vector product). The last expression clearly states that this represents a rotation of $\mathbf{s}$ about the axis in the direction coinciding with the vector $\hat{\theta}$ by an infinitesimal angle $|\hat{\theta}|$.

---

[14]The tensor constants C_{jkl} in the commutation relations, $[e_j, e_k] = C_{jkl} e_l$, are called the *structure constants*. If $C_{jkl} = 0$ for all indices such as the case of T_k, the commutation relation is called *Abelian*. If $C_{jkl} \neq 0$, the relation is *non-Abelian*.

12.5.3. Gauge transformation (rotation symmetry)

When we consider local rotation of a fluid element about an arbitrary reference point x_0 within fluid, attention must be paid to a local transformation δs at a neighboring point $x_0 + s$ (for a small s). In this regard, it is remarked that the previous translational transformation is concerned with *pointwise* transformation, while the rotational transformation here is concerned with local transformation in the *neighborhood* of x_0.

Rotational gauge transformation of a point s in a frame F (an inertial frame, say) to s' in a noninertial frame F' is expressed as

$$s'(s,t) = s + \delta s, \quad t' = t, \tag{12.51}$$

instead of (12.30) for the translation. The variation δs is defined by an infinitesimal rotation $\theta = \theta^k e_k$ applied to the displacement vector s:

$$\delta s \equiv \eta(s,t) = \theta s = \hat{\theta} \times s. \tag{12.52}$$

Corresponding transformation of velocity ($v = (d/dt)s_a$) is as follows:

$$v'(s') = v(s) + (d/dt)\eta, \tag{12.53}$$

$$\delta v = (d/dt)\eta \equiv \partial_t \eta(s,t) + (v \cdot \nabla_s)\eta(s,t) = \partial_t \theta s + \theta v. \tag{12.54}$$

instead of (12.31) and (12.32) where $\nabla_s = (\partial/\partial s_i)$. As before (Sec. 12.4.3), it is noted that in this transformation the points s and s' are the same with respect to the frame F. This means that the coordinate frame F' is rotated by an angle $-\hat{\theta}$ with respect to F with a common origin x_0.

According to the scenario of covariant derivative of Sec. 12.3.2, a new covariant derivative D_t and a new gauge field $\Omega = \Omega^k e_k$ are introduced with respect to the rotational symmetry. We have the following:

$$D_t s = d_t s + \Omega s, \quad d_t s = u, \quad \Omega s = \Omega_{ij} s_j = \hat{\Omega} \times s, \tag{12.55}$$

where d_t and ∂_t of (12.36) are replaced by D_t and d_t respectively. The gauge field Ω is defined by $\Omega = \Omega^k e_k$ and $\Omega_{ij} = -\Omega_{ji}$ where

$\Omega_{ij} \equiv (\Omega^k e_k)_{ij}$. Its axial vector counterpart is $\hat{\Omega} = (\Omega^1, \Omega^2, \Omega^3)$. The operator d_t denotes a time derivative when local rotation is absent. Then, $d_t s$ may be represented as $a_{ij} s_j + O(|\mathbf{s}|^2)$ by using a symmetric tensor a_{ij} (see the next section for the details).

In an analogous way to the translational transformations, variations of s and u are written as

$$\delta s = \theta^k e_k s, \qquad (12.56)$$

$$\delta u_i = \theta^k e_k u, \qquad (12.57)$$

for the rotational transformations, where the first expression is equivalent to (12.52). Corresponding to the expression (12.38), variation of Ω is defined by[15]

$$\delta \Omega^k = \partial_t \theta^k + \varepsilon_{klm} \theta^l \, \Omega^m, \qquad (12.58)$$

where the structure constant ε_{klm} of (12.50) is used as a rule in the non-Abelian symmetry [Wnb95]. Then, we can show

$$\delta(\mathrm{D}_t s) = \partial_t \theta s + \theta \, \mathrm{D}_t s. \qquad (12.59)$$

In fact, we obtain

$$
\begin{aligned}
\delta(\mathrm{D}_t s) &= \delta u + \delta \Omega^k e_k s + \Omega^k e_k \delta s = \theta^k e_k u \\
&\quad + (\varepsilon_{klm} \theta^l \, \Omega^m - \partial_t \theta^k) e_k s + \Omega^m e_m \theta^l e_l s \\
&= \partial_t \theta s + \theta \, \mathrm{D}_t s,
\end{aligned}
\qquad (12.60)
$$

where we have used the following equality from the relation (12.50):

$$e_m e_l + \varepsilon_{lmk} e_k = e_l e_m,$$

where $\varepsilon_{lmk} = \varepsilon_{klm}$. Therefore, the covariant derivative $\mathrm{D}_t s$ transforms like $\mathbf{v}$ of (12.54).

Thus, it is found that the expression (12.55) gives a gauge-covariant definition of velocity:

$$v := \mathrm{D}_t s = d_t s + \Omega s = u + \Omega_{ij} s_j, \qquad (12.61)$$

where $\Omega_{ij} = (\Omega^k e_k)_{ij}$ is a skew-symmetric tensor.

---

[15]The vector form is $\delta \hat{\Omega} = \hat{\theta} \times \hat{\Omega} + \partial_t \hat{\theta}$.

12.5.4. *Significance of local rotation and the gauge field*

In the translation symmetry of Sec. 12.4.4, we obtained the expression of gauge-covariant velocity (12.42), i.e. $u = d_t x$. Using s in place of x, we have

$$d_t s = u(\mathbf{x}_0 + \mathbf{s}) = u(\mathbf{x}_0) + \Delta u \approx u_0 + s_j \partial_j u,$$
$$(d_t s)_i = (u_0)_i + a_{ij} s_j, \tag{12.62}$$

where $u_0 = u(\mathbf{x}_0)$, and the deviation δu is expressed as $a_{ij} s_j$. The coefficient matrix a_{ij} is assumed to be symmetric ($a_{ij} = a_{ji}$), because a skew-symmetric part (if any) can be absorbed into Ω_{ij} of (12.61). The symmetric part δu is also expressed as

$$\Delta u = \nabla_s \phi(s) = a_{ij} s_j, \quad \phi(\mathbf{s}) = \frac{1}{2} a_{ij} s_i s_j,$$

where $\phi(\mathbf{s})$ represents a velocity potential of an irrotational field. Thus, the expressions (12.61) and (12.62) combine to

$$D_t s = d_t s + \Omega s = u_0 + a_{ij} s_j + \Omega_{ij} s_j$$
$$= u_0 + \nabla_s \phi(s) + \hat{\Omega} \times \mathbf{s}. \tag{12.63}$$

Suppose that a velocity field $\mathbf{v}(\mathbf{x}_0 + \mathbf{s})$ is given. Then, we have an analogous local expression. In fact, relative deviation $\Delta \mathbf{v}$ is defined by

$$\Delta \mathbf{v} = \mathbf{v}(\mathbf{x}_0 + \mathbf{s}) - \mathbf{v}(\mathbf{x}_0) = (\mathbf{s} \cdot \nabla)\mathbf{v}(\mathbf{x}_0) + O(|\mathbf{s}|^2). \tag{12.64}$$

As we did in Sec. 1.4 of Chapter 1, to the first order of $|\mathbf{s}|$, we have

$$\Delta v_k = (\mathbf{s} \cdot \nabla)v_k(\mathbf{x}_0) = s_j \partial_j v_k = s_j(e_{jk} + w_{jk}), \tag{12.65}$$

where the term $\partial_j v_k$ is decomposed as $e_{jk} + w_{jk}$, defined by

$$e_{jk} = \frac{1}{2}(\partial_j v_k + \partial_k v_j), \quad w_{jk} = \frac{1}{2}(\partial_j v_k - \partial_k v_j) = \frac{1}{2}\varepsilon_{jkl}(\boldsymbol{\omega})_l.$$

where $\omega = \varepsilon_{ijk} 2w_{jk} = \operatorname{curl} \mathbf{v}$ (*vorticity*). Thus, we obtain

$$
\begin{aligned}
\mathbf{v}(\mathbf{x}_0 + \mathbf{s}) &= \mathbf{v}(\mathbf{x}_0) + e_{jk}s_j + w_{jk}s_j \\
&= \mathbf{v}(\mathbf{x}_0) + \nabla_s \Phi(\mathbf{s}) + \frac{1}{2}\omega \times \mathbf{s}, \quad \Phi(\mathbf{s}) = \frac{1}{2}e_{jk}(\mathbf{x}_0)\, s_j s_k.
\end{aligned}
\tag{12.66}
$$

Comparing this with (12.63), it is seen that u_0 and $\phi(s)$ correspond to $\mathbf{v}(\mathbf{x}_0)$ and $\Phi(\mathbf{s})$ of (12.66), respectively. Moreover we find that $\hat{\Omega} = \frac{1}{2}\omega$. So far, $\mathbf{x}_0 + \mathbf{s}$ was a local position vector around a fixed point $\mathbf{x}_0$. In general, using $\mathbf{x}$ in place of $\mathbf{x}_0 + \mathbf{s}$, we arrive at a definition of velocity. Namely, identifying as $u_0 = \mathbf{v}(\mathbf{x}_0)$, $\phi(s) = \Phi(\mathbf{s})$ and $\hat{\Omega} = \frac{1}{2}\omega$, we obtain the definition:

$$
\mathbf{v}(\mathbf{x}, t) = D_t \mathbf{x}. \tag{12.67}
$$

It is remarkable that the gauge field $2\hat{\Omega}$ is equivalent to the vorticity ω. Namely, *the vorticity ω is nothing but the gauge field*:

$$
\omega = 2\hat{\Omega}. \tag{12.68}
$$

12.5.5. *Lagrangian associated with the rotation symmetry*

According to the scenario of the gauge principle, an additional Lagrangian (called a *kinetic* term [Fr97]) is to be defined in connection with the rotation symmetry in order to obtain a nontrivial equation for the gauge field ω. A *possible* type of the Lagrangian is proposed as

$$
L_A = -\int_M \rho \langle \mathbf{v}, \mathbf{b} \rangle \mathrm{d}^3 \mathbf{x}, \tag{12.69}
$$

where $\mathbf{b}$ is a vector associated with the gauge field ω (considered in Appendix F.2). It is almost obvious from the argument of Sec. 12.6.1 that the Lagrangian (12.69) is invariant under local rotational transformations of $SO(3)$. An explicit form of $\mathbf{b}$ is proposed in Appendix F.2(b) so as to conform to the translation symmetry.

12.6. Variational formulation for flows of an ideal fluid

12.6.1. *Covariant derivative (in summary)*

Now we define the operator D_t of covariant derivative by

$$D_t \equiv \partial_t + (\mathbf{v} \cdot \nabla). \tag{12.70}$$

It is deduced from the analysis of previous sections that the covariant derivative $D_t \mathbf{x}$ represents the particle velocity. Namely, $D_t \mathbf{x} = \mathbf{v}$ and $\mathbf{v}(\mathbf{x}_a, t) = (d/dt)\mathbf{x}_a(t)$.

12.6.2. *Particle velocity*

The velocity $\mathbf{v}$ can be defined by the covariant derivative $D_t \mathbf{x}$. Additional remarks are given as follows. The Lagrange particle coordinates $\mathbf{a} = (a_k)$ are regarded as functions of $\mathbf{x}$ and t and satisfy the following equations,

$$D_t a_k = \partial_t a_k + (\mathbf{v} \cdot \nabla)a_k = 0, \quad k = 1, 2, 3, \tag{12.71}$$

because the particle of the name tag (a_k) moves with the velocity $\mathbf{v}$ by definition. Setting as $\mathbf{x} = \mathbf{X}(\mathbf{a}, t)$ for the particle position, we have

$$\mathbf{v} = D_t \mathbf{X}(\mathbf{a}(\mathbf{x}, t), t) = \partial_t \mathbf{X}(\mathbf{a}, t) + D_t \mathbf{a} \cdot \nabla_a \mathbf{X} = \partial_t \mathbf{X}(\mathbf{a}, t), \tag{12.72}$$

by using (12.71), where $(\nabla_a \mathbf{X}) = (\partial X^k / \partial a^l)$. The right-hand side defines the velocity $\mathbf{v}_a$ of a particle $\mathbf{a}$. Thus we have the equality relation, $D_t \mathbf{x} = \mathbf{v}$.

On the other hand, regarding $\mathbf{x}$ as a field variable, we have

$$D_t \mathbf{x} = (\partial_t + \mathbf{v} \cdot \nabla)\mathbf{x} = \mathbf{v}. \tag{12.73}$$

Thus, the $D_t \mathbf{x}$ signifies the velocity of a material particle $\mathbf{a}$.

12.6.3. *Action principle*

According to the previous sections, the full Lagrangian is defined by

$$L_T = L_f + L_\epsilon + L_\phi + L_\psi + L_A$$

$$= \int_M d^3\mathbf{x} \, \Lambda_T[\mathbf{v}, \rho, s, \phi, \psi, \mathbf{b}],$$

$$\Lambda_T \equiv \frac{1}{2}\rho\langle\mathbf{v}, \mathbf{v}\rangle - \rho\epsilon(\rho, s) - \rho D_t\phi - \rho s D_t\psi - \rho\langle\mathbf{v}, \mathbf{b}\rangle, \quad (12.74)$$

where $\mathbf{v}$, ρ and ϵ are the velocity, density and internal energy (per unit mass) of the fluid, and $\phi(\mathbf{x}, t)$ and $\psi(\mathbf{x}, t)$ are scalar functions, and $\mathbf{b}$ is a vector gauge field, and $D_t = \partial_t + \mathbf{v} \cdot \nabla$. The action is defined by $I = \int_{t_1}^{t_2} \int_M \Lambda_T dt \, d^3\mathbf{x}$. The action principle is given by

$$\delta I = \delta \int_{t_1}^{t_2} \int_M dt \, d^3\mathbf{x} \, \Lambda_T = 0. \quad (12.75)$$

Usually, in the variational formulation (of the Eulerian representation of independent variables $\mathbf{x}$ and t), the Euler's equation of motion is derived under the constraints of the continuity equation and isentropic equation. In the present analysis, the variational principle based on the gauge principle gives us the continuity equation and the isentropic equation as *outcomes* of variations of the Lagrangian $L_T[\mathbf{v}, \rho, s, \phi, \psi, \mathbf{b}]$ with respect to variations of the gauge potentials ϕ and ψ. The potentials have intrinsic physical significance in the framework of the gauge theory.

Here, we have to take into consideration a certain thermodynamic property that the fluid is an ideal fluid in which there is no mechanism of dissipation of kinetic energy into heat. That is, there is no heat production within fluid. By thermodynamics, change of the internal energy ϵ and enthalpy $h = \epsilon + p/\rho$ can be expressed in terms of changes of density $\delta\rho$ and entropy δs as

$$\delta\epsilon = \left(\frac{\partial\epsilon}{\partial\rho}\right)_s \delta\rho + \left(\frac{\partial\epsilon}{\partial s}\right)_\rho \delta s = \frac{p}{\rho^2}\delta\rho + T\delta s, \quad (12.76)$$

$$\delta h = \frac{1}{\rho}\delta p + T\delta s, \quad (12.77)$$

where $(\partial\epsilon/\partial\rho)_s = p/\rho^2$ and $(\partial\epsilon/\partial s)_\rho = T$ with p the fluid pressure and T the temperature, $(\cdot)_s$ denoting the change with s kept fixed. If there is no heat production, we have $T\delta s = 0$. Then,

$$\delta\epsilon = (\delta\epsilon)_s = \left(\frac{\partial\epsilon}{\partial\rho}\right)_s \delta\rho = \frac{p}{\rho^2}\delta\rho, \quad \delta h = \frac{1}{\rho}\delta p. \tag{12.78}$$

12.6.4. *Outcomes of variations*

Writing Λ_T as

$$\Lambda_T = \Lambda_T[\mathbf{u}, \rho, s, \phi, \psi, \mathbf{b}] \equiv \frac{1}{2}\rho\langle\mathbf{v}, \mathbf{v}\rangle - \rho\epsilon(\rho, s) - \rho(\partial_t + \mathbf{v}\cdot\nabla)\phi$$
$$- \rho s(\partial_t + \mathbf{v}\cdot\nabla)\psi - \rho\langle\mathbf{v}, \mathbf{b}\rangle, \tag{12.79}$$

we take variations of the field variables $\mathbf{v}$, ρ, s and potentials ϕ and ψ. (Variation of $\mathbf{b}$ is considered separately in Appendix F.2.) Independent variations are taken for those variables. Substituting the variations $\mathbf{v}+\delta\mathbf{v}$, $\rho+\delta\rho s+\delta s$, $\phi+\delta\phi$ and $\psi+\delta\psi$ into $\Lambda_T[\mathbf{v}, \rho, s, \phi, \psi, \mathbf{b}]$ and writing its variation as $\delta\Lambda_T$, we obtain

$$\delta\Lambda_T = \delta\mathbf{v}\cdot\rho(\mathbf{v} - \nabla\phi - s\nabla\psi - \mathbf{w}) - \delta s\rho\mathrm{D}_t\psi \tag{12.80}$$
$$+ \delta\rho\left(\frac{1}{2}u^2 - h - \mathrm{D}_t\phi - s\mathrm{D}_t\psi - \mathbf{v}\cdot\mathbf{b}\right)$$
$$+ \delta\phi(\partial_t\rho + \nabla\cdot(\rho\mathbf{v})) - \partial_t(\rho\delta\phi) - \nabla\cdot(\rho\mathbf{v}\delta\phi)$$
$$+ \delta\psi(\partial_t(\rho s) + \nabla\cdot(\rho s\mathbf{v})) - \partial_t(\rho s\delta\psi) - \nabla\cdot(\rho s\mathbf{v}\delta\psi), \tag{12.81}$$

where

$$\mathbf{w} = (w_i), \quad w_i = (\partial/\partial v^i)\langle\mathbf{v}, \mathbf{b}\rangle, \tag{12.82}$$

and h is the specific enthalpy defined by $h = \epsilon + \rho(\partial\epsilon/\partial\rho)_s = \epsilon + p/\rho$.

Thus, the variational principle, $\delta I = 0$ for independent arbitrary variations $\delta\mathbf{v}$, $\delta\rho$ and δs, results in

$$\delta\mathbf{v}: \quad \mathbf{v} = \nabla\phi + s\nabla\psi + \mathbf{w}, \tag{12.83}$$

$$\delta\rho: \quad \frac{1}{2}v^2 - h - \mathrm{D}_t\phi - s\mathrm{D}_t\psi - \mathbf{v}\cdot\mathbf{b} = 0, \tag{12.84}$$

$$\delta s: \quad \mathrm{D}_t\psi \equiv \partial_t\psi + \mathbf{v}\cdot\nabla\psi = 0. \tag{12.85}$$

From the variations of $\delta\phi$ and $\delta\psi$, we obtain

$$\delta\phi: \quad \partial_t\rho + \nabla \cdot (\rho\mathbf{v}) = 0, \tag{12.86}$$

$$\delta\psi: \quad \partial_t(\rho s) + \nabla \cdot (\rho s\mathbf{v}) = 0. \tag{12.87}$$

Using (12.86), the second equation can be rewritten as

$$\partial_t s + \mathbf{v} \cdot \nabla s = D_t s = 0. \tag{12.88}$$

i.e. the fluid flow is *adiabatic*. Thus, the continuity equation (12.86) and the entropy equation (12.88) have been derived from the variational principle. These must be supplemented by the equation of particle motion (12.73). The equation of isentropy, (12.88), is consistent with (12.78).

Equation (12.83) gives the velocity by

$$\mathbf{v} = \nabla\phi + s\nabla\psi + \mathbf{w}. \tag{12.89}$$

The vorticity $\boldsymbol{\omega} = \nabla \times \mathbf{v}$ is represented by

$$\boldsymbol{\omega} = \nabla s \times \nabla\psi + \nabla \times \mathbf{w}. \tag{12.90}$$

12.6.5. *Irrotational flow*

For a homentropic fluid in which the entropy s is a uniform constant s_0 at all points, we have $e = e(\rho)$, and

$$\mathrm{d}e = \frac{p}{\rho^2}\mathrm{d}\rho, \quad \mathrm{d}h = \frac{1}{\rho}\mathrm{d}p. \tag{12.91}$$

from (12.76) and (12.77) since $\delta s = 0$. In addition, assuming that the vector gauge field $\mathbf{b}$ is absent, the motion becomes *irrotational*. In fact, from (12.89), we have

$$\mathbf{v} = \nabla\Phi, \quad \Phi = \phi + s_0\psi, \tag{12.92}$$

(since $\mathbf{w} = 0$), i.e. the velocity field has a potential Φ, and $\boldsymbol{\omega} = 0$ from (12.90). Since we have

$$D_t\phi + s_0 D_t\psi = \partial_t\Phi + \mathbf{v} \cdot \nabla\Phi = \partial_t\Phi + v^2.$$

Equation (12.84) becomes

$$\frac{1}{2}v^2 + h + \partial_t \Phi = 0, \tag{12.93}$$

which is equivalent to (5.28) with $\chi = 0$. Taking its gradient reduces to the Euler's equation (5.24):

$$\partial_t \mathbf{v} + \nabla\left(\frac{1}{2}v^2\right) = -\nabla h, \quad \text{where } \nabla h = \frac{1}{\rho}\nabla p. \tag{12.94}$$

Note that the left-hand side is the *material* time derivative for the potential velocity $u_k = \partial_k \Phi$. In fact, using $\partial_i(u^2/2) = u_k\partial_i u_k = (\partial_k\Phi)\partial_i\partial_k\Phi = (\partial_k\Phi)\partial_k\partial_i\Phi = u_k\partial_k u^i$, we obtain

$$\partial_t \mathbf{v} + \nabla\left(\frac{1}{2}v^2\right) = \partial_t\mathbf{v} + (\mathbf{v}\cdot\nabla)\mathbf{v} = D_t\mathbf{v}. \tag{12.95}$$

It is interesting to find that the fluid motion is driven by the velocity potential Φ, where $\Phi = \phi + s_0\psi$ with ϕ, ψ the gauge potentials.

We recall that the flow of a superfluid in the degenerate ground state, which is characterized with zero entropy, is represented by a velocity potential in Chapter 11. There, it is shown that such a fluid of macroscopic number of bosons is represented by a single wave function in the degenerate ground state, where the quantum-mechanical current is described by a potential function (phase of the wave function). Therefore the corresponding velocity is irrotational. In this case, local rotation would not be captured.

12.6.6.   *Clebsch solution*

If the entropy s is not uniform (i.e. a function of points and time) but in case that the vector field $\mathbf{b}$ is still absent, the present solution is equivalent to the classical Clebsch solution owing to the property $D_t\psi = 0$. From (12.89) and (12.90), the velocity and vorticity are

$$\mathbf{v} = \nabla\phi + s\nabla\psi, \quad \boldsymbol{\omega} = \nabla s \times \nabla\psi. \tag{12.96}$$

In this case, the vorticity is connected with nonuniformity of entropy. In addition, Eq. (12.84) can be written as

$$\frac{1}{2}v^2 + h + \partial_t\phi + s\partial_t\psi = 0, \tag{12.97}$$

because we have

$$D_t\phi + sD_t\psi = \partial_t\phi + s\partial_t\psi + \mathbf{v}\cdot(\nabla\phi + s\nabla\psi) = \partial_t\phi + s\partial_t\psi + \mathbf{v}^2.$$

It is shown in Problem 12.1 that Euler's equation of motion,

$$\partial_t\mathbf{v} + \boldsymbol{\omega}\times\mathbf{v} = -\nabla\left(\frac{1}{2}v^2 + h\right), \tag{12.98}$$

is satisfied by Eq. (12.97) owing to the conditions (12.85) and (12.88), under the definition (12.96) and the barotropic relation $h(p) = \int^p dp'/\rho(p')$. In this case, the helicity vanishes (Problem 12.1).

Equation (12.98) can be written also as

$$\partial_t\mathbf{v} + (\mathbf{v}\cdot\nabla)\mathbf{v} = -\frac{1}{\rho}\mathrm{grad}\,p, \tag{12.99}$$

because of the identity: $\boldsymbol{\omega}\times\mathbf{v} = (\mathbf{v}\cdot\nabla)\mathbf{v} - \nabla(\frac{1}{2}v^2)$ and the relation (12.77) under (12.88).

12.7.   Variations and Noether's theorem

Differential form of momentum conservation results from the Noether theorem associated with *local* translational symmetry. Here, variations are taken with respect to translational transformations only, with the gauge potentials fixed. As before, the action I is defined by

$$I = \int_{t_1}^{t_2} dt \int_M d^3\mathbf{x}\left[\Lambda_f + \Lambda_\epsilon + \Lambda_\phi + \Lambda_\psi + \Lambda_A\right], \tag{12.100}$$

$$\Lambda_f = \frac{1}{2}\rho\langle\mathbf{v},\mathbf{v}\rangle, \quad \Lambda_\epsilon = -\rho\epsilon(\rho, s)$$

and $\Lambda_\phi = -\rho D_t\phi$, $\Lambda_\psi = -\rho s D_t\psi$, $\Lambda_A = -\rho\langle\mathbf{v},\mathbf{b}\rangle$.

12.7.1. *Local variations*

We consider the following infinitesimal coordinate transformation:

$$\mathbf{x}'(\mathbf{x}, t) = \mathbf{x} + \boldsymbol{\xi}(\mathbf{x}, t). \tag{12.101}$$

In Sec. 12.4, we considered transformation properties of d_t and $d_t q$ under this local transformation. By this transformation, a volume element $d^3\mathbf{x}$ is changed to

$$d^3\mathbf{x}' = \frac{\partial(\mathbf{x}')}{\partial(\mathbf{x})} d^3\mathbf{x} = (1 + \partial_k \xi_k) d^3\mathbf{x},$$

where $\partial(\mathbf{x}')/\partial(\mathbf{x}) = 1 + \partial_k \xi_k$ (where $\partial_k \xi_k = \operatorname{div} \boldsymbol{\xi}$) is the Jacobian of the transformation (up to the first order terms). From (12.31), (12.32) and (12.70), the velocity $\mathbf{v}(\mathbf{x})$ is transformed locally as

$$\mathbf{v}'(\mathbf{x}') = \mathbf{v}(\mathbf{x}) + D_t \boldsymbol{\xi},$$

where the points $\mathbf{x}$ and $\mathbf{x}'$ are the same points with respect to the inertial space F (see Sec. 12.4.3). We denote this transformation by $\Delta\mathbf{v}$:

$$\Delta\mathbf{v} = \mathbf{v}'(\mathbf{x}') - \mathbf{v}(\mathbf{x}) = D_t \boldsymbol{\xi}. \tag{12.102}$$

According to the change of volume element $d^3\mathbf{x}$, there is change of density ρ. The invariance of mass is

$$\rho(\mathbf{x})d^3\mathbf{x}(\mathbf{x}) = \rho'(\mathbf{x}')d^3\mathbf{x}'(\mathbf{x}'), \quad \text{thus } \Delta(\rho d^3\mathbf{x}) = 0. \tag{12.103}$$

Hence, we obtain $\rho(\mathbf{x}) = (1 + \operatorname{div} \boldsymbol{\xi})\rho'(\mathbf{x}')$. Therefore,

$$\Delta\rho = \rho'(\mathbf{x}') - \rho(\mathbf{x}) = -\rho \operatorname{div} \boldsymbol{\xi} = -\rho \partial_k \xi_k, \tag{12.104}$$

to the first order of $|\boldsymbol{\xi}|$. The invariance of entropy $s\rho d^3\mathbf{x}(\mathbf{x}) = s'\rho'd^3\mathbf{x}'(\mathbf{x}')$ is expressed by

$$\Delta s = s'(\mathbf{x}') - s(\mathbf{x}) = 0. \tag{12.105}$$

The gauge fields $D_t \phi$ and $D_t \phi$ remain unvaried:

$$\Delta(D_t \phi) = 0, \quad \Delta(D_t \psi) = 0.$$

Combining with (12.103), these imply

$$\Delta(\Lambda_\phi d^3\mathbf{x}) = 0, \quad \Delta(\Lambda_\psi d^3\mathbf{x}) = 0. \tag{12.106}$$

Correspondingly, $\Delta(\Lambda_A d^3\mathbf{x}) = 0$ is assumed.

The variation field $\boldsymbol{\xi}(\mathbf{x}, t)$ is constrained so as to vanish on the boundary surface S of $M \subset E^3$, as well as at both ends of time t_1, t_2 for the action I (where M is chosen arbitrarily):

$$\boldsymbol{\xi}(\mathbf{x}_S, t) = 0, \quad {}^\forall t, \quad \text{for } \mathbf{x}_S \in S = \partial M, \tag{12.107}$$

$$\boldsymbol{\xi}(\mathbf{x}, t_1) = 0, \quad \boldsymbol{\xi}(\mathbf{x}, t_2) = 0, \quad \text{for } {}^\forall \mathbf{x} \in M. \tag{12.108}$$

When we consider the symmetry with respect to global transformation, we take the limit:

$$\boldsymbol{\xi}(\mathbf{x}, t) \to \boldsymbol{\xi}_0 \text{ (a uniform constant vector)}.$$

12.7.2. *Invariant variation*

It is required that, under the infinitesimal variations (12.101)–(12.106), the action I is invariant (as it should be), i.e.

$$0 = \Delta I \equiv I' - I = \int dt \int_{M'} d^3\mathbf{x}' \big[\Lambda_f(\mathbf{v}', \rho') + \Lambda_\epsilon(\rho', s') \big] (\mathbf{x}', t)$$

$$- \int dt \int_M d^3\mathbf{x} \, [\Lambda_f(\mathbf{v}, \rho) + \Lambda_\epsilon(\rho, s)](\mathbf{x}, t).$$

The integration space M' is the same as M with respect to the frame F, although different expressions are given. Using the original variables in the first integral, we have

$$\Delta I = \int dt \int_M d^3\mathbf{x} \frac{\partial(\mathbf{x}')}{\partial(\mathbf{x})} \big(\Lambda_f(\mathbf{v} + \Delta\mathbf{v}, \rho + \Delta\rho) + \Lambda_\epsilon(\rho + \Delta\rho, s + \Delta s) \big)$$

$$- \int dt \int_M d^3\mathbf{x} \, [\Lambda_f(\mathbf{v}, \rho) + \Lambda_\epsilon(\rho, s)] = 0, \tag{12.109}$$

where $[\partial(\mathbf{x}')/\partial(\mathbf{x})] d^3 x = (1 + \partial_k \xi_k) d^3 \mathbf{x}$. Thus we obtain

$$\Delta I = \int dt \int_M d^3\mathbf{x} \left\{ \frac{\partial \Lambda_f}{\partial \mathbf{v}} \Delta\mathbf{v} + \frac{\partial \Lambda_f}{\partial \rho} \Delta\rho + \left(\frac{\partial \Lambda_\epsilon}{\partial \rho} \right)_s \Delta\rho + \left(\frac{\partial \Lambda_\epsilon}{\partial s} \right)_\rho \Delta s \right.$$

$$\left. + \big[\Lambda_f(\mathbf{v}, \rho) + \Lambda_\epsilon(\rho, s) \big] \partial_k \xi_k \right\},$$

to the first order of variations. Substituting (12.102), (12.104) and (12.105), we have

$$\Delta I = \int dt \int_M d^3\mathbf{x} \left\{ \frac{\partial \Lambda_f}{\partial \mathbf{v}} D_t \boldsymbol{\xi} + \left(\frac{\partial \Lambda_f}{\partial \rho} + \left(\frac{\partial \Lambda_\epsilon}{\partial \rho} \right)_s \right) (-\rho \partial_k \xi_k) \right.$$

$$\left. + [\Lambda_f + \Lambda_\epsilon] \partial_k \xi_k \right\} = 0, \qquad (12.110)$$

where $\Lambda_f + \Lambda_\epsilon = \frac{1}{2}\rho v^2 - \rho \epsilon$, and

$$\frac{\partial \Lambda_f}{\partial \mathbf{v}} = \rho \mathbf{v}, \quad \frac{\partial \Lambda_f}{\partial \rho} = \frac{1}{2}v^2, \quad \left(\frac{\partial \Lambda_\epsilon}{\partial \rho} \right)_s = -h. \qquad (12.111)$$

It is immediately seen that the second and third terms of (12.110) can be combined by using (12.111) as

$$-\rho \left(\frac{\partial \Lambda_f}{\partial \rho} + \left(\frac{\partial \Lambda_\epsilon}{\partial \rho} \right)_s \right) (\partial_k \xi_k) + [\Lambda_f + \Lambda_\epsilon] \partial_k \xi_k$$

$$= \rho \left(-\frac{1}{2}v^2 + h + \frac{1}{2}v^2 - e \right) \partial_k \xi_k$$

$$= \rho(h - e)\partial_k \xi_k = \rho \frac{p}{\rho} \partial_k \xi_k = p \partial_k \xi_k, \qquad (12.112)$$

where $h = \epsilon + p/\rho$. Hence, the second and third terms of (12.110) are reduced to the single term $p(\partial_k \xi_k)$, which can be expressed further as $\partial_k(p\,\xi_k) - (\partial_k p)\xi_k$.

12.7.3. *Noether's theorem*

We now consider the outcome obtained from the arbitrary variation of $\boldsymbol{\xi}$. We write (12.110) as

$$\Delta I = \int dt \int_M d^3\mathbf{x} F[\xi_k, \partial_\alpha \xi_k] = 0. \qquad (12.113)$$

By using (12.112), the integrand $F[\xi_k, \partial_\alpha \xi_k]$ reduces to

$$F[\xi_k, \partial_\alpha \xi_k] = \frac{\partial \Lambda_f}{\partial \mathbf{v}} D_t \boldsymbol{\xi} + p(\partial_k \xi_k). \qquad (12.114)$$

In view of the definitions $D_t \xi_k = \partial_t \xi_k + v_l \partial_l \xi_k$, this can be rewritten as

$$F[\xi_k, \partial_\alpha \xi_k] = \xi_k \left[-\partial_t \left(\frac{\partial \Lambda_f}{\partial v_k} \right) - \partial_l \left(v_l \frac{\partial \Lambda_f}{\partial v_k} \right) - (\partial_k p) \right] + \text{Div},$$

(12.115)

where the divergence terms are collected in the term Div:

$$\text{Div} = \partial_t \left(\frac{\partial \Lambda_f}{\partial v_k} \xi_k \right) + \partial_l \left(v_l \frac{\partial \Lambda_f}{\partial v_k} \xi_k \right) + \partial_k (p \xi_k).$$

(12.116)

Using (12.111), the expression (12.115) with (12.116) becomes

$$F[\xi_k, \partial_\alpha \xi_k] = \xi_k [-\partial_t (\rho v_k) - \partial_l (\rho v_l v_k) - \partial_k p]$$
$$+ \partial_t (\rho v_k \xi_k) + \partial_l (v_l \rho v_k \xi_k) + \partial_l (p \delta_k^l \xi_k).$$

(12.117)

Substituting (12.117) into (12.113), the variational principle (12.113) can be written as

$$0 = \Delta I = \int dt \int_M d^3 \mathbf{x} \, \xi_k \left(-\partial_t (\rho v_k) - \partial_l (\rho v_l v_k) - \partial_k p \right)$$

$$+ \int dt \left[\frac{d}{dt} \int_M d^3 \mathbf{x} (\rho v_k \xi_k) \right] + \int dt \int_S dS n_l (\rho v_l v_k + p \delta_{lk}) \xi_k,$$

(12.118)

where S is the boundary surface of M and $(n_l) = \mathbf{n}$ is a unit outward normal to S. The terms on the second line are integrated terms, which came from the second line of (12.117). These vanish owing to the imposed conditions (12.107) and (12.108).

Thus, the invariance of I for arbitrary variation of ξ_k satisfying the conditions (12.107) and (12.108) results in

$$\partial_t (\rho v_k) + \partial_l (\rho v_l v_k + p \delta_{lk}) = 0.$$

(12.119)

This is the *conservation equation of momentum*. If we use the continuity equation (12.86), we obtain the Euler's equation of motion:

$$\partial_t v_k + v_l \partial_l v_k + \frac{1}{\rho} \partial_k p = 0.$$

(12.120)

These are equivalent to (3.21) and (3.14) with $f_i = 0$ respectively.

Now, we can consider the outcome of global gauge invariance with respect to a global translation of ξ_k = const., without the conditions (12.107) and (12.108). Using Eq. (12.119) obtained from the variational principle, the first line of (12.118) vanishes. Thus, for ξ_k = const., we obtain from (12.118),

$$\xi_k \left[\frac{d}{dt} \int_M d^3\mathbf{x}(\rho v_k) + \int_S dS\, n_l(\rho v_l v_k + p\delta_{lk}) \right] = 0, \qquad (12.121)$$

taking the constant ξ_k out of the integral signs. For arbitrary ξ_k ($k = 1, 2, 3$), the expression within [ ] must vanish. Therefore,

$$\frac{d}{dt} \int_M d^3\mathbf{x}(\rho v_k) = -\int_S dS\, n_l(\rho v_l v_k + p\delta_{lk}), \qquad (12.122)$$

for $k = 1, 2, 3$. This is the conservation of total momentum, equivalent to (3.23) without the external force f_i.

12.8. Additional notes

12.8.1. *Potential parts*

From the equation of motion (12.120), the vorticity equation (7.1) can be derived by taking its curl:

$$\partial_t \boldsymbol{\omega} + \mathrm{curl}(\boldsymbol{\omega} \times \mathbf{v}) = 0, \qquad (7.1)$$

for a homentropic fluid (of uniform entropy). Suppose that this is solved and a velocity field $\mathbf{v}(\mathbf{x}, t)$ is found. Then, the velocity $\mathbf{v}$ should satisfy the following equation:

$$\partial_t \mathbf{v} + \boldsymbol{\omega} \times \mathbf{v} = \nabla H_*$$

for a scalar function $H_*(\mathbf{x}, t)$. Obviously, taking curl of this equation reduces to (7.1). On the other hand, the equation of motion (12.120) was transformed to (3.30) for a homentropic fluid in Sec. 3.4.1:

$$\partial_t \mathbf{v} + \boldsymbol{\omega} \times \mathbf{v} = -\mathrm{grad}\left(\frac{\mathbf{v}^2}{2} + h\right) \qquad (3.30)$$

(neglecting the force potential $\chi = 0$). Thus, we must have

$$\frac{\mathbf{v}^2}{2} + h + H_* = f(t), \qquad (12.123)$$

where $f(t)$ is an arbitrary function of t.

This is equivalent to (12.84). In fact, Eq. (12.84) is

$$0 = \frac{1}{2}v^2 - h - \mathrm{D}_t\phi - s\mathrm{D}_t\psi - \mathbf{v} \cdot \mathbf{b}$$

$$= \frac{1}{2}v^2 - h - \partial_t\phi - s\partial_t\psi - \mathbf{v} \cdot (\nabla\phi + s\nabla\psi + \mathbf{b})$$

$$= -\frac{1}{2}v^2 - h - \partial_t\phi - s\partial_t\psi - \mathbf{v} \cdot (\mathbf{b} - \mathbf{w}) \qquad (12.124)$$

by using the definition (12.89) of $\mathbf{v}$. Comparing (12.123) and (12.124), it is found that both expressions are equivalent if $H_* - f(t) = \partial_t\phi + s\partial_t\psi + \mathbf{v} \cdot (\mathbf{b} - \mathbf{w})$.

12.8.2. *Additional note on the rotational symmetry*

A brief but important remark must be given to the Lagrangian L_A of (12.69) associated with the rotation symmetry: $L_A = -\int_M \rho\langle\mathbf{b}, \mathbf{v}\rangle\mathrm{d}^3\mathbf{x}$. In Appendix F.2, the same Lagrangian is represented as

$$L_A = -\int_M \langle \mathcal{L}_X A^1, \boldsymbol{\omega}\rangle\mathrm{d}^3\mathbf{x} = -\int_M \langle \nabla \times (\mathcal{L}_X A^1), \mathbf{v}\rangle\,\mathrm{d}^3\mathbf{x}. \quad (12.125)$$

where A^1 is a gauge potential 1-form (defined there), and X is a tangent vector defined by $\partial_t + v^k\partial_k$, which is equivalent to the operator of convective derivative D_t as far as its form is concerned.[16] Thus, the vector field $\mathbf{b}$ is derived from a vector gauge potential A^1 by the relation $\rho\mathbf{b} = \nabla \times (\mathcal{L}_X A^1)$, where $\mathcal{L}_X A^1$ denotes the Lie derivative of

---

[16]For a scalar function f (0-form), the Lie derivative of f is given by $\mathcal{L}_X f = Xf = \mathrm{D}_t f$. However, if applied to other forms or vectors F (say), $\mathcal{L}_X F$ is different from $\mathrm{D}_t F = \partial_t F + v^k\partial_k F$, where D_t is a simple differential operator D_t. In this regard, concerning the terms $\mathrm{D}_t\phi$ and $\mathrm{D}_t\psi$ of (12.74), it is consistent and more appropriate to write $\mathcal{L}_X\phi$ and $\mathcal{L}_X\psi$, which are equivalent to $\mathrm{D}_t\phi$ and $\mathrm{D}_t\psi$ since ϕ and ψ are scalar functions.

the form A^1 along the flow generated by X. In the Appendix, differential calculus are carried out in order to elucidate the significance of the gauge potential A^1 associated with the rotation symmetry.

It is remarkable that the action principle for arbitrary variations of A^1 results in the vorticity equation (Appendix F.2):

$$\partial_t \boldsymbol{\omega} + \mathrm{curl}(\boldsymbol{\omega} \times \mathbf{v}) = 0.$$

Thus, it is found that *the vorticity equation is an equation of a gauge field.*

12.9. Problem

Problem 12.1 Clebsch solution

Verify that the Euler's equation of motion for a barotropic fluid of $p = p(\rho)$,

$$\partial_t \mathbf{v} + \boldsymbol{\omega} \times \mathbf{v} = -\nabla \left(\frac{1}{2} v^2 + \int^{\mathbf{x}} \frac{dp}{\rho} \right), \qquad (12.126)$$

can be solved in general by

$$\mathbf{v} = \nabla\phi + s\nabla\psi, \qquad (12.127)$$

$$\frac{1}{2} v^2 + h + \partial_t \phi + s\partial_t \psi = 0, \quad h = \int \frac{dp}{\rho}, \qquad (12.128)$$

$$D_t s = 0, \quad D_t \psi = 0, \quad D_t = \partial_t + \mathbf{v} \cdot \nabla, \qquad (12.129)$$

where s, ϕ, ψ are scalar functions of $\mathbf{x}$ [Lamb32, Sec. 167].

In addition, show that the helicity H vanishes where H is defined by an integral of $\mathbf{v} \cdot \boldsymbol{\omega}$ over unbounded space, when $\boldsymbol{\omega}$ vanishes at infinity.

Appendix A

Vector analysis

A.1. Definitions

We use the notation of vectors as

$$\mathbf{A} = (A_1, A_2, A_3), \quad \mathbf{B} = (B_1, B_2, B_3), \quad \mathbf{C} = (C_1, C_2, C_3),$$

or simply as $\mathbf{A} = (A_i)$, $\mathbf{B} = (B_i)$, $\mathbf{C} = (C_i)$ in the cartesian frame of reference. Kronecker's delta δ_{ij} is a second-order tensor, and ε_{ijk} is a third-order skew-symmetric tensor, which are defined by

$$\delta_{ij} = \begin{cases} 1 \ (i = j) \\ 0 \ (i \neq j) \end{cases} \tag{A.1}$$

$$\varepsilon_{ijk} = \begin{cases} 1, & (1,2,3) \rightarrow (i,j,k) : \text{even permutation} \\ -1, & (1,2,3) \rightarrow (i,j,k) : \text{odd permutation} \\ 0 & \text{otherwise} : \text{(with repeated indices)}. \end{cases} \tag{A.2}$$

It is useful to introduce the following vectorial differential operators,

$$\nabla = (\partial_x, \partial_y, \partial_z), \tag{A.3}$$

$$\operatorname{grad} f := (\partial_x, \partial_y, \partial_z) f, \tag{A.4}$$

where $f(x, y, z)$ is a scalar function.

A.2. Scalar product

A *scalar product* of two vectors $\mathbf{A}$ and $\mathbf{B}$ is defined by

$$\mathbf{A} \cdot \mathbf{B} = A_k B_k = A_1 B_1 + A_2 B_2 + A_3 B_3 = |\mathbf{A}|\,|\mathbf{B}| \cos\theta, \qquad \text{(A.5)}$$

where $|\mathbf{A}|$ denotes the magnitude of $\mathbf{A}$, i.e. $|\mathbf{A}| = \sqrt{A_1^2 + A_2^2 + A_3^2}$, and θ is the angle between $\mathbf{A}$ and $\mathbf{B}$.

If $\mathbf{A}$ is the differential operator ∇, we have the definition of *divergence* div:

$$\operatorname{div}\mathbf{B} := \nabla \cdot \mathbf{B} = \partial_k B_k = \partial_1 B_1 + \partial_2 B_2 + \partial_3 B_3. \qquad \text{(A.6)}$$

If $\mathbf{B}$ is the differential operator ∇, then we have the definition of $\mathbf{A} \cdot \operatorname{grad}$:

$$\mathbf{A} \cdot \operatorname{grad} := \mathbf{A} \cdot \nabla = A_k \partial_k = A_1 \partial_1 + A_2 \partial_2 + A_3 \partial_3. \qquad \text{(A.7)}$$

A.3. Vector product

A *vector product* of $\mathbf{A}$ and $\mathbf{B}$ is defined by

$$\mathbf{A} \times \mathbf{B} = (A_2 B_3 - A_3 B_2, A_3 B_1 - A_1 B_3, A_1 B_2 - A_2 B_1) \qquad \text{(A.8)}$$

$$= (A_2 B_3 - A_3 B_2)\mathbf{i} + (A_3 B_1 - A_1 B_3)\mathbf{j}$$
$$+ (A_1 B_2 - A_2 B_1)\mathbf{k} \qquad \text{(A.9)}$$

$$= \begin{vmatrix} \mathbf{i} & \mathbf{j} & \mathbf{k} \\ A_1 & A_2 & A_3 \\ B_1 & B_2 & B_3 \end{vmatrix}, \qquad \text{(A.10)}$$

where $\mathbf{i}$, $\mathbf{j}$, $\mathbf{k}$ are unit vectors in the directions of x, y, z axes respectively, i.e. $|\mathbf{i}| = |\mathbf{j}| = |\mathbf{k}| = 1$, $\mathbf{i} \cdot \mathbf{j} = \mathbf{j} \cdot \mathbf{k} = \mathbf{k} \cdot \mathbf{i} = 0$. Using the angle θ between $\mathbf{A}$ and $\mathbf{B}$, the magnitude of $\mathbf{A} \times \mathbf{B}$ is given by

$$|\mathbf{A} \times \mathbf{B}| = |\mathbf{A}||\mathbf{B}| \sin\theta. \qquad \text{(A.11)}$$

$\mathbf{A} \times \mathbf{B}$ is perpendicular to the plane spanned by $\mathbf{A}$ and $\mathbf{B}$ and directed in right-handed way when rotated from $\mathbf{A}$ to $\mathbf{B}$. Using the third-order skew-symmetric tensor ε_{ijk}, the vector product is written compactly as

$$(\mathbf{A} \times \mathbf{B})_i = \varepsilon_{ijk} A_j B_k, \quad (i = 1, 2, 3). \tag{A.12}$$

In fact, we obtain $(\mathbf{A} \times \mathbf{B})_1 = A_2 B_3 - A_3 B_2$, etc.

If $\mathbf{A}$ is the differential operator ∇, we have the definition of *rotation* curl:

$$\operatorname{curl} \mathbf{B} := \nabla \times \mathbf{B} = \varepsilon_{ijk} \partial_j B_k \tag{A.13}$$

$$= (\partial_2 B_3 - \partial_3 B_2, \partial_3 B_1 - \partial_1 B_3, \partial_1 B_2 - \partial_2 B_1) \tag{A.14}$$

$$= \begin{vmatrix} \mathbf{i} & \mathbf{j} & \mathbf{k} \\ \partial_1 & \partial_2 & \partial_3 \\ B_1 & B_2 & B_3 \end{vmatrix}. \tag{A.15}$$

A.4. Triple products

There are two kinds of triple product: scalar and vector. Using three vectors $\mathbf{A}$, $\mathbf{B}$, $\mathbf{C}$, the *scalar triple product* is defined by a scalar product of $\mathbf{A}$ and $\mathbf{B} \times \mathbf{C}$, which is represented as

$$\mathbf{A} \cdot (\mathbf{B} \times \mathbf{C}) = A_1(B_2 C_3 - B_3 C_2) + A_2(B_3 C_1 - B_1 C_3)$$
$$+ A_3(B_1 C_2 - B_2 C_1) \tag{A.16}$$

$$= \begin{vmatrix} A_1 & A_2 & A_3 \\ B_1 & B_2 & B_3 \\ C_1 & C_2 & C_3 \end{vmatrix}. \tag{A.17}$$

As is evident from the definition of the determinant, we have the equality, $\mathbf{A} \cdot (\mathbf{B} \times \mathbf{C}) = \mathbf{B} \cdot (\mathbf{C} \times \mathbf{A}) = \mathbf{C} \cdot (\mathbf{A} \times \mathbf{B})$. This is equal to the volume of a hexahedron composed of three vectors $\mathbf{A}$, $\mathbf{B}$, $\mathbf{C}$.

A *vector triple product* is defined by

$$[\mathbf{A} \times (\mathbf{B} \times \mathbf{C})]_i = A_k B_i C_k - A_k B_k C_i, \tag{A.18}$$

which is an ith component, and summation with respect to k is understood. If $\mathbf{A}$, $\mathbf{B}$, $\mathbf{C}$ are ordinary vectors, then rearranging the order, we have

$$\mathbf{A} \times (\mathbf{B} \times \mathbf{C}) = (\mathbf{A} \cdot \mathbf{C})\mathbf{B} - (\mathbf{A} \cdot \mathbf{B})\mathbf{C}. \qquad \text{(A.19)}$$

A particularly useful expression is obtained when we set $\mathbf{A} = \mathbf{C} = \mathbf{v}$ and $\mathbf{B} = \nabla = (\partial_1, \partial_2, \partial_3)$ in (A.18), which is

$$[\mathbf{v} \times (\nabla \times \mathbf{v})]_i = v_k \partial_i v_k - v_k \partial_k v_i = \partial_i \left(\frac{1}{2} v^2 \right) - (\mathbf{v} \cdot \nabla) v_i, \quad \text{(A.20)}$$

where $v^2 = v_1^2 + v_2^2 + v_3^2$. Thus we obtain the following vector identity:

$$\mathbf{v} \times (\nabla \times \mathbf{v}) = \nabla \left(\frac{1}{2} v^2 \right) - (\mathbf{v} \cdot \nabla)\mathbf{v}. \qquad \text{(A.21)}$$

From (A.18), we can derive an useful identity between the second-order tensor δ_{ij} of (A.1) and the third-order tensor ε_{ijk} of (A.2). Applying the formula (A.12) two times, the left-hand side of (A.18) is

$$\varepsilon_{ijk} A_j (\varepsilon_{klm} B_l C_m) = \varepsilon_{kij} \varepsilon_{klm} A_j B_l C_m,$$

where the equality $\varepsilon_{ijk} = \varepsilon_{kij}$ is used. The right-hand side is written as

$$\delta_{jm} \delta_{il} A_j B_l C_m - \delta_{jl} \delta_{im} A_j B_l C_m.$$

Thus, equating the coefficients of $A_j B_l C_m$, we obtain the identity,

$$\varepsilon_{kij} \varepsilon_{klm} = \delta_{il} \delta_{jm} - \delta_{im} \delta_{jl}, \qquad \text{(A.22)}$$

where summation with respect to $k = 1, 2, 3$ is understood on the left-hand side.

When $\mathbf{A}$ is a differential operator ∇ in Eq. (A.19), ∇ acts on $\mathbf{C}$ as well as on $\mathbf{B}$. Thus, Eq. (A.19) is replaced by

$$\nabla \times (\mathbf{B} \times \mathbf{C}) = (\nabla \cdot \mathbf{C})\mathbf{B} + (\mathbf{C} \cdot \nabla)\mathbf{B} - (\nabla \cdot \mathbf{B})\mathbf{C} - (\mathbf{B} \cdot \nabla)\mathbf{C}. \quad \text{(A.23)}$$

This is rewritten as

$$\text{curl}(\mathbf{B} \times \mathbf{C}) = (\text{div}\,\mathbf{C})\mathbf{B} + (\mathbf{C} \cdot \text{grad})\mathbf{B} - (\text{div}\,\mathbf{B})\mathbf{C} - (\mathbf{B} \cdot \text{grad})\mathbf{C}.$$
$$\text{(A.24)}$$

A.5. Differential operators

Identities:

$$\nabla \times \nabla f = \text{curl}(\text{grad } f) \equiv 0, \tag{A.25}$$

$$\nabla \cdot (\nabla \times \mathbf{v}) = \text{div}(\text{curl } \mathbf{v}) \equiv 0, \tag{A.26}$$

$$\nabla \times (\nabla \times \mathbf{v}) = \nabla(\nabla \cdot \mathbf{v}) - \nabla^2 \mathbf{v} = \text{grad}(\text{div } \mathbf{v}) - \nabla^2 \mathbf{v}. \tag{A.27}$$

Nabla operator:

$$\nabla = (\partial_x, \partial_y, \partial_z) \quad \text{[cartesian coordinates } (x, y, z)] \tag{A.28}$$

$$= \left(\frac{\partial}{\partial x}, \frac{\partial}{\partial \sigma}, \frac{1}{\sigma} \frac{\partial}{\partial \phi} \right) \quad \text{[cylindrical coordinates } (x, \sigma, \phi)] \tag{A.29}$$

$$= \left(\frac{\partial}{\partial r}, \frac{1}{r} \frac{\partial}{\partial \theta}, \frac{1}{r \sin \theta} \frac{\partial}{\partial \varphi} \right) \quad \text{[spherical coordinates } (r, \theta, \varphi)] \tag{A.30}$$

Laplacian operator:

$$\Delta = \frac{\partial^2}{\partial x^2} + \frac{\partial^2}{\partial y^2} + \frac{\partial^2}{\partial z^2} = \nabla^2$$

$$\text{[cartesian coordinates } (x, y, z)] \tag{A.31}$$

$$= \frac{\partial^2}{\partial x^2} + \frac{1}{\sigma} \frac{\partial}{\partial \sigma} \left(\sigma \frac{\partial}{\partial \sigma} \right) + \frac{1}{\sigma^2} \frac{\partial^2}{\partial \phi^2}$$

$$\text{[cylindrical coordinates } (x, \sigma, \phi)] \tag{A.32}$$

$$= \frac{1}{r^2} \frac{\partial}{\partial r} \left(r^2 \frac{\partial}{\partial r} \right) + \frac{1}{r^2 \sin \theta} \frac{\partial}{\partial \theta} \left(\sin \theta \frac{\partial}{\partial \theta} \right) + \frac{1}{r^2 \sin^2 \theta} \frac{\partial^2}{\partial \varphi^2}$$

$$\text{[spherical coordinates } (r, \theta, \varphi)] \tag{A.33}$$

A.6. Integration theorems

(i) *Line integral:*

$$\int_{\mathbf{x}_1 \, (\gamma)}^{\mathbf{x}_2} (\text{grad } f) \cdot d\mathbf{l} = f(\mathbf{x}_2) - f(\mathbf{x}_1), \tag{A.34}$$

where $f(\mathbf{x})$ is a scalar function, and $d\mathbf{l}$ a line-element of a curve γ from $\mathbf{x}_1$ to $\mathbf{x}_2$.

(ii) **Stokes's theorem** for a vector field $\mathbf{B}(\mathbf{x})$:

$$\int_A (\text{curl}\,\mathbf{B}) \cdot \mathbf{n}\,dA = \oint_C \mathbf{B} \cdot d\mathbf{l}, \qquad (A.35)$$

where A is an open surface bounded by a closed curve C of a line-element $d\mathbf{l}$, and $\mathbf{n}$ is a unit normal to the surface element dA.

(iii) **Gauss's theorem** for a vector field $\mathbf{A}(\mathbf{x})$:

$$\int_V \text{div}\,\mathbf{A}\,dV = \oint_S \mathbf{A} \cdot \mathbf{n}\,dS, \qquad (A.36)$$

where S is a closed surface bounding a volume V, and $\mathbf{n}$ is a unit outward normal to S.

A.7. δ function

The delta function $\delta(x)$ is defined as follows. For an interval D of a variable x including $x = 0$, suppose that the following integral formula is valid always,

$$\int_D f(x)\,\delta(x)\,dx = f(0), \qquad (A.37)$$

for any continuous function $f(x)$ of a real variable x, then the function $\delta(x)$ is called the **delta function**, or Dirac's delta function.

The delta function is defined variously as follows:

$$\delta(x) = \lim_{a \to 0} \frac{1}{\sqrt{\pi}a}\, e^{-x^2/a^2} \qquad (A.38)$$

$$= \lim_{a \to 0} \frac{a}{\pi(x^2 + a^2)} \qquad (A.39)$$

$$= \lim_{a \to 0} \frac{1}{2a}\, \text{sech}^2 \frac{x}{a} \qquad (A.40)$$

$$= \lim_{k \to \infty} \frac{\sin kx}{\pi x} \qquad (A.41)$$

$$= \frac{1}{2\pi} \int_{-\infty}^{\infty} e^{ikx}\,dk. \qquad (A.42)$$

The delta functions in 2D-space and 3D-space are defined as

$$2D : \delta_2(\mathbf{r}) = \delta(x)\,\delta(y), \qquad \mathbf{r} = (x, y) \qquad (A.43)$$
$$3D : \delta_3(\mathbf{r}) = \delta(x)\,\delta(y)\,\delta(z), \quad \mathbf{r} = (x, y, z). \qquad (A.44)$$

The nth order derivative of the delta function $\delta^{(n)}(x)$ is defined by the following integral relation:

$$\int_D f(x)\delta^{(n)}(x)\mathrm{d}x = (-1)^n f^{(n)}(0).$$

Appendix B

Velocity potential, stream function

B.1. Velocity potential

If a vector field $\mathbf{v}(\mathbf{x})$ is *irrotational* at all points $\mathbf{x} = (x, y, z)$ of an open *simply-connected* domain D under consideration, $\mathbf{v}$ satisfies the equation, $\mathrm{curl}\,\mathbf{v} = 0$. Then there exists a scalar function $\Phi(\mathbf{x})$ for $\mathbf{x} \in D$ such that $\mathbf{v}(\mathbf{x})$ is represented as

$$\mathbf{v} = \mathrm{grad}\,\Phi = \nabla\Phi = (\Phi_x, \Phi_y, \Phi_z). \qquad (\text{B.1})$$

In fact, if $\mathrm{curl}\,\mathbf{v} = 0$, we have from Stokes's theorem (A.35) of Appendix A,

$$\oint_C \mathbf{v} \cdot \mathrm{d}\mathbf{l} = 0. \qquad (\text{B.2})$$

Taking two points O and P on the oriented closed curve C (Fig. B.1), the curve C is divided into two parts C_1 (O to P, say) and $C - C_1$ (P to O). Reversing the orientation of $C - C_1$ and writing its reversed curve as C_2, the above equation becomes

$$\int_{C_1} \mathbf{v} \cdot \mathrm{d}\mathbf{l} - \int_{C_2} \mathbf{v} \cdot \mathrm{d}\mathbf{l} = 0.$$

(The minus sign of the second term corresponds to interchanging both ends $\mathbf{x}_1$ and $\mathbf{x}_2$ in the line integral (A.34).) This can be written as

$$\int_{O\,(C_1)}^{P} \mathbf{v} \cdot \mathrm{d}\mathbf{l} = \int_{O\,(C_2)}^{P} \mathbf{v} \cdot \mathrm{d}\mathbf{l} := \Phi(O, P). \qquad (\text{B.3})$$

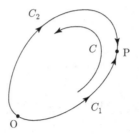

Fig. B.1.   Contour curves C_1 and C_2.

The closed curve C can be chosen arbitrarily within the domain of irrotationality. Hence in the above expression, the integral path from O to P can be taken arbitrarily in D, and the integral is regarded as a function of the end points O and P only, which is defined by a function $\Phi(O, P)$ in (B.3). If the point O is fixed as the coordinate origin, Φ is a function of P only. Thus, writing P as $\mathbf{x} = (x, y, z)$, we obtain a function $\Phi(\mathbf{x})$ defined by

$$\Phi(\mathbf{x}) = \int_O^{\mathbf{x}} \mathbf{v} \cdot d\mathbf{l}. \tag{B.4}$$

Taking its derivative, we have

$$d\Phi = \mathbf{v} \cdot d\mathbf{x} = u \, dx + v \, dy + w \, dz,$$

where $\mathbf{v} = (u, v, w)$ and $d\mathbf{x} = (dx, dy, dz)$. On the other hand, we have $d\Phi = \Phi_x \, dx + \Phi_y \, dy + \Phi_z \, dz$. Thus, we find that

$$u = \Phi_x, \quad v = \Phi_y, \quad w = \Phi_z. \tag{B.5}$$

This is equivalent to the expression (B.1).

     The function $\Phi(\mathbf{x})$ defined by (B.1) for velocity $\mathbf{v}$ is termed the **velocity potential**.

B.2.   Stream function (2D)

Suppose that we have a two-dimensional flow of an incompressible fluid in the cartesian (x, y)-plane. Writing the velocity as $\mathbf{v} = (u, v)$, the condition of incompressibility is given by

$$\mathrm{div}\,\mathbf{v} = \partial_x u + \partial_y v = 0. \tag{B.6}$$

A stream function $\Psi(x,y)$ is defined by

$$u = \Psi_y = \partial_y \Psi, \quad v = -\Psi_x = -\partial_x \Psi. \tag{B.7}$$

It is seen that the above continuity equation (B.6) is satisfied identically by this definition of velocities.

In the two-dimensional (x,y) space, the Gauss's theorem for a vector field (A_x, A_y) is described by (instead of (A.36) for 3D)

$$\int_D (\partial_x A_x + \partial_y A_y)\,\mathrm{d}x\,\mathrm{d}y = \oint_C (-A_y\,\mathrm{d}x + A_x\,\mathrm{d}y), \tag{B.8}$$

where C is an anti-clockwise closed curve bounding a two-dimensional domain D.

When (A_x, A_y) is replaced by (u, v), the left-hand side vanishes owing to (B.6). Hence, we have

$$0 = \oint_C (-v\,\mathrm{d}x + u\,\mathrm{d}y) = \oint_C \mathbf{w} \cdot \mathrm{d}\mathbf{l}, \quad \mathbf{w} := (-v, u), \tag{B.9}$$

where a new vector field $\mathbf{w}$ is defined.

Obviously, there is analogy between the above (B.9) and Eq. (B.2) in the previous section. By the same reasoning as before, it can be shown that there exists a function $\Psi(x,y)$ defined by

$$\Psi(x,y) = \int_O^{\mathbf{X}} \mathbf{w} \cdot \mathrm{d}\mathbf{l} = \int_O^{\mathbf{X}} (-v\,\mathrm{d}x + u\,\mathrm{d}y). \tag{B.10}$$

Thus, we find that $u = \partial_y \Psi$ and $v = -\partial_x \Psi$, which are equivalent to (B.7).

The contours defined by $\Psi(x,y) = \mathrm{const.}$ in the (x,y)-plane coincide with stream-lines, because we have

$$\mathrm{d}\Psi = \partial_x \Psi\,\mathrm{d}x + \partial_y \Psi\,\mathrm{d}y = -v\,\mathrm{d}x + u\,\mathrm{d}y = 0,$$

which reduces to the equation of stream-lines: $\mathrm{d}x/u = \mathrm{d}y/v$.

The function $\Psi(x,y)$ defined by (B.10) or (B.7) is termed the **stream function** for two-dimensional flows.

B.3.   Stokes' stream function (axisymmetric)

An axisymmetric flow without swirl is defined by the velocity,

$$\mathbf{v}(r, \theta) = (v_r, v_\theta, 0),$$

in the spherical polar coordinates (r, θ, ϕ). Stokes' stream function $\Psi(r, \theta)$ can be introduced for such axisymmetric flows of an incompressible fluid by

$$v_r = \frac{1}{r^2 \sin\theta} \frac{\partial \Psi}{\partial \theta}, \qquad v_\theta = -\frac{1}{r \sin\theta} \frac{\partial \Psi}{\partial r}. \tag{B.11}$$

Using the formula (D.20) in Appendix D.2, it is readily seen that the above expressions satisfy the continuity equation $\operatorname{div}\mathbf{v} = 0$ identically.

The contours defined by $\Psi(r, \theta) = \text{const.}$ in the (r, θ) plane coincide with stream-lines, because we have

$$0 = \mathrm{d}\Psi = \partial_r \Psi \,\mathrm{d}r + \partial_\theta \Psi \,\mathrm{d}\theta = r \sin\theta \,(-v_\theta \mathrm{d}r + v_r \, r\mathrm{d}\theta),$$

which reduces to the equation of stream-lines: $\mathrm{d}r/v_r = r\,\mathrm{d}\theta/v_\theta$.

Appendix C

Ideal fluid and ideal gas

It is a common approach in physics to try to simplify representation of material properties as much as possible. It is called *modeling*. For example, *an ideal fluid* means a fluid with vanishing diffusion coefficients of both viscosity and thermal conduction. This is useful to represent fluid motions in which transfer phenomena due to molecular motion can be neglected. This enables separation of macroscopic motions from microscopic irreversible processes due to molecular thermal motion. In an ideal fluid, the motion is adiabatic and the stress tensor is isotropic, i.e. stress force being normal to any surface element. As a result, many useful expressions can be given concisely (Chapters 5–8).

An ideal gas, or equivalently *a perfect gas*, is one which is governed by the equation of state (for a unit mass):

$$pV = \frac{R}{\mu_m} T \quad \text{or} \quad p = \frac{R}{\mu_m} \rho T, \tag{C.1}$$

where $R = 8.314 \times 10^7$ [erg/deg·mol] is the gas constant, and p, ρ, T, V are respectively the pressure, density, temperature and specific volume $(V = 1/\rho)$ of the gas, and μ_m is the molecular mass. In an ideal gas, the interaction force between molecules is so weak to be negligible.

For a gram-molecule of an ideal gas,

$$pV = RT.$$

The sound speed c in an ideal gas is given by

$$c^2 = \gamma \frac{p}{\rho} = \gamma \frac{RT}{\mu_m},$$ (C.2)

where $\gamma = C_p/C_v$ is the specific heat ratio, called the *adiabatic index*. C_p and C_v are specific heats (i.e. per unit mass) with constant pressure and volume, respectively. For *monoatomic gases*, $\gamma = 5/3$, and *diatomic gases*, $\gamma = 7/5$ at ordinary temperatures. However, for a *polytropic gas*, γ is a certain constant n (see the last paragraph).

The gas constant is related to the difference of C_p and C_v by the relation,

$$\frac{R}{\mu_m} = C_p - C_v = (\gamma - 1)\, C_v.$$ (C.3)

Adiabatic transition from (p_0, ρ_0) to (p, ρ) is connected by the relation,

$$\frac{p}{p_0} = \left(\frac{\rho}{\rho_0}\right)^\gamma.$$ (C.4)

The internal energy e and enthalpy h per unit mass of an ideal gas are given by

$$e = C_v T = \frac{pV}{\gamma - 1} = \frac{c^2}{\gamma(\gamma - 1)},$$ (C.5)

$$h = e + \frac{1}{\rho}p = C_p T = \gamma \frac{pV}{\gamma - 1} = \frac{c^2}{(\gamma - 1)}.$$ (C.6)

The name *polytropic* is used after *polytropic processes*. This is a process in which the pressure varies in proportion to some power of density, i.e. $p \propto \rho^n$. For a gas with constant specific heat, such a process may be either isothermal ($p \propto \rho$), or adiabatic with $p \propto \rho^\gamma$ with γ as the specific heat ratio.

Appendix D

Curvilinear reference frames: Differential operators

D.1. Frenet–Serret formula for a space curve

Let a space curve C be defined by $\mathbf{x}(s)$ in $\mathbb{R}^3$ with s as the arc-length parameter. Then, the unit tangent vector is given by

$$\mathbf{t} = \frac{\mathrm{d}\mathbf{x}}{\mathrm{d}s}, \quad \|\mathbf{t}\| = \langle \mathbf{t},\, \mathbf{t} \rangle^{1/2} = 1.$$

Differentiating the equation $\langle \mathbf{t},\, \mathbf{t} \rangle = 1$ with respect to s, we have $\langle \mathbf{t},\, (\mathrm{d}\mathbf{t}/\mathrm{d}s) \rangle = 0$. Hence, $\mathrm{d}\mathbf{t}/\mathrm{d}s$ is orthogonal to $\mathbf{t}$ and also to curve C, and the vector $\mathrm{d}\mathbf{t}/\mathrm{d}s$ defines a unique direction (if $\mathrm{d}\mathbf{t}/\mathrm{d}s \neq 0$) in a plane normal to C at $\mathbf{x}$ called the direction of principal normal, represented by

$$\frac{\mathrm{d}\mathbf{t}}{\mathrm{d}s} = \frac{\mathrm{d}^2\mathbf{x}}{\mathrm{d}s^2} = k(s)\,\mathbf{n}(s), \quad \|\mathbf{n}\| = \langle \mathbf{n},\, \mathbf{n} \rangle^{1/2} = 1, \qquad \text{(D.1)}$$

where $\mathbf{n}$ is the vector of unit *principal normal* and $k(s)$ is the curvature. Then, we can define the unit *binormal* vector $\mathbf{b}$ by the equation, $\mathbf{b} = \mathbf{t} \times \mathbf{n}$, which is normal to the osculating plane spanned by $\mathbf{t}$ and $\mathbf{n}$. Thus, we have a local right-handed orthonormal frame $(\mathbf{t}, \mathbf{n}, \mathbf{b})$ at each point $\mathbf{x}$.

Analogously to $\mathrm{d}\mathbf{t}/\mathrm{d}s \perp \mathbf{t}$, the vector $\mathrm{d}\mathbf{n}/\mathrm{d}s$ is orthogonal to $\mathbf{n}$. Hence, it may be written as $\mathbf{n}' = \alpha\mathbf{t} - \tau\mathbf{b}$, where the prime denotes $\mathrm{d}/\mathrm{d}s$, and $\alpha, \tau \in \mathbb{R}$. Differentiating $\mathbf{b}$, we obtain

$$\frac{\mathrm{d}\mathbf{b}}{\mathrm{d}s} = \mathbf{t} \times \mathbf{n}' + \mathbf{t}' \times \mathbf{n} = -\tau\,\mathbf{t} \times \mathbf{b} = \tau(s)\,\mathbf{n}(s), \qquad \text{(D.2)}$$

where (D.1) is used, and τ is the *torsion* of curve C at s. Since $\mathbf{n} = \mathbf{b} \times \mathbf{t}$, we have

$$\frac{d\mathbf{n}}{ds} = \mathbf{b} \times \mathbf{t}' + \mathbf{b}' \times \mathbf{t} = -k\mathbf{t} - \tau\mathbf{b}. \tag{D.3}$$

where (D.1) and (D.2) are used.

Collecting the three equations (D.1)–(D.3), we obtain a set of differential equations for $(\mathbf{t}(s), \mathbf{n}(s), \mathbf{b}(s))$:

$$\frac{d}{ds} \begin{pmatrix} \mathbf{t} \\ \mathbf{n} \\ \mathbf{b} \end{pmatrix} = \begin{pmatrix} 0 & k & 0 \\ -k & 0 & -\tau \\ 0 & \tau & 0 \end{pmatrix} \begin{pmatrix} \mathbf{t} \\ \mathbf{n} \\ \mathbf{b} \end{pmatrix}. \tag{D.4}$$

This is called the *Frenet–Serret equations* for a space curve.

D.2. Cylindrical coordinates

In the reference frame of cylindrical coordinates (r, θ, z) with z being the cylindrical axis, and (r, θ) the polar coordinate in the (x, y) plane (θ the azimuthal angle with respect to the x axis), we define three unit vectors $\mathbf{e}_r$, $\mathbf{e}_\theta$, $\mathbf{e}_z$ in each direction of coordinate axes. With respect to the cartesian reference frame (x, y, z), the unit vectors are represented as

$$\mathbf{e}_r = (\cos\theta, \sin\theta, 0), \quad \mathbf{e}_\theta = (-\sin\theta, \cos\theta, 0), \quad \mathbf{e}_z = (0, 0, 1), \tag{D.5}$$

which are characterized by the orthonormality:

$$\begin{array}{ccc} \mathbf{e}_r \cdot \mathbf{e}_\theta = 0, & \mathbf{e}_\theta \cdot \mathbf{e}_z = 0, & \mathbf{e}_\theta \cdot \mathbf{e}_z = 0, \\ |\mathbf{e}_r| = 1, & |\mathbf{e}_\theta| = 1, & |\mathbf{e}_z| = 1. \end{array} \tag{D.6}$$

The velocity vector $\mathbf{v}$ and ∇ operator are written as

$$\mathbf{v} = \mathbf{e}_r v_r + \mathbf{e}_\theta v_\theta + \mathbf{e}_z v_z, \quad \nabla = \mathbf{e}_r \partial_r + \mathbf{e}_\theta \frac{1}{r}\partial_\theta + \mathbf{e}_z \partial_z. \tag{D.7}$$

Derivatives of the three unit vectors with respect to θ are given by

$$\partial_\theta \mathbf{e}_r = \mathbf{e}_\theta, \quad \partial_\theta \mathbf{e}_\theta = -\mathbf{e}_r, \quad \partial_\theta \mathbf{e}_z = 0. \tag{D.8}$$

The other derivatives with respect to r and z are all zero.

Using (D.7) and (D.6), div **v** is given by

$$\nabla \cdot \mathbf{v} = \left(\mathbf{e}_r \partial_r + \mathbf{e}_\theta \frac{1}{r} \partial_\theta + \mathbf{e}_z \partial_z \right) \cdot \left(\mathbf{e}_r v_r + \mathbf{e}_\theta v_\theta + \mathbf{e}_z v_z \right)$$

$$= \partial_r v_r + \frac{1}{r} v_r + \frac{1}{r} \partial_\theta v_\theta + \partial_z v_z. \tag{D.9}$$

Note that $\partial_r v_r + (1/r) v_r = (1/r) \partial_r (r v_r)$.

Similarly, in view of $\nabla^2 f = \operatorname{div}(\nabla f)$ for a scalar function f, we obtain

$$\nabla^2 f = \left(\mathbf{e}_r \partial_r + \mathbf{e}_\theta \frac{1}{r} \partial_\theta + \mathbf{e}_z \partial_z \right) \cdot \left(\mathbf{e}_r \partial_r f + \mathbf{e}_\theta \frac{1}{r} \partial_\theta f + \mathbf{e}_z \partial_z f \right)$$

$$= \left(\partial_r^2 + \frac{1}{r} \partial_r + \frac{1}{r^2} \partial_\theta^2 + \partial_z^2 \right) f. \tag{D.10}$$

From this we can define the ∇^2 operator in the cylindrical system by

$$\nabla^2 \equiv \partial_r^2 + \frac{1}{r} \partial_r + \frac{1}{r^2} \partial_\theta^2 + \partial_z^2 = \frac{1}{r} \partial_r (r \partial_r) + \frac{1}{r^2} \partial_\theta^2 + \partial_z^2. \tag{D.11}$$

Applying the ∇^2 operator on a scalar function f gives immediately $\nabla^2 f$ of (D.10). However, applying the same ∇^2 operator on a vector field **v** must be considered carefully because the derivatives of $\mathbf{e}_r$, $\mathbf{e}_\theta$ are not zero. In fact, we obtain the following:

$$\nabla^2 \mathbf{v} = \nabla^2 \left(\mathbf{e}_r v_r + \mathbf{e}_\theta v_\theta + \mathbf{e}_z v_z \right) = \mathbf{e}_r \left(\nabla^2 v_r - \frac{v_r}{r^2} - \frac{2}{r^2} \partial_\theta v_\theta \right)$$

$$+ \mathbf{e}_\theta \left(\nabla^2 v_\theta + \frac{2}{r^2} \partial_\theta v_r - \frac{v_\theta}{r^2} \right) + \mathbf{e}_z \left(\nabla^2 v_z \right), \tag{D.12}$$

The scalar product of **v** and ∇ is defined by

$$(\mathbf{v} \cdot \nabla) \equiv v_r \partial_r + v_\theta \frac{1}{r} \partial_\theta + v_z \partial_z. \tag{D.13}$$

Then the convection term $(\mathbf{v} \cdot \nabla) \mathbf{v}$ is

$$(\mathbf{v} \cdot \nabla) \mathbf{v} = (\mathbf{v} \cdot \nabla) \cdot \left(\mathbf{e}_r v_r + \mathbf{e}_\theta v_\theta + \mathbf{e}_z v_z \right)$$

$$= \mathbf{e}_r \left((\mathbf{v} \cdot \nabla) v_r - \frac{1}{r} v_\theta^2 \right) + \mathbf{e}_\theta \left((\mathbf{v} \cdot \nabla) v_\theta + \frac{1}{r} v_r v_\theta \right)$$

$$+ \mathbf{e}_z (\mathbf{v} \cdot \nabla) v_z. \tag{D.14}$$

Operating curl, we have

$$\nabla \times \mathbf{v} = \mathbf{e}_z \left(\frac{1}{r} \frac{\partial(r v_\theta)}{\partial r} - \frac{1}{r} \frac{\partial v_r}{\partial \theta} \right) + \mathbf{e}_r \left(\frac{1}{r} \frac{\partial v_z}{\partial \theta} - \frac{\partial v_\theta}{\partial z} \right)$$

$$+ \mathbf{e}_\theta \left(\frac{\partial v_r}{\partial z} - \frac{\partial v_z}{\partial r} \right), \tag{D.15}$$

D.3. Spherical polar coordinates

In the reference frame of spherical polar coordinates (r, θ, ϕ) with r being the radial coordinate, θ the polar angle and ϕ the azimuthal angle about the polar axis $\theta = 0$ (coinciding with the z axis), we define three unit vectors $\mathbf{e}_r$, $\mathbf{e}_\theta$, $\mathbf{e}_\phi$ in each direction of coordinate axes. With respect to the cartesian reference frame (x, y, z), the unit vectors are represented as

$$\mathbf{e}_r = (\sin\theta \cos\phi, \sin\theta \sin\phi, \cos\theta),$$
$$\mathbf{e}_\theta = (\cos\theta \cos\phi, \cos\theta \sin\phi, -\sin\theta), \tag{D.16}$$
$$\mathbf{e}_\phi = (-\sin\phi, \cos\phi, 0).$$

These satisfy the orthonormality: $\mathbf{e}_r \cdot \mathbf{e}_\theta = 0$, $\mathbf{e}_\theta \cdot \mathbf{e}_\phi = 0$, $\mathbf{e}_\theta \cdot \mathbf{e}_\phi = 0$, and $|\mathbf{e}_r| = 1$, $|\mathbf{e}_\theta| = 1$, $|\mathbf{e}_\phi| = 1$.

The velocity vector $\mathbf{v}$ and ∇ operator are written as

$$\mathbf{v} = \mathbf{e}_r v_r + \mathbf{e}_\theta v_\theta + \mathbf{e}_\phi v_\phi, \quad \nabla = \mathbf{e}_r \partial_r + \mathbf{e}_\theta \frac{1}{r} \partial_\theta + \mathbf{e}_\phi \frac{1}{r \sin\theta} \partial_\phi. \tag{D.17}$$

Derivatives of the three unit vectors with respect to θ and ϕ are given by

$$\partial_\theta \mathbf{e}_r = \mathbf{e}_\theta, \quad \partial_\theta \mathbf{e}_\theta = -\mathbf{e}_r, \quad \partial_\theta \mathbf{e}_\phi = 0, \tag{D.18}$$

$$\partial_\phi \mathbf{e}_r = \sin\theta \, \mathbf{e}_\phi, \quad \partial_\phi \mathbf{e}_\theta = \cos\theta \, \mathbf{e}_\phi, \quad \partial_\phi \mathbf{e}_\phi = -\sin\theta \, \mathbf{e}_r - \cos\theta \, \mathbf{e}_\theta. \tag{D.19}$$

The derivatives with respect to r are all zero.

Using (D.17), div **v** is given by

$$\nabla \cdot \mathbf{v} = \left(\mathbf{e}_r \partial_r + \mathbf{e}_\theta \frac{1}{r} \partial_\theta + \mathbf{e}_\phi \frac{1}{r \sin \theta} \partial_\phi \right) \cdot (\mathbf{e}_r v_r + \mathbf{e}_\theta v_\theta + \mathbf{e}_\phi v_\phi)$$

$$= \frac{1}{r^2} \partial_r (r^2 v_r) + \frac{1}{r \sin \theta} \partial_\theta (\sin \theta \, v_\theta) + \frac{1}{r \sin \theta} \partial_\phi v_\phi. \quad \text{(D.20)}$$

Similarly, in view of $\nabla^2 f = \text{div}(\nabla f)$ for a scalar function f, we obtain

$$\nabla^2 f = \left(\mathbf{e}_r \partial_r + \mathbf{e}_\theta \frac{1}{r} \partial_\theta + \mathbf{e}_\phi \frac{1}{r \sin \theta} \partial_\phi \right)$$

$$\times \left(\mathbf{e}_r \partial_r f + \mathbf{e}_\theta \frac{1}{r} \partial_\theta f + \mathbf{e}_\phi \frac{1}{r \sin \theta} \partial_\phi f \right)$$

$$= \frac{1}{r^2} \partial_r (r^2 \partial_r f) + \frac{1}{r^2 \sin \theta} \partial_\theta (\sin \theta \, \partial_\theta f) + \frac{1}{r^2 \sin^2 \theta} \partial_\phi^2 f. \quad \text{(D.21)}$$

From this, we can define ∇^2 operator in the spherical polar system by

$$\nabla^2 \equiv \frac{1}{r^2} \partial_r (r^2 \partial_r) + \frac{1}{r^2 \sin \theta} \partial_\theta (\sin \theta \, \partial_\theta) + \frac{1}{r^2 \sin^2 \theta} \partial_\phi^2 \quad \text{(D.22)}$$

$$= \partial_r^2 + \frac{2}{r} \partial_r + \frac{1}{r^2 \sin \theta} \partial_\theta (\sin \theta \, \partial_\theta) + \frac{1}{r^2 \sin^2 \theta} \partial_\phi^2. \quad \text{(D.23)}$$

Applying ∇^2 operator on a scalar function f gives immediately $\nabla^2 f$ of (D.22). However, applying the same ∇^2 operator on a vector field **v** must be considered carefully because the derivatives of $\mathbf{e}_r$, $\mathbf{e}_\theta$, $\mathbf{e}_\phi$ are not zero. In fact, we obtain the following:

$$\nabla^2 \mathbf{v} = \nabla^2 (\mathbf{e}_r v_r + \mathbf{e}_\theta v_\theta + \mathbf{e}_\phi v_\phi)$$

$$= \mathbf{e}_r \left(\nabla^2 v_r - \frac{2 v_r}{r^2} - \frac{2}{r^2 \sin \theta} \partial_\theta (v_\theta \sin \theta) - \frac{2}{r^2 \sin \theta} \partial_\phi v_\phi \right)$$

$$+ \mathbf{e}_\theta \left(\nabla^2 v_\theta + \frac{2}{r^2} \partial_\theta v_r - \frac{v_\theta}{r^2 \sin^2 \theta} - \frac{2 \cos \theta}{r^2 \sin^2 \theta} \partial_\phi v_\phi \right)$$

$$+ \mathbf{e}_\phi \left(\nabla^2 v_\phi + \frac{2}{r^2 \sin \theta} \partial_\phi v_r - \frac{v_\phi}{r^2 \sin^2 \theta} + \frac{2 \cos \theta}{r^2 \sin^2 \theta} \partial_\phi v_\theta \right).$$

$$\text{(D.24)}$$

The scalar product of $\mathbf{v}$ and ∇ is defined by

$$(\mathbf{v} \cdot \nabla) \equiv v_r \partial_r + v_\theta \frac{1}{r}\partial_\theta + v_\phi \frac{1}{r \sin \theta}\partial_\phi. \qquad \text{(D.25)}$$

Then the convection term $(\mathbf{v} \cdot \nabla)\mathbf{v}$ is

$$
\begin{aligned}
(\mathbf{v} \cdot \nabla)\mathbf{v} &= (\mathbf{v} \cdot \nabla) \cdot \left(\mathbf{e}_r v_r + \mathbf{e}_\theta v_\theta + \mathbf{e}_\phi v_\phi\right) \\
&= \mathbf{e}_r\left((\mathbf{v} \cdot \nabla)v_r - \frac{1}{r}v_\theta^2 - \frac{1}{r}v_\phi^2\right) \\
&\quad + \mathbf{e}_\theta\left((\mathbf{v} \cdot \nabla)v_\theta + \frac{1}{r}v_r v_\theta - \frac{1}{r}v_\phi^2 \cot\theta\right) \\
&\quad + \mathbf{e}_\phi\left((\mathbf{v} \cdot \nabla)v_\phi + \frac{1}{r}v_r v_\phi + \frac{1}{r}v_\theta v_\phi \cot\theta\right). \qquad \text{(D.26)}
\end{aligned}
$$

Operating curl, we have

$$
\begin{aligned}
\nabla \times \mathbf{v} &= \mathbf{e}_r \frac{1}{r \sin\theta}\left(\partial_\theta(v_\phi \sin\theta) - \partial_\phi v_\theta\right) \\
&\quad + \mathbf{e}_\theta \frac{1}{r}\left(\frac{1}{\sin\theta}\partial_\phi v_r - \partial_r(r v_\phi)\right) \\
&\quad + \mathbf{e}_\phi \frac{1}{r}\left(\partial_r(r v_\theta) - \partial_\theta v_r\right). \qquad \text{(D.27)}
\end{aligned}
$$

Rate-of-strain tensors are

$$
e_{rr} = \partial_r v_r \quad e_{\theta\theta} = \frac{1}{r}\partial_\theta v_\theta + \frac{v_r}{r},
$$
$$
e_{\phi\phi} = \frac{1}{r \sin\theta}\partial_\phi v_\phi + \frac{v_r}{r} + \frac{1}{r}v_\theta \cot\theta, \qquad \text{(D.28)}
$$

$$
e_{r\theta} = \frac{r}{2}\partial_r\left(\frac{v_\theta}{r}\right) + \frac{1}{2r}\partial_\theta v_r, \quad e_{\phi r} = \frac{1}{2r \sin\theta}\partial_\phi v_r + \frac{r}{2}\partial_r\left(\frac{v_\phi}{r}\right),
$$
$$
e_{\theta\phi} = \frac{\sin\theta}{2r}\partial_\theta\left(\frac{v_\phi}{\sin\theta}\right) + \frac{1}{2r \sin\theta}\partial_\phi v_\theta. \qquad \text{(D.29)}
$$

Appendix E

First three structure functions

According to Secs. 10.4.4 and 10.4.5, in the frame of the principal axes (s_1, s_2, s_3) of the rate of strain tensor e_{ij} with the principal values σ_1, σ_2, σ_3, the longitudinal velocity difference is given by

$$\Delta v_l(\mathbf{s}) = (\sigma_1 s_1, \sigma_2 s_2, \sigma_3 s_3) \cdot (s_1, s_2, s_3)/s$$
$$= s^{-1}(\sigma_1 s_1^2 + \sigma_2 s_2^2 + \sigma_3 s_3^2)$$
$$= s(\sigma_1 \alpha^2 + \sigma_2 \beta^2 + \sigma_3 \gamma^2),$$

where

$$\mathbf{s}/s = (\alpha, \beta, \gamma) = (\sin\theta\cos\phi, \sin\theta\sin\phi, \cos\phi),$$

with θ as the polar angle and ϕ the azimuthal angle to denote a point on a unit sphere. The average $\langle \cdot \rangle_{sp}$ is replaced with the integral

$$\frac{1}{4\pi}\int d\Omega, \quad \text{where } d\Omega = \sin\theta\, d\theta d\phi.$$

Therefore, we have

$$F_p(s) = \langle (\Delta v_l(\mathbf{s}))^p \rangle_{sp} = \frac{s^p}{4\pi}\int (\sigma_1 \alpha^2 + \sigma_2 \beta^2 + \sigma_3 \gamma^2)^p\, d\Omega.$$

In particular,

$$\langle \alpha^2 \rangle_{sp} = \langle \beta^2 \rangle_{sp} = \langle \gamma^2 \rangle_{sp} = \frac{1}{3}.$$

Using the above expressions, we immediately find

$$F_1(s) = s\left(\sigma_1\langle\alpha^2\rangle_{sp} + \sigma_2\langle\beta^2\rangle_{sp} + \sigma_3\langle\gamma^2\rangle_{sp}\right) = \frac{1}{3}(\sigma_1 + \sigma_2 + \sigma_3)s.$$

In the present case of incompressible flow, we have $\sigma_1 + \sigma_2 + \sigma_3 = 0$.

For the second-order structure function $F_2(s)$, we note that

$$\langle\alpha^4\rangle_{sp} = \langle\beta^4\rangle_{sp} = \langle\gamma^4\rangle_{sp} = \frac{1}{5}.$$

$$\langle\alpha^2\beta^2\rangle_{sp} = \langle\beta^2\gamma^2\rangle_{sp} = \langle\gamma^2\alpha^2\rangle_{sp} = \frac{1}{15}.$$

Using these together with $(\sigma_1 + \sigma_2 + \sigma_3)^2 = 0$, we obtain

$$F_2(s) = \frac{2}{15}\left(\sigma_1^2 + \sigma_2^2 + \sigma_3^2\right)s^2.$$

For the third-order structure function $F_3(s)$, we note first that

$$(\sigma_1 + \sigma_2 + \sigma_3)^3 = 0, \quad \sigma_1^2\sigma_2 + \sigma_1\sigma_2^2 = \sigma_1\sigma_2(\sigma_1 + \sigma_2) = -\sigma_1\sigma_2\sigma_3.$$

From these and similar relations, we obtain

$$\sigma_1^3 + \sigma_2^3 + \sigma_3^3 = 3\sigma_1\sigma_2\sigma_3.$$

Note that

$$\langle\alpha^6\rangle_{sp} = \langle\beta^6\rangle_{sp} = \langle\gamma^6\rangle_{sp}, \quad \langle\alpha^2\beta^4\rangle_{sp} = \langle\beta^2\gamma^4\rangle_{sp} = \cdots$$

Using these relations, we obtain

$$\frac{F_3}{s^3} = \sigma_1\sigma_2\sigma_3\left[3\langle\alpha^6\rangle_{sp} - 9\langle\alpha^2\beta^4\rangle_{sp} + 6\langle\alpha^2\beta^2\gamma^2\rangle_{sp}\right]$$

By the calculation, we have $\langle\alpha^6\rangle_{sp} = 1/7$, $\langle\alpha^2\beta^4\rangle_{sp} = 1/35$ and $\langle\alpha^2\beta^2\gamma^2\rangle_{sp} = 1/105$. Substituting these, we finally obtain

$$\frac{F_3}{s^3} = \frac{8}{35}\sigma_1\sigma_2\sigma_3.$$

Appendix F

Lagrangians

F.1. Galilei invariance and Lorentz invariance

F.1.1. *Lorentz transformation*

Lorentz transformation is a transformation of space-time from one frame F to another F' moving with a relative space-velocity $\mathbf{U}$. When the relative velocity $U = |\mathbf{U}|$ is toward the x-direction, the transformation law is expressed as follows (where $\beta = U/c$, and c is the light speed):

$$t' = \frac{1}{1-\beta^2}\left(t - \frac{Ux}{c^2}\right),$$

$$x' = \frac{1}{1-\beta^2}(x - Ut), \quad y' = y, \quad z' = z. \tag{F.1}$$

In the Lorentz transformation with the variables $(x^\mu) = (ct, \mathbf{x})$,[1] the length $|\mathrm{d}s|$ of a line-element $\mathrm{d}s = (c\,\mathrm{d}t, \mathrm{d}\mathbf{x})$ is a *scalar*, namely a Lorentz-invariant. With the Minkowski metric $g_{\mu\nu} = \mathrm{diag}(-1, 1, 1, 1)$,[2] its square is given by

$$|\mathrm{d}s|^2 \equiv g_{\mu\nu}\,\mathrm{d}x^\mu \mathrm{d}x^\nu$$

$$= -c^2\mathrm{d}t^2 + \langle \mathrm{d}\mathbf{x}, \mathrm{d}\mathbf{x}\rangle = -c^2(\mathrm{d}t')^2 + \langle \mathrm{d}\mathbf{x}', \mathrm{d}\mathbf{x}'\rangle, \tag{F.2}$$

i.e. *invariant* under the transformation between $(t, \mathbf{x})$ and $(t', \mathbf{x}')$.

---

[1] Spatial components are denoted by bold letters: $\mathbf{x} = (x, y, z)$.
[2] The metric $g_{\mu\nu}$ is defined by $\mathrm{d}s^2 = g_{\mu\nu}\,\mathrm{d}x^\mu \mathrm{d}x^\nu$.

Scalar product of a 4-momentum $P = (E/c, \mathbf{p})$ of a particle of mass m with the line-element ds is also an invariant (by the definition of scalar),

$$(P, ds) = -\frac{E}{c} c\,dt + \mathbf{p} \cdot d\mathbf{x} = (\mathbf{p} \cdot \dot{\mathbf{x}} - E)dt \quad (= \Lambda\,dt)$$

$$= m(v^2 - c^2)\,dt = -m_0 c^2\,d\tau, \tag{F.3}$$

where m_0 is the rest mass, $\mathbf{v}$ and $\mathbf{p} = m\mathbf{v}$ are 3-velocity and 3-momentum of the particle respectively [Fr97; LL75], and

$$E = mc^2, \quad m = \frac{m_0}{(1 - \beta^2)^{1/2}}, \quad \beta = \frac{v}{c},$$

$$d\mathbf{x} = \mathbf{v}dt, \quad d\tau = (1 - \beta^2)^{1/2}\,dt \quad \text{(proper time)}.$$

The expression on the left-most side (P, ds) of (F.3) is a scalar product so that it is invariant with respect to the Lorentz transformation, while the factor $d\tau$ on the right-most side is the proper time so that it is invariant. Hence all the five expressions of (F.3) are equivalent and invariant with respect to the Lorentz transformation.

F.1.2.   *Lorentz-invariant Galilean Lagrangian*

The factor $\Lambda = \mathbf{p} \cdot \dot{\mathbf{x}} - E$ in the middle of (F.3) is what is called the Lagrangian in *Mechanics*. Hence it is found that either of the five expressions of (F.3), denoted as Λdt, might be taken as the integrand of the action I.

Next, we consider a *Lorentz-invariant* Lagrangian $\Lambda_L^{(0)}$ in the limit as $v/c \to 0$, and seek its appropriate counterpart Λ_G in the Galilean system. In this limit, the mass m and energy mc^2 are approximated by m_0 and $m_0(c^2 + \frac{1}{2}v^2 + \epsilon)$ respectively in a macroscopic fluid system ([LL87], Sec. 133), where ϵ is the internal energy per unit fluid mass. The first expression of the second line of (F.3) is, then asymptotically,

$$mv^2 - mc^2 \Rightarrow m_0 v^2 - m_0 \left(c^2 + \frac{1}{2}v^2 + \epsilon\right) = (\rho\,d^3x)\left(\frac{1}{2}v^2 - \epsilon - c^2\right),$$

where m_0 is replaced by $\rho(x)\,\mathrm{d}^3x$. Thus, the Lagrangian $\Lambda_\mathrm{L}^{(0)}$ would be defined by

$$\Lambda_\mathrm{L}^{(0)}\mathrm{d}t = \int_M \mathrm{d}^3x\rho(x)\left(\frac{1}{2}\langle v(x), v(x)\rangle - \epsilon - c^2\right)\mathrm{d}t. \qquad \text{(F.4)}$$

The third $-c^2\mathrm{d}t$ term is necessary so as to satisfy the Lorentz-invariance ([LL75], Sec. 87). The reason is as follows. Obviously, the term $\langle v(x), v(x)\rangle$ is not invariant with the Galilei transformation, $\mathbf{v} \mapsto \mathbf{v}' = \mathbf{v} - \mathbf{U}$. Using the relations $\mathrm{d}\mathbf{x} = \mathbf{v}\mathrm{d}t$ and $\mathrm{d}\mathbf{x}' = \mathbf{v}'\mathrm{d}t' = (\mathbf{v} - \mathbf{U})\mathrm{d}t'$, the invariance (F.2) leads to

$$c^2\mathrm{d}t' = c^2\mathrm{d}t + \left(-\langle \mathbf{v}, \mathbf{U}\rangle + \frac{1}{2}U^2\right)\left(1 + O((v/c)^2)\right)\mathrm{d}t \qquad \text{(F.5)}$$

The second term on the right makes the Lagrangian $\Lambda_\mathrm{L}^{(0)}\,\mathrm{d}t$ Lorentz-invariant exactly in the $O((v/c)^0)$ terms in the limit as $v/c \to 0$.

When we consider a fluid flow as a Galilean system, the following prescription is applied. Suppose that the flow is investigated in a finite domain M in space. Then the $c^2\mathrm{d}t$ term gives a constant $c^2\mathcal{M}\mathrm{d}t$ to $\Lambda_\mathrm{L}^{(0)}\mathrm{d}t$, where $\mathcal{M} = \int_M \mathrm{d}^3x\,\rho(x)$ is the total mass in the domain M. In carrying out variation of the action I, the total mass $\mathcal{M}$ is fixed at a constant.

Now, keeping this in mind implicitly, we define the Lagrangian Λ_G of a fluid motion in the Galileian system by

$$\Lambda_\mathrm{G}\,\mathrm{d}t = \int_M \mathrm{d}^3x\rho(x)\left(\frac{1}{2}\langle v(x), v(x)\rangle - \epsilon\right)\mathrm{d}t. \qquad \text{(F.6)}$$

Only when we need to consider its Galilei invariance, we use the Lagrangian $\Lambda_\mathrm{L}^{(0)}$ of (F.4). In the variational formulation in the main text (Sec. 12.6), local conservation of mass is derived from the action principle. As a consequence, the mass is conserved globally. Thus, the use of Λ_G will not cause serious problem except in the case requiring its Galilei invariance.[3]

---

[3]Second term of (F.5) is written in the form of total time derivative since $\langle \mathbf{v}, \mathbf{U}\rangle - \frac{1}{2}U^2 = (\mathrm{d}/\mathrm{d}t)(\langle \mathbf{x}(t), \mathbf{U}\rangle - \frac{1}{2}U^2 t)$. Therefore it is understood in Newtonian mechanics that this term does not play any role in the variational formulation.

F.2.   Rotation symmetry

We consider here the significance of the Lagrangian L_A associated with the rotation symmetry and its outcome derived from the variational principle. To that end, the differential calculus is helpful (see [Ka04] for exterior differentials and products).

(a) **Representations with exterior algebra**
To begin with, the flow field of velocity $\mathbf{v} = (v^k)$ is represented by a tangent vector X:

$$X = X^\mu \partial_\mu = \partial_t + v^1 \partial_1 + v^2 \partial_2 + v^3 \partial_3 = \partial_t + v^k \partial_k, \qquad (F.7)$$

where the greek letter μ denotes $0, 1, 2, 3$. The time component (X^0) is set to be unity: $X^0 = 1$ and $X^k = v^k$ (a roman letter k denotes $1, 2, 3$). This is also equivalent with the differential operator D_t of covariant derivative.

Next, we define a velocity 1-form (one-form) V^1 and potential 1-form A^1 by

$$\begin{aligned} V^1 &= v_k \, dx^k = v_1 \, dx^1 + v_2 \, dx^2 + v_3 \, dx^3, \\ A^1 &= A_k \, dx^k = A_1 \, dx^1 + A_2 \, dx^2 + A_3 \, dx^3, \end{aligned} \qquad (F.8)$$

and vorticity 2-form Ω^2 by

$$\Omega^2 = dV^1 = \omega_1 \, dx^2 \wedge dx^3 + \omega_2 \, dx^3 \wedge dx^1 + \omega_3 \, dx^1 \wedge dx^2, \qquad (F.9)$$

where $\boldsymbol{\omega} = (\omega_1, \omega_2, \omega_3) = \nabla \times \mathbf{v}$. In the flat space, both components are equal, $v^k = v_k$ and $A^k = A_k$. Significance of the vector potential $\mathbf{A} = (A^1, A^2, A^3)$ will be made clear below.

If f is a scalar function (also called a 0-form), we shall define the Lie derivative of f with respect to X by

$$\mathcal{L}_X f = X^\mu \partial_\mu f \equiv D_t f.$$

The Lie derivative of a 1-form A^1 is defined by

$$\mathcal{L}_X A^1 := (X^\mu \partial_\mu A_i + A_k \partial_i X^k) \, dx^i = \Psi_i \, dx^i, \qquad (F.10)$$

$$\Psi_i := (\mathcal{L}_X A^1)_i = X^\mu \partial_\mu A_i + A_k \partial_i X^k \qquad (F.11)$$

where the $\mu = 0$ term drops in the second term of Ψ^i since $X^0 = 1$. Given a vector field Y, the value of $\mathcal{L}_X A^1[Y]$ measures the derivative of the value of A^1 (as one moves along the orbit of X, i.e. the particle path) evaluated on a vector field Y frozen to the flow generated by X.

Let us take an exterior product of $\mathcal{L}_X A^1$ and Ω^2, which gives a volume form $\mathrm{d}V^3 \equiv \mathrm{d}x^1 \wedge \mathrm{d}x^2 \wedge \mathrm{d}x^3$ with a coefficient of scalar product $\langle \Psi, \boldsymbol{\omega} \rangle$ where $\Psi = (\Psi^i)$:

$$(\mathcal{L}_X A^1) \wedge \Omega^2 = \langle \Psi, \boldsymbol{\omega} \rangle \, \mathrm{d}V^3 = \langle \mathcal{L}_X A^1, \boldsymbol{\omega} \rangle \, \mathrm{d}V^3 = \mathcal{L}_X A^1[\boldsymbol{\omega}] \mathrm{d}V^3.$$

(b) **Lagrangian L_A and its variation with respect to A**

Lagrangian L_A associated with the rotation symmetry is proposed as

$$L_A = -\int_M \mathcal{L}_X A^1[\boldsymbol{\omega}] \mathrm{d}^3\mathbf{x} = -\int_M \langle \Psi, \boldsymbol{\omega} \rangle \mathrm{d}^3\mathbf{x} = -\int_M \langle \nabla \times \Psi, \mathbf{v} \rangle \mathrm{d}^3\mathbf{x},$$
(F.12)

Comparing this with (12.69), we find that

$$\rho \mathbf{b} = \nabla \times \Psi = \nabla \times (\mathcal{L}_X A^1).$$
(F.13)

Using the definition (F.7) for X, the expression of Ψ is given by

$$\begin{aligned}
\Psi = \mathcal{L}_X A &= \partial_t A + v^k \partial_k A + A_k \nabla v^k \\
&= \partial_t A + v^k \partial_k A - v^k \nabla A_k + \nabla (A_k v^k) \\
&= \partial_t A + (\nabla \times A) \times \mathbf{v} + \nabla (A_k v^k)
\end{aligned}$$

Therefore, defining $\mathbf{B}$ by $\nabla \times A$, we have

$$\rho \mathbf{b} = \nabla \times \Psi = [\partial_t \mathbf{B} + \nabla \times (\mathbf{B} \times \mathbf{v})].$$

Variation δL_A of L_A with respect to the variation $\delta \mathbf{v}$ is given by

$$\begin{aligned}
\delta L_A &= -\int_M [\langle \nabla \times \Psi, \delta \mathbf{v} \rangle + \langle \delta \Psi, \boldsymbol{\omega} \rangle] \mathrm{d}^3\mathbf{x} \\
&= -\int_M \langle (\nabla \times \Psi + \boldsymbol{\omega} \times \mathbf{B}), \delta \mathbf{v} \rangle \mathrm{d}^3\mathbf{x}.
\end{aligned}$$

This leads to the following expression of $\mathbf{w}$ defined by (12.82)

$$\rho \mathbf{w} = \nabla \times \Psi + \boldsymbol{\omega} \times \mathbf{B} = \mathcal{L}_X \mathbf{B} + \boldsymbol{\omega} \times \mathbf{B},$$
(F.14)

where $\mathcal{L}_X\mathbf{B} \equiv \partial_t\mathbf{B} + \nabla \times (\mathbf{B} \times \mathbf{v})$. For flows of a homentropic fluid, the velocity is given by (12.89): $\mathbf{v} = \nabla\phi + \mathbf{w}$. Therefore, the helicity is

$$H = \int_V \boldsymbol{\omega} \cdot \mathbf{v}\, d^3\mathbf{x} = \int_V \boldsymbol{\omega} \cdot \mathbf{w}\, d^3\mathbf{x} = \int_V \rho^{-1}(\boldsymbol{\omega} \cdot \mathcal{L}_X\mathbf{B})d^3\mathbf{x}.$$

Thus it is found that the term $\mathcal{L}_X\mathbf{B}$ generates the helicity.

Density of the Lagrangian L_A is defined by $\Lambda_A := -\langle \boldsymbol{\Psi}, \boldsymbol{\omega} \rangle$:

$$\Lambda_A := -\langle \boldsymbol{\Psi}, \boldsymbol{\omega} \rangle = -\omega^i \partial_t A_i - \omega^i v^k \partial_k A_i - \omega^i A_k \partial_i v^k.$$

Rearranging the first two terms, we obtain

$$\Lambda_A = \left(\partial_t \omega^i + \partial_k(v^k \omega^i) - \omega^k \partial_k v^i\right)A_i - \partial_t(\omega^i A_i) - \partial_k(\omega^i v^k A_i)$$

Next, we consider the action $I_A = \int_{t_1}^{t_2} dt \int_M \Lambda_A\, d^3\mathbf{x}$ and its variation with respect to the variation $\delta\mathbf{A}$. Applying the same boundary conditions (12.107) and (12.108) to δA_i (used for the variations of $\boldsymbol{\xi}$), we obtain

$$\delta I_A = \int_{t_1}^{t_2} dt \int_M \left[\partial_t \omega^i + \partial_k(v^k \omega^i) - \omega^k \partial_k v^i\right]\delta A_i d^3\mathbf{a}.$$

Invariance of I_A for arbitrary variations δA_i implies

$$\partial_t \omega^i + v^k \partial_k \omega^i + \omega^i \partial_k v^k - \omega^k \partial_k v^i = 0.$$

Thus, we have found the equation for the *gauge field* $\boldsymbol{\omega}$. In vector notation, this is written as

$$\partial_t \boldsymbol{\omega} + (\mathbf{v} \cdot \nabla)\boldsymbol{\omega} + \boldsymbol{\omega}(\nabla \cdot \mathbf{v}) - (\boldsymbol{\omega} \cdot \nabla)\mathbf{v} = 0. \tag{F.15}$$

This is nothing but the vorticity equation, which is also written as

$$\partial_t \boldsymbol{\omega} + \text{curl}(\boldsymbol{\omega} \times \mathbf{v}) = 0. \tag{F.16}$$

Solutions

Problem 1

1.1 : This is not the particle path, since the pattern is a snap-shot. It is not the stream-line either because there is no physical reason (no physical constraint) that the fluid is moving in the tangent direction of the line. The ink-particles composing the lines are not the family that passed at a fixed point at successive times. Therefore the pattern is not the streak-line.

The pattern is interpreted as a diffeomorphism mapping from the compact domain occupied by the initial ink distribution on the surface onto the present domain of streaky pattern which is formed by the straining surface deformation (stretching in one direction and compression in the perpendicular direction) and folding of the elongated domain. Thus initial drop-like domain is mapped in diffeomorphic way to a streaky domain by the motion of the water surface with eddies.

1.2 : Consider a fluid particle located at $\mathbf{x}$ at a time t, and denote its subsequent position at a time $t + \delta t$ by $T(\mathbf{x})$ which is given by $\mathbf{x} + \mathbf{v}\delta t + O(\delta t^2)$. Suppose that the line element $\delta\mathbf{x}$ between two neighboring points at $\mathbf{x}$ and $\mathbf{x} + \delta\mathbf{x}$ is represented by $\delta\mathbf{X}$ at a subsequent time $t + \delta t$, where $\delta\mathbf{x} = (\delta x, \delta y, \delta z)^T$ with ^{T} denoting the transpose. The new line element $\delta\mathbf{X} = (\delta X, \delta Y, \delta Z)^T$ will be given by

$$
\begin{aligned}
\delta\mathbf{X} &= T(\mathbf{x} + \delta\mathbf{x}) - T(\mathbf{x}) \\
&= (\mathbf{x} + \delta\mathbf{x} - \mathbf{x}) + (\mathbf{v}(\mathbf{x} + \delta\mathbf{x}) - \mathbf{v}(\mathbf{x}))\delta t + O(\delta t^2) \\
&= \delta\mathbf{x} + \left(\frac{\partial \mathbf{v}}{\partial x_k} \delta x_k + O(|\delta\mathbf{x}|^2) \right)\delta t + O(\delta t^2) \\
&:= A\,\delta\mathbf{x} + O(|\delta\mathbf{x}|^2\delta t) + O(\delta t^2)
\end{aligned}
$$

343

where

$$A\,\delta\mathbf{x} = \begin{pmatrix} 1 + \partial_x u\,\delta t & \partial_y u\,\delta t & \partial_z u\,\delta t \\ \partial_x v\,\delta t & 1 + \partial_y v\,\delta t & \partial_z v\,\delta t \\ \partial_x w\,\delta t & \partial_y w\,\delta t & 1 + \partial_z w\,\delta t \end{pmatrix} \begin{pmatrix} \delta x \\ \delta y \\ \delta z \end{pmatrix}$$

Denoting the volume of the parallelpiped constructed with $(\delta x, \delta y, \delta z)$ by V_0, and the volume constructed with $(\delta X, \delta Y, \delta Z)$ by V_1, the ratio V_1/V_0 is given by

$$\frac{V_1}{V_0} = \left| \frac{\partial X_i}{\partial x_j} \right| = |A|$$

where $|\cdot|$ denotes the determinant and

$$|A| = 1 + \operatorname{div}\mathbf{v}\,\delta t + O(\delta t^2),$$

where $\operatorname{div}\mathbf{v} := \partial_x u + \partial_y v + \partial_z w$. Thus, we obtain

$$\frac{1}{V}\frac{dV}{dt} = \lim_{\delta t \to 0}\frac{1}{V_0}\frac{V_1 - V_0}{\delta t} = \operatorname{div}\mathbf{v}.$$

1.3 : Setting $Q = x$, y, and z in (1.14), we obtain $Dx/Dt = u$, $Dy/Dt = v$, and $Dz/Dt = w$, where the particle position coordinates x, y and z are regarded as field variables (rather than independent variables). If we set $Q = u$, we obtain $Du/Dt = \partial_t u + (\mathbf{v}\cdot\nabla)u$. Similarly, we obtain the expressions for Dv/Dt and Dw/Dt.

Problem 2

2.1 : We obtain $u'(0) = du/dy|_{y=0} = 4U/d$ and $u'(d) = -4U/d$. The viscous friction exerted by the fluid on the wall at $y = 0$ and d (per unit length along the channel) is given by $\sigma_f = \sum \sigma_{xy}^{(v)} n_y = n_y\,\mu\,du/dy|_{y=0} + n_y\,\mu\,du/dy|_{y=d}$, where $n_y(0) = 1$ and $n_y(d) = -1$ by the definition (2.18). Thus, we obtain $F = 2 \times 4\mu U/d = 8\mu U/d$ (per unit depth to the third direction, z). The factor 2 means two contributions from the upper and lower walls.

2.2 : The governing equation is $d^2T/dz^2 = 0$ from (2.6). Its general solution is given by $T(z) = a + bz$ (a, b: constants). From the boundary values $T(0) = T_1$ and $T(d) = T_2$, we obtain $a = T_1$ and $b = (T_2 - T_1)/d$.

2.3 : Writing the integrand of (2.26) (except the function $u_0(\xi)$ of a parameter ξ) as

$$f(x,t) = \frac{1}{2\sqrt{\pi\lambda t}}\, e^{-X^2/4\lambda t}, \quad X = x - \xi,$$

we have $\partial_x f = -(X/(2\lambda t))\, f$, and

$$\partial_t f = -\frac{1}{2t}\, f + \frac{X^2}{4\lambda t^2}\, f, \quad \lambda \partial_x^2 f = -\frac{1}{2t}\, f + \frac{X^2}{4\lambda t^2}\, f.$$

Thus, $f(x,t)$ satisfies Eq. (2.24), therefore $u(x,t)$ does so, as far as the integration with respect to ξ converges. As t tends to $+0$ (from positive side), $f(x,t)$ tends to the delta function $\delta(X)$ (see (A.38) of Appendix A). Hence, $u(x,t) \to \int_{-\infty}^{\infty} u_0(\xi)\,\delta(x - \xi)\mathrm{d}\xi = u_0(x)$, satisfying the initial condition.

2.4 : Taking the coordinate origin at the center of the cube of side a, we consider the equation of the angular momentum of the cube. Denoting the moment of inertia of the cube around the z axis by I_z, and the angular velocity around it by Ω_z which is assumed uniform over the cube, the z component angular momentum is given by $M_z = I_z\Omega_z$, where $I_z = \int(x^2 + y^2)\rho\,\mathrm{d}x\mathrm{d}y\mathrm{d}z$. The dynamical equation is given by $(\mathrm{d}/\mathrm{d}t)(I_z\Omega_z) = N_z$, where N_z is the moment of the stress force over the faces from the outside fluid. To the first approximation, the total surface forces exerted on the faces perpendicular to the x axis are $f_x^{(+)} = (\sigma_{xx}, \sigma_{yx}, \sigma_{zx})\, a^2$ for the face of positive x, and $f_x^{(-)} = -(\sigma_{xx}, \sigma_{yx}, \sigma_{zx})\, a^2$ for the face of negative x. Hence, the z component of the moment of the stress forces is given by $N_z^{(x)} = \frac{1}{2}a\left(f_x^{(+)}\right)_y - \frac{1}{2}a\left(f_x^{(-)}\right)_y = \sigma_{yx}\, a^3$.

Similarly, the surface forces exerted on the faces perpendicular to the y axis are $(\sigma_{xy}, \sigma_{yy}, \sigma_{zy})\, a^2$ for the face of positive y, and $-(\sigma_{xy}, \sigma_{yy}, \sigma_{zy})\, a^2$ for the face of negative y. Hence, the z component of the moment of the stress forces is given by $N_z^{(y)} = -\sigma_{xy}\, a^3$. Therefore, the total z component moment is given by $N_z^{(y)} = (\sigma_{yx} - \sigma_{xy})\, a^3$. Since a is assumed to be infinitesimal, we have $I_z = O(a^5)$, while Ω is independent of a. The equation $(\mathrm{d}/\mathrm{d}t)I_z\,\Omega = N_z$ divided by a^3 reduces to $\sigma_{yx} - \sigma_{xy} = 0$ in the limit as $a \to 0$. Thus, we obtain the symmetry $\sigma_{xy} = \sigma_{yx}$. Similarly, we obtain $\sigma_{yz} = \sigma_{zy}$ and $\sigma_{zx} = \sigma_{xz}$.

2.5 : Using (2.27), we obtain

$$\sigma_{ij}^{(v)} = (a\delta_{ij}\delta_{kl} + b\delta_{ik}\delta_{jl} + c\delta_{il}\delta_{jk})e_{kl} = a\delta_{ij}D + be_{ij} + ce_{ji}.$$

where $D = e_{kk}$. The symmetry $\sigma_{ij}^{(v)} = \sigma_{ji}^{(v)}$ leads to $b = c$. Writing $b = \mu$, $a = \zeta - (2/3)\mu$, and $e_{ij} = D_{ij} + (1/3)D\delta_{ij}$ of (2.15), we obtain the expression (2.14).

Problem 3

3.1 : From (3.7), (3.21) and (3.37) with $f_i = 0$, we obtain
Mass conservation equation: $\rho_t + (\rho u)_x = 0$.
Momentum conservation equation: $(\rho u)_t + (\rho u^2)_x + p_x = 0$.
Energy conservation equation: $[\rho(\frac{1}{2}u^2 + e)]_t + [\rho u(\frac{1}{2}u^2 + h)]_x = 0$.

Problem 4

4.1 : (i) Using the formula of ∇^2 of (D.11), we obtain

$$\nabla^2 u = (\partial_x^2 + \partial_y^2 + \partial_z^2)u = \partial_x^2 u + \frac{1}{r}\partial_r(r\partial_r)u + \frac{1}{r^2}\partial_\theta^2 u,$$

with ∂_z^2 replaced by ∂_x^2. Setting $\partial_x^2 u = 0$, we have

$$\partial_y^2 u + \partial_z^2 u = \frac{1}{r}\partial_r(r\partial_r)u + \frac{1}{r^2}\partial_\theta^2 u.$$

Thus, Eq. (4.39) reduces to (4.77).
(ii) The x velocity $u(r)$ is governed by Eq. (4.77) with $\partial_\theta^2 u = 0$:

$$(ru')' = -(P/\mu)r, \quad u' = du/dr.$$

Solving this ordinary differential equation, we obtain $u = -(P/4\mu)r^2 + \text{const}$. Applying the boundary condition $u(a) = 0$, this results in (4.78).
(iii) The total rate of flow Q per unit time is given by

$$Q = \int_0^a u_{\text{HP}}(r)\,2\pi r\,dr = \int_0^a \frac{P}{4\mu}(a^2 - r^2)\,2\pi r\,dr = \frac{\pi a^4 P}{8\mu}.$$

4.2 : Substituting $u = f(y)\,e^{i\omega t}$ into (4.44), we obtain $f_{yy} - i(\omega/\nu)f = 0$. Hence, $f(y) = C\exp[\pm(1 + i)\sqrt{\omega/2\nu}\,y]$. The decaying condition at infinity requires the lower negative sign, and the boundary condition at $y = 0$ requires $C = U$. Thus we obtain

$$u(y, t) = \Re\left[U \exp\left(-(1+i)\sqrt{\frac{\omega}{2\nu}}\, y\right) e^{i\omega t}\right].$$

Its real part yields the solution (4.79). The representative thickness δ of the oscillating boundary layer will be given by the y-value corresponding to the place where the amplitude decays to $1/e$ of the boundary value. Thus, $\delta = \sqrt{2\nu/\omega}$.

4.3 : The solution is given by $v(r) = ar + b/r$. This must satisfy $v(r_1) = r_1\Omega_1$ and $v(r_2) = r_2\Omega_2$. Solving for a and b, we obtain

$$a = \frac{r_2^2\Omega_2 - r_1^2\Omega_1}{r_2^2 - r_1^2}, \quad b = (\Omega_1 - \Omega_2)\frac{r_1^2 r_2^2}{r_2^2 - r_1^2}.$$

4.4 : (i) The continuity equation (4.33) leads to $v = -\int_0^y \partial_x u \, dy$. From this, we have $v = O(U(\delta/l))$ according to the estimates below (4.33). Then all the terms of (4.29), uu_x, vu_y, $(1/\rho)\, p_x$, νu_{xx} and νu_{yy} (neglecting u_t because of the steady problem) are estimated respectively by the order of magnitudes as follows:

$$\frac{U^2}{l}, \quad \frac{U^2}{l}, \quad \frac{\Delta p}{\rho l}, \quad \frac{U^2}{l}\frac{1}{R_e}, \quad \frac{U^2}{l}\frac{1}{R_e}\frac{l^2}{\delta^2}.$$

where $R_e = Ul/\nu$ and Δp is the order of the pressure difference in the distance l. If $\Delta p = O(\rho U^2)$ and $(l/\delta)^2 \approx R_e$, then all the terms are of the order U^2/l except the term νu_{xx} which is of the order $(U^2/l)\, R_e^{-1}$. If $R_e \gg 1$, this term may be neglected and we obtain (4.32).

Regarding Eq. (4.30), all the terms of uv_x, vv_y, $(1/\rho)\, p_y$, νv_{xx} and νv_{yy} are estimated respectively as follows:

$$\frac{U^2}{l}\frac{\delta}{l}, \quad \frac{U^2}{l}\frac{\delta}{l}, \quad \frac{\Delta p}{\rho\delta}, \quad \frac{U^2}{l}\frac{\delta}{l}\frac{1}{R_e}, \quad \frac{U^2}{l}\frac{\delta}{l}\frac{1}{R_e}\frac{l^2}{\delta^2}.$$

All are of the order $(U^2/l)(\delta/l)$ or smaller under the above assumptions of $\Delta p = O(\rho U^2)$ and $(l/\delta)^2 \approx R_e$, except the pressure term $\Delta p/(\rho\delta) = (U^2/\delta) \gg (U^2/l)(\delta/l)$. Thus, all the other terms are smaller than the term $(1/\rho)p_y$ by R_e^{-1}, and we obtain (4.80) from (4.30).

(ii) By the definition (B.7) of stream function of Appendix B.2, we have

$$u = \partial_y \psi = U f'(\eta), \quad v = -\partial_x \psi = \frac{1}{2} (\nu U/x)^{\frac{1}{2}} (\eta f' - f),$$

where $f' = df/d\eta$. The boundary condition (4.28) requires $f = f' = 0$ at $\eta = 0$, while the condition (4.27) requires $f' \to 1$ ($f \to \eta$) as $\eta \to \infty$.

Furthermore, $u_x = -\frac{1}{2} U(\eta/x) f''$, $u_y = U(\eta/y) f''$ and $u_{yy} = U(\eta/y)^2 f'''$. Substituting these into Eq. (4.32) with $p_x = 0$, we obtain (4.82).

4.5 : (i) By the definition (B.7) of stream function, we have

$$u = \partial_y \psi = A x^{-\frac{1}{3}} f'(\zeta),$$

$$v = -\partial_x \psi = \frac{1}{3} A x^{-2/3} (2\zeta f' - f),$$

$$u_y = A x^{-1} f'', \quad u_x = -\frac{1}{3} A x^{-4/3} (f' + 2\zeta f''),$$

$$u_{yy} = A x^{-5/3} f''',$$

where $f' = df/d\zeta$. The boundary condition (4.84) requires $f = f'' = 0$ at $\zeta = 0$. As $\zeta \to \pm\infty$, both of u and v must tend to 0. Therefore,

$$f' = O(\zeta^{-1-\alpha}) \to 0, \quad f = O(\zeta^{-\alpha}) \to 0, \quad (\alpha > 0).$$

Substituting the above derivatives into Eq. (4.32) with $p_x = 0$, one can derive the equation,

$$-\frac{1}{3} A (f')^2 - \frac{1}{3} A f f'' = \nu f'''.$$

If we set $A = 3\nu$, we obtain (4.86).

(ii) The first and second terms of (4.32) can be rewritten as $u u_x = (\frac{1}{2} u^2)_x$, and

$$v u_y = (vu)_y - v_y u = (vu)_y + u_x u = (vu)_y + \left(\frac{1}{2} u^2\right)_x.$$

The equality $v_y = -u_x$ was used from (4.33). Therefore, Eq. (4.32) can be written as

$$(u^2)_x = (\nu u_y - uv)_y.$$

Integrating both sides with respect to y from $-\infty$ to ∞, we obtain

$$\frac{\mathrm{d}}{\mathrm{d}x}\int_{-\infty}^{\infty} u^2(x,y)\,\mathrm{d}y = [\nu u_y - uv]_{-\infty}^{\infty} = 0,$$

because $f(\zeta)$, $f'(\zeta) \to 0$ as $\zeta \to \pm\infty$. Thus, $\int_{-\infty}^{\infty} u^2\mathrm{d}y$ is invariant with respect to x. This integral multiplied with ρ denotes a physically significant quantity, i.e. the total momentum flux M across a fixed x position. Invariance of M means that there is no external momentum source in the x direction under uniform p.

(iii) Equation (4.86) is written as $f''' + (ff')' = 0$. This can be integrated once, yielding $f'' + ff' = 0$ by using the boundary conditions $f = f'' = 0$ at $\zeta = 0$. This can be integrated once more, yielding

$$f' + \frac{1}{2}f^2 = \frac{1}{2}c^2, \quad c = f(\infty) = \text{const.} \ (> 0).$$

Solving this using $c = 2b$, we obtain

$$f(\zeta) = 2b\tanh(b\zeta), \quad b = \left(\frac{M}{48\,\rho\nu^2}\right)^{\frac{1}{3}},$$

$$u(x,y) = 3\nu x^{-\frac{1}{3}}f'(\zeta) = ax^{-\frac{1}{3}}\operatorname{sech}^2(b\zeta),$$

$$a = 6\nu b^2 = \left(\frac{3M^2}{32\,\rho^2\nu}\right)^{\frac{1}{3}},$$

$$M = \rho\int_{-\infty}^{\infty} u^2(x,y)\,\mathrm{d}y = \rho a^2 x^{-2/3}\int_{-\infty}^{\infty}\operatorname{sech}^4(b\zeta)\,\mathrm{d}y$$

$$= 2\frac{\rho a^2}{b}\int_0^{\infty}\operatorname{sech}^4 Z\,\mathrm{d}Z = \frac{4}{3}\frac{\rho a^2}{b} = 48\,\rho\nu^2\,b^3.$$

The jet profile $u(\zeta) = \operatorname{sech}^2\zeta$ is called the *Bickley jet*. The last expression defines the relation between M and b, where the following integration formula is used: $\int_0^{\infty}\operatorname{sech}^4 Z\,\mathrm{d}Z = \int_0^{\infty}\operatorname{sech}^2 Z\operatorname{sech}^2 Z\,\mathrm{d}Z = \int_0^1(1 - T^2)\,\mathrm{d}T = \frac{2}{3}$, with $T \equiv \tanh Z$ and $\mathrm{d}T = \operatorname{sech}^2 Z\,\mathrm{d}Z$.

4.6 : (i) Introducing a variable τ defined by x/U, Eq. (4.88) takes the form of a diffusion equation with a "time" variable τ: $\partial_\tau u = \nu u_{yy}$.

In unbounded space of y, the solution satisfying the boundary condition (4.89) is well-known and given by

$$u(x, y) = Q \sqrt{\frac{U}{4\pi\nu x}} \exp\left(-\frac{Uy^2}{4\nu x}\right), \tag{A}$$

where Q is a constant whose meaning will become clear just below. Carrying out the integral of $u(x, y)$ with respect to y, we obtain

$$\int_{-\infty}^{\infty} u(x, y)\mathrm{d}y$$

$$= Q \sqrt{\frac{U}{4\pi\nu x}} \int_{-\infty}^{\infty} \exp\left(-\frac{Uy^2}{4\nu x}\right)\mathrm{d}y = Q,$$

where the integration formula $\int_{-\infty}^{\infty} \exp(-\eta^2)\,\mathrm{d}\eta = \sqrt{\pi}$ is used. Thus, it is found that the mass flux defect at a fixed x position is a constant Q independent of x.

(ii) We shall use the momentum equation in the integral form (3.23) where $P_{ik} = \rho v_i v_k + p\delta_{ik}$. For the *control* surface A_0, we choose a large rectangular box with sides parallel and perpendicular to the undisturbed uniform stream, including the body inside. The sides S of A_0 parallel to the stream are located far enough from the body to lie entirely outside the wake. The fluid within A_0 is acted on by forces on the surrounding surface A_0, and by forces at the body surface whose resultant should be $-D$ (sink) in the x direction.

Under the steady flow assumption, the drag D is given by

$$D = \int_F (p_1 + \rho u_1^2 - p_2 - \rho u_2^2)\mathrm{d}F - \rho \int_S uv_n\mathrm{d}S, \tag{B}$$

where u_1, p_1 and u_2, p_2 are the x-velocity and the pressure at the upstream and downstream faces F, respectively. The second integral term is the x-momentum flux out of the side surface S (so that $A_0 = F + S$), where the pressure effect cancels out on both sides. All the viscous stresses on A_0 are assumed to be small, and so neglected.

The mass flux across A_0 must be exactly zero, which is given by

$$\int_F (u_2 - u_1)\mathrm{d}F + \int_S v_n \mathrm{d}S = 0. \qquad \text{(C)}$$

We may put $u = U$ in the second term of (B). With the aid of (C), we have

$$D = \int_F (p_1 + \rho u_1(u_1 - U) - p_2 - \rho u_2(u_2 - U))\mathrm{d}F. \qquad \text{(D)}$$

At the far downstream face, the difference from the far upstream lies in the existence of the wake, regarded as an "in-flow" towards the body. The volume flux of this "in-flow" is equal to Q calculated in (i). This in-flow in the wake must be compensated by an equal volume flux away from the body outside the wake where the flow is practically irrotational. Thus, the presence of the wake is associated with a source-like flow (Fig. 4.12), with its strength being Q. At large distance r from the body, the source-flow velocity $\mathbf{v}$ falls off as r^{-1} in two-dimensions (see (5.63) of Sec. 5.8.1). In this same region, the Bernouli's theorem (5.12) of Sec. 5.1 is applicable so that we have $p_1 - p_2 = -\frac{1}{2}\rho((\mathbf{v}_1)^2 - (\mathbf{v}_2)^2)$.

Thus, far from the body, the flow is a superposition of a uniform stream and the source-like flow outside the wake, and within the wake, we have the velocity (A) obtained in (i). In this situation, most terms in the expression (D) cancel, and the only remaining contribution is obtained from the wake, given by $D = \rho U \int (U - u_2)\mathrm{d}F = \rho U Q$, where $U - \mathcal{U}_2 = \mathcal{U}(x,y)$. Thus the drag D is interpreted as the momentum flux associated with dragging of the volume flux Q.

4.7 : The force F_i exerted by a sphere S_r is given by the integral (4.70) together with stress tensor $\sigma_{ij} = -p\delta_{ij} + 2\mu e_{ij}$, where $F_i = D_i$ if $n_r = -1$. The rate-of-strain tensors e_{ij} in the spherical polar coordinates are defined by (D.28) and (D.29) in Appendix D.3. Using (4.63) and setting $r = a$, we obtain the rate-of-strain tensors in the frame (r, θ, ϕ) as follows:

$$e_{rr} = -\frac{2}{a^2}\cos\theta, \quad e_{\theta\theta} = e_{\phi\phi} = \frac{1}{a^2}\cos\theta,$$

$$e_{r\theta} = e_{\theta\phi} = e_{\phi r} = 0.$$

The pressure is given by (4.65) as $p_S = (2\mu/a^2)\cos\theta$. Hence, the stress tensors are

$$\sigma_{rr} = -\frac{6\mu}{a^2}\cos\theta, \quad \sigma_{\theta\theta} = \sigma_{\phi\phi} = 0,$$

$$\sigma_{r\theta} = \sigma_{\theta\phi} = \sigma_{\phi r} = 0.$$

Thus, we obtain nonzero force (on the fluid external to S_r) along the polar axis x:

$$F_x = -\int_{r=a} \sigma_{rr}\cos\theta\,dS = \int \frac{6\mu}{a^2}\cos^2\theta\,2\pi a^2\sin\theta\,d\theta = 8\pi\mu.$$

4.8 : An irrotational incompressible flow ($\boldsymbol{\omega} = 0$ and div $\mathbf{v} = 0$) with $p = $ const. constitutes a solution of the Stokes equation (4.60) with $\mathbf{F} = 0$ (see Sec. 4.8.1). In fact, the force F_i on a closed surface S is given by

$$F_i = \oint_S \sigma_{ij}n_j dS = \oint_S (-p\delta_{ij} + 2\mu e_{ij})n_j dS$$

$$= \int_V (-\partial_i p + \mu\nabla^2 v_i)dV = \int_V (-\nabla p - \mu\nabla\times\boldsymbol{\omega})_i dV = 0,$$

where i, j denote the cartesian x, y, z components. Transformation from the surface integral on S to the volume integral in the space V out of S was carried out to obtain the second line. This is assured by the assumption that there is no singularity out of S. In addition, the following equalities $\partial_j(2e_{ij}) = \partial_j(\partial_i v_j + \partial_j v_i) = \nabla^2 v_i = -(\nabla\times\boldsymbol{\omega})_i = 0$ are used.

Problem 5

5.1 :
$$dF = \partial_x F\,dx + \partial_y F\,dy = (\Phi_x + i\Psi_x)dx + (\Phi_y + i\Psi_y)dy$$
$$= (u - iv)dx + (v + iu)dy = (u - iv)(dx + idy)$$
$$= (u - iv)dz$$

If $dy = 0$, then $dz = dx$. If $dx = 0$, then $dz = idy$.

5.2 : The equation of stream-lines $dx/u = dy/v$ is given by $dx/Ax = dy/(-Ay)$, which reduces to $ydx + xdy = d(xy) = 0$. Therefore, the stream-lines are given by $xy = $ const., i.e. $\Psi = xy$.

Regarding $y = 0$ as a wall and considering the upper half space ($y > 0$), the flow impinges on the wall $y = 0$ from above.

5.3 : The normals to each family of curves $\Phi(x, y) = $ const. and $\Psi(x, y) = $ const. are given by $\operatorname{grad} \Phi = (\Phi_x, \Phi_y)$ and $\operatorname{grad} \Psi = (\Psi_x, \Psi_y)$. Their scalar product is

$$\operatorname{grad} \Phi \cdot \operatorname{grad} \Psi = \Phi_x \Psi_x + \Phi_y \Psi_y = 0,$$

owing to (5.52). This verifies the orthogonal intersection of the two families of curves.

5.4 : Because $F(z)$ is differentiable (since $F(z)$ is an analytic function), $dZ = F'(z)dz$, and we have

$$Z_1 - Z_0 = |F'(z_0)|e^{i\phi}(z_1 - z_0), \quad Z_2 - Z_0 = |F'(z_0)|e^{i\phi}(z_2 - z_0),$$

where $F'(z_0) = |F'(z_0)|e^{i\phi}$. The two infinitesimal segments $z_1 - z_0$ and $z_2 - z_0$ are rotated by the same angle ϕ. Hence, the intersection angle between $Z_1 - Z_0$ and $Z_2 - Z_0$ is unchanged. This is represented by the following relation:

$$\arg\left[\frac{Z_2 - Z_0}{Z_1 - Z_0}\right] = \arg\left[\frac{z_2 - z_0}{z_1 - z_0}\right] = \theta.$$

5.5 : Substituting $\zeta = \sigma e^{i\phi}$ into (5.88), we have

$$z = x + iy = \left(\sigma + \frac{a^2}{\sigma}\right)\cos\phi + i\left(\sigma - \frac{a^2}{\sigma}\right)\sin\phi. \qquad (A)$$

(i) Setting $\sigma = a$, we obtain $x = 2a\cos\phi$, and $y = 0$. For $\phi \in [0, 2\pi]$, we have $x \in [-2a, 2a]$.

The exterior of the circle $\sigma = a$ is expressed by $\sigma > a$ and $\phi \in [0, 2\pi]$, which corresponds to the whole z-plane with the cut L. According to (A), counter-clockwise rotation around $\zeta = 0$ corresponds to the same counter-clockwise rotation around L.

The interior of the circle $\sigma = a$ corresponds to the whole z plane with the cut L, again. But this time, the counter-clockwise rotation around $\zeta = 0$ corresponds to the clockwise rotation around L, according to (A).

(ii) As $\zeta \to \infty$, $F_\alpha \to Ue^{-i\alpha}\zeta$. Hence, the flow tends to an inclined uniform flow with an angle α counter-clockwise with the real axis of the ζ-plane. The second term, which was neglected at infinity, is a dipole $Ua^2 e^{i\alpha}/\zeta$ with its axis inclined by an angle $\pi + \alpha$ with the positive real axis of ξ. Substituting $\zeta = \sigma e^{i\phi}$ into (5.89), we obtain

$$F_\alpha = \left(\sigma + \frac{a^2}{\sigma}\right)\cos(\phi - \alpha) + i\left(\sigma - \frac{a^2}{\sigma}\right)\sin(\phi - \alpha).$$

This means that the circle $\sigma = a$ is a stream-line. Thus, F_α represents a uniform flow around a circle of radius a, and the flow is inclined at an angle α with respect to the positive real axis of ζ.

(iii) By the transformation (5.88), the circle $\sigma = a$ collapses to the double segments of L representing a flat plate of length $4a$. As $\zeta \to \infty$, $z \approx \zeta$. Therefore, the flow at infinity of the z-plane is also an inclined uniform flow with an angle α counter-clockwise with the real axis of z.

(iv) If $\alpha = \pi/2$, the potential $F_\alpha(\zeta)$ reduces to the potential $F_\perp(\zeta)$ of (5.90). Hence, $F_\perp(\zeta)$ represents a vertical uniform flow from below (upward) in the ζ-plane. Since $z \approx \zeta$ as $\zeta \to \infty$, the flow in the z-plane is the same vertical uniform flow from below impinging on the horizontal flat plate perpendicularly. Because $Z = -iz = ze^{-i\pi/2}$, the Z-plane is obtained by rotating the z-plane by 90° clockwise. Thus, $F_\perp$ represents a flow impinging on a vertical flat plate at right angles from left in the Z-plane.

5.6 : If $n \neq -1$,

$$\oint_C z^n dz = \frac{1}{n+1}[z^{n+1}]_C = 0.$$

The cases of $n = -2, -3, \ldots$ (negative integers $\neq -1$) are multipoles. If $n = -1$ (simple pole), using $z = re^{i\theta}$,

$$\oint_C \frac{1}{z} dz = [\log z]_C = [\log r + i\theta]_C = 2\pi i.$$

5.7 : Note that the unit normal to C is written as $\mathbf{n} = (n_x, n_y)dl = (dy, -dx)$, according to the footnote to Sec. 5.7. We can write

$$X - iY = \oint_C (-pn_x)dl - i\oint_C (-pn_y)dl = -\oint_C p(dy + idx).$$

Note that $dy + idx = i(dx - idy) = id\bar{z}$. Furthermore, the quantity $(\overline{dF/dz})d\bar{z}$ is real, because the fluid velocity must be tangential, i.e. along the body contour C with $dz = ds\,e^{i\phi}$, $\arg[dF/dz]$ should be $-\phi$ (parallel), or $\pi - \phi$ (anti-parallel). Therefore, we have

$(\overline{\mathrm{d}F/\mathrm{d}z})\mathrm{d}\bar{z} = \overline{(\mathrm{d}F/\mathrm{d}z)\mathrm{d}z}$. Thus, using (5.57), we obtain (5.92), since $\oint_C p_0 \, \mathrm{d}\bar{z} = \overline{p_0 \oint_C \mathrm{d}z} = 0$ (the residue is zero, see Problem 5.6).

5.8 : (i) The first term $U(z + a^2/z)$ describes the uniform flow around a cylinder of radius a, while the second term $-(\gamma/2\pi i)\log z$ represents a clockwise vortex at $z = 0$ ($\gamma > 0$). There is no singularity except at $z = 0$. The stream function is given by

$$\Psi = \Im[F_\gamma(z)] = (r - (a^2/r))\sin\theta + (\gamma/2\pi)\log r,$$

for $z = re^{i\theta}$ (see (5.70)). Obviously, for $r = a$, $\Psi = (\gamma/2\pi)\log a = $ const. Thus, the circle $r = a$ is a stream-line and regarded as a cylinderical surface. The complex velocity is

$$w(z) = \frac{\mathrm{d}F_\gamma}{\mathrm{d}z} = U\left(1 - \frac{a^2}{z^2}\right) - \frac{\gamma}{2\pi i z}, \qquad \text{(B)}$$

which tends to the uniform flow $(U, 0)$ as $z \to \infty$.

Thus, the complex potential $F_\gamma(z)$ represents a uniform flow (of velocity U) around a circular cylinder of radius a (Fig. 5.13). Integrating $w(z)$ along the counter-clockwise closed contour C, we obtain

$$\oint_C w(z)\mathrm{d}z = -\frac{\gamma}{2\pi i} \oint_C \frac{1}{z}\mathrm{d}z = -\frac{\gamma}{2\pi i} 2\pi i = -\gamma.$$

Thus, the velocity circulation $\Gamma(C)$ around the cylinder is $-\gamma$ (i.e. clockwise), according to (5.59) and (5.91).

(ii) Using (B) of (i), we obtain

$$X - iY = \frac{1}{2}i\rho_0 \oint_C \left(\frac{\mathrm{d}F}{\mathrm{d}z}\right)^2 \mathrm{d}z$$

$$= \frac{1}{2}i\rho_0 \oint_C \left(\cdots - \frac{U\gamma}{\pi i z} + \cdots\right)\mathrm{d}z$$

$$= -\frac{1}{2}i\rho_0 \frac{U\gamma}{\pi i} 2\pi i = -i\rho_0 U\gamma$$

(see Problem 5.6). Thus, $X = 0$ and $Y = \rho_0 U\gamma$.

5.9 : The integral $(1/4\pi) \int K \, d\Omega$ means the angle average $\bar{K}$ over all directions. Hence, the left-hand side of (5.79) is written as

$$\frac{1}{4\pi} \int A_i e_i \, B_k e_k \, d\Omega = A_i B_k \overline{e_i e_k}$$

By using the expression (5.94), it is not difficult to obtain $\overline{e_i e_k} = \frac{1}{3}\delta_{ik}$. In particular, for the diagonal elements we have

$$e_x^2 + e_y^2 + e_z^2 = 1, \qquad \overline{e_x^2} = \overline{e_y^2} = \overline{e_z^2}, \; = \frac{1}{3}.$$

Thus we obtain $A_i B_k \overline{e_i e_k} = A_i B_k \frac{1}{3}\delta_{ik} = \frac{1}{3}A_i B_i$.

5.10 : At the end of Sec. 5.9, the induced mass of a sphere is given by $m_{ij} = \frac{1}{2}(m_{\text{fluid}}) \delta_{ij}$, a half of the displaced mass. Hence the equation of motion (5.86) reduces to

$$\left(m_s + \frac{1}{2}m_{\text{fluid}} \right) \frac{dU_i}{dt} = f_i.$$

In the bubble problem, we may set $m_s = 0$. Taking the z axis in the vertically upward direction, we have an upward buoyancy force corresponding to the displaced mass of water, given by $f_z = m_{\text{fluid}} \, g$ with $f_x = f_y = 0$. Thus we obtain $dU_z/dt = 2g$ (and $dU_x/dt = dU_y/dt = 0$).

Problem 6

6.1 : (i) The continuity equation (6.47) becomes $\partial_t p + u\partial_x p + \rho c^2 \partial_x u = 0$. Multiplying this by $1/(\rho c)$ and adding Eq. (6.48), we obtain the first equation (6.78). Subtracting (6.48) results in the second (6.79).

(ii) Let us define the two functions $F_\pm$ by

$$F_\pm := u \pm \int_{p_0}^{p} \frac{dp}{\rho c} \qquad (p_0 : \text{const.}).$$

Their t-derivative and x-derivative are

$$\partial_t F_\pm = \partial_t u \pm \frac{1}{\rho c}\partial_t p, \qquad \partial_x F_\pm = \partial_x u \pm \frac{1}{\rho c}\partial_x p.$$

F_+ satisfies (6.80) and F_- satisfies (6.81). This verifies that F_+ and F_- are equivalent to J_+ and J_-, respectively.

(iii) Suppose that we have a family of curves C_+ on the (x, t) plane defined by the equation $dx/dt = u + c$. Taking differential of the function J_+ along C_+, we have

$$dJ_+ = \partial_t J_+ \, dt + \partial_x J_+ \, dx = (\partial_t J_+ + \partial_x J_+ \, (dx/dt)) \, dt$$
$$= (\partial_t J_+ + (u + c) \, \partial_x J_+) \, dt = 0,$$

by (6.80). Thus, the function $J_+(x, t)$ is invariant along the family of curves C_+. Likewise, the function $J_-(x, t)$ is invariant along the family of curves C_- defined by $dx/dt = u - c$. The curves $C_\pm$ are called the *characteristics* in the theory of first-order partial differential equations like (6.80) or (6.81).

6.2 : Defining $X = x - c_0 t$, Eq. (6.82) becomes $-c_0 u_X + (u^2/2)_X = \nu u_{XX}$. Integrating this with respect to X, we obtain

$$u_X = \frac{1}{2\nu}((u - c_0)^2 - C^2), \quad (C : \text{a constant}).$$

Setting $v = u - c_0$, this becomes $dv/(v^2 - C^2) = dX/2\nu$. This can be integrated to give $v = -C \tanh[(C/2\nu)(X - X_0)]$, namely

$$u = c_0 - C \tanh[(C/2\nu)(x - c_0 t - X_0)],$$

where C, X_0 are constants. From the given boundary conditions at infinity, we obtain $c_0 = (c_1 + c_2)/2$ and $C = (c_1 - c_2)/2$. Thus, finally, we find the solution:

$$u = \frac{c_1 + c_2}{2} - \frac{c_1 - c_2}{2} \tanh\left[\frac{c_1 - c_2}{4\nu}(x - c_0 t - X_0)\right].$$

6.3 : (i) Using (6.84), we have

$$kx - \omega t = k_0 x - \omega(k_0)t + (k - k_0)(x - c_g t).$$

Substituting this into the integral (6.83), we obtain

$$\zeta(x, t) = e^{i[k_0 x - \omega(k_0)t]} \int A(k) \, e^{i(x - c_g t)(k - k_0)} \, dk$$
$$= F(x - c_g t)e^{i[k_0 x - \omega(k_0)t]}, \quad F(\xi) \equiv \int A(k)e^{i\,\xi(k - k_0)} \, dk$$

(ii) We substitute $A(k) = A_0 \exp[-a(k - k_0)^2]$ into the Fourier form of $F(\xi)$ given in (i). Using $\chi = k - k_0$, we obtain

$$F(\xi) = A_0 \int e^{-a\chi^2 + i\xi\chi} d\chi$$

$$= A_0 \int \exp\left[-a(\chi - i\xi/2a)^2 - \xi^2/4a \right] d\chi$$

$$= A_0 e^{-\xi^2/4a} \int e^{-a\eta^2} d\eta$$

$$= A_0 \sqrt{\frac{\pi}{a}} e^{-\xi^2/4a}, \quad \left(\int_{-\infty}^{\infty} e^{-y^2} dy = \sqrt{\pi} \right).$$

Thus, we obtain the wave packet:

$$\zeta(x, t) = A_0 \sqrt{\frac{\pi}{a}} \exp\left[-\frac{(x - c_g t)^2}{4a} \right] e^{i[k_0 x - \omega(k_0)t]}.$$

Problem 7

7.1 : Using $\mathbf{v} = \nabla \times \mathbf{A}$, the relation $\nabla \times \mathbf{v} = \boldsymbol{\omega}$ is transformed to

$$\nabla \times (\nabla \times \mathbf{A}) = \nabla(\nabla \cdot \mathbf{A}) - \nabla^2 \mathbf{A} = \boldsymbol{\omega}.$$

If $\nabla \cdot \mathbf{A} = 0$ is satisfied, we have $\nabla^2 \mathbf{A} = -\boldsymbol{\omega}$. This (Poisson's equation) can be integrated to yield $\mathbf{A}(\mathbf{x}) = (1/4\pi) \int_D \boldsymbol{\omega}(\mathbf{y}) \, d^3\mathbf{y}/|\mathbf{x} - \mathbf{y}|$, which is (7.5a). The property $\nabla \cdot \mathbf{A} = 0$ can be shown as follows:

$$\nabla_x \mathbf{A}(\mathbf{x}) = \frac{1}{4\pi} \int_D d^3\mathbf{y} \, \boldsymbol{\omega}(\mathbf{y}) \cdot \nabla_x \frac{1}{|\mathbf{x} - \mathbf{y}|}$$

$$= -\frac{1}{4\pi} \int_D d^3\mathbf{y} \, \nabla_y \cdot \frac{\boldsymbol{\omega}(\mathbf{y})}{|\mathbf{x} - \mathbf{y}|}$$

since $\nabla_y \cdot \boldsymbol{\omega}(\mathbf{y}) = 0$, and $\nabla_x = -\nabla_y$ if operated to $1/|\mathbf{x} - \mathbf{y}|$. The last integral is transformed to a surface integral which vanishes by (7.4a).

7.2 : (a) We take time derivative of the impulse $\mathbf{P}$ of (7.9) and use the equation (7.1). Then we obtain

$$\frac{d}{dt}\mathbf{P} = \frac{1}{2} \int_D \mathbf{x} \times \partial_t \boldsymbol{\omega} \, d^3\mathbf{x} = \frac{1}{2} \int_D \mathbf{x} \times (\nabla \times \mathbf{q}) \, d^3\mathbf{x}$$

where $\mathbf{q} = -\boldsymbol{\omega} \times \mathbf{v}$. Using (7.5e) with v_i replaced by q_i, the right-hand side reduces to $\int q_i \, d^3\mathbf{x} = \int (\mathbf{v} \times \boldsymbol{\omega})_i d^3\mathbf{x}$, the other integrals vanishing. The integrand can be written as

$$q_i = (\mathbf{v} \times \boldsymbol{\omega})_i = (\mathbf{v} \times (\nabla \times \mathbf{v}))_i = \partial_i \left(\frac{1}{2} v^2 \right) - \partial_k (v_k v_i), \qquad \text{(A)}$$

since $\partial_k v_k = 0$. Integration of these terms are transformed to vanishing surface integrals. Thus we obtain $(\mathrm{d}/\mathrm{d}t)\mathbf{P} = 0$, stating invariance of $\mathbf{P}$.

(b) Regarding the angular impulse $\mathbf{L}$,

$$\frac{\mathrm{d}}{\mathrm{d}t}\mathbf{L} = \frac{1}{3} \int_D \mathbf{x} \times (\mathbf{x} \times \partial_t \boldsymbol{\omega}) \, \mathrm{d}^3 \mathbf{x}$$

$$= \frac{1}{3} \int_D \mathbf{x} \times [\mathbf{x} \times (\nabla \times \mathbf{q})] \, \mathrm{d}^3 \mathbf{x}, \qquad \text{(B)}$$

where $\mathbf{q} = \mathbf{v} \times \boldsymbol{\omega}$. We have the following identity,

$$\mathbf{x} \times [\mathbf{x} \times (\nabla \times \mathbf{q})] = 3\,\mathbf{x} \times \mathbf{q} + \mathbf{C},$$

where $\mathbf{C}$ consists of two divergence terms of the component $C_i = \varepsilon_{jkl} \partial_k (x_i x_j q_l) - \varepsilon_{ikl} \partial_k (|\mathbf{x}|^2 q_l)$. Furthermore, $\mathbf{q}$ can be written as $q_i = \partial_k V_{ki}$ where $V_{ki} = \frac{1}{2} v^2 \delta_{ki} - v_k v_i$ from the above (A). Hence, the ith component of integrand of (B), $\frac{1}{3}\mathbf{x} \times [\mathbf{x} \times (\nabla \times \mathbf{q})]$, can be written as

$$(\mathbf{x} \times \mathbf{q})_i = \varepsilon_{ijk}\, x_j \, \partial_l V_{lk} = \varepsilon_{ijk}\, \partial_l (x_j V_{lk}),$$

since $\varepsilon_{ijk} V_{jk} = 0$. Thus, the integral of (B) can be converted to vanishing surface integrals, and we obtain invariance of $\mathbf{L}$: $(\mathrm{d}/\mathrm{d}t)\mathbf{L} = 0$.

(c) Regarding the helicity H, let us consider the following identity for two arbitrary vectors $\mathbf{A}$ and $\mathbf{B}$: $\nabla \cdot (\mathbf{A} \times \mathbf{B}) = (\nabla \times \mathbf{A}) \cdot \mathbf{B} - \mathbf{A} \cdot (\nabla \times \mathbf{B})$. When $\mathbf{A}$ and $\mathbf{B}$ decay at infinity like the vector $\mathbf{v}$, we have the equality:

$$\int \mathbf{A} \cdot (\nabla \times \mathbf{B}) \mathrm{d}V = \int (\nabla \times \mathbf{A}) \cdot \mathbf{B} \, \mathrm{d}V. \qquad \text{(C)}$$

Setting as $\mathbf{A} = \mathbf{v}$ and $\mathbf{B} = \partial_t \mathbf{v}$, we obtain

$$\int \mathbf{v} \cdot (\nabla \times \partial_t \mathbf{v}) \mathrm{d}V = \int \partial_t \mathbf{v} \cdot (\nabla \times \mathbf{v}) \mathrm{d}V.$$

Therefore, we have $dH/dt = 2 \int \mathbf{v} \cdot \partial_t \boldsymbol{\omega} dV$. Using (7.1), we obtain

$$\frac{d}{dt} H = -2 \int \mathbf{v} \cdot \left[\nabla \times (\boldsymbol{\omega} \times \mathbf{v}) \right] dV$$

$$= -2 \int \boldsymbol{\omega} \cdot (\boldsymbol{\omega} \times \mathbf{v}) dV = 0,$$

by the orthogonarity of the two factors, where (C) is used for the second equality. Thus, we find $dH/dt = 0$.

7.3 : Differentiating (7.31) and using (7.27), we have

$$\frac{dR^2}{dt} = \iint_D (x^2 + y^2) \, \partial_t \omega \, dx dy$$

$$= -\iint_D (x^2 + y^2) \left(u \, \partial_x \omega + v \, \partial_y \omega \right) dx dy$$

$$= -\int_{\partial D} (x^2 + y^2) \, u\omega \, dy - \int_{\partial D} (x^2 + y^2) \, v\omega \, dx$$

$$+ 2 \iint_D (xu\omega + yv\omega) \, dx dy$$

First two contour integrals along ∂D vanish by (7.21). Using (7.33) for u and (7.34) for v,

$$\frac{dR^2}{dt} = \frac{1}{\pi} \iint_D \iint_{D'} \frac{xy' - yx'}{(x' - x)^2 + (y' - y)^2}$$

$$\times \, \omega(x, y) \, \omega(x', y') \, dx dy \, dx' dy'$$

The value of the last integral should not be changed by exchanging the pairs of integration variables, (x, y) and (x', y'). However, it is observed that the integrand changes its sign. This means that the integral itself must vanish: therefore $dR^2/dt = 0$. Thus, R^2 is invariant.

7.4 : The formula (5.20) becomes

$$\Gamma[C] = \oint_C \mathbf{v} \cdot d\mathbf{l} = \int_S \omega_z \, dS.$$

For the closed curve C, we take a rectangle $\overrightarrow{abcda}$ where two longer sides $\overrightarrow{ab}$ and $\overrightarrow{cd}$ of length l are parallel to x axis with negative y-value $-\varepsilon$ and positive y-value $+\varepsilon$, respectively. Shorter sides are $\overrightarrow{bc}$ and $\overrightarrow{da}$

with their lengths being 2ε. Orientation of C is counter-clockwise. Then, the circulation Γ is

$$\Gamma[C] = \int_a^b u\,dx + \int_c^d u\,dx = \frac{1}{2}Ul - \left(-\frac{1}{2}U\right)l = Ul.$$

For the vorticity $\boldsymbol{\omega} = (0, 0, \omega)$ integral is taken over the area of $S = 2\varepsilon l$ of the rectangle $abcd$:

$$\Gamma[C] = \int_0^l dx \int_{-\varepsilon}^{\varepsilon} dy\, \omega_z(x, y) = Ul.$$

Since this holds for any l and ε, this implies that $\int_{-\varepsilon}^{\varepsilon} dy\omega_z(x, y) = U$ must hold for any ε. This leads to $\omega_z = U\,\delta(y)$. If the contour C does not cut the x axis, we have $\Gamma[C] = 0$. This is also satisfied by $\omega_z = U\delta(y)$.

7.5 : Taking $s = 0$ at the origin O, i.e. $\mathbf{y}(0) = 0$, the position $\mathbf{y}(s)$ on the curve $\mathcal{F}$ near O is expanded in the frame K as

$$\mathbf{y}(s) = \mathbf{y}'(0)\,s + \frac{1}{2}\mathbf{y}''(0)s^2 + O(s^3) = s\mathbf{t} + \frac{1}{2}k_0 s^2\,\mathbf{n} + O(s^3),$$

where k_0 is the curvature of the curve $\mathcal{F}$ at $s = 0$ (see Appendix D.1). Substituting this, the integrand of (7.67) is found to be written as

$$\frac{z\mathbf{n} - y\mathbf{b}}{(r^2 + s^2)^{3/2}} \left(1 + \frac{3}{2}k_0 \frac{ys^2}{r^2 + s^2}\right) - k_0 \frac{\frac{1}{2}s^2\mathbf{b} + yst}{(r^2 + s^2)^{3/2}} + O(k_0^2). \quad \text{(D)}$$

To determine the behavior of $\mathbf{u}(\mathbf{x})$ as $r \to 0$, we substitute (D) into (7.67) and evaluate contribution to $\mathbf{u}$ from the above nearby portion of the filament $(-\lambda < s < \lambda)$. Changing the variable from s to $\sigma = s/r$ and taking limit $\lambda/r \to \infty$, it is found that the Biot–Savart integral (7.67) is expressed in the frame K as

$$\mathbf{u}(\mathbf{x}) = \frac{\gamma}{2\pi} \left(\frac{y}{r^2}\mathbf{b} - \frac{z}{r^2}\mathbf{n}\right) + \frac{\gamma}{4\pi}k_0 \left(\log\frac{\lambda}{r}\right)\mathbf{b} + \text{(b.t.)},$$

where (b.t.) denotes remaining bounded terms.

7.6 : Substituting (7.81) into (7.72), the left hand side is $a\omega(\sin\theta, -\cos\theta, \lambda)$, whereas the right hand side is $a^2 k^3 h(\sin\theta, -\cos\theta, 1/h)$. Thus, the equation (7.72) is satisfied if $\omega = ak^3h$ and $h\lambda = 1$. Requiring that s is an arc-length parameter, i.e. $ds^2 = dx^2 + dy^2 + dz^2$, we must have $a^2 k^2(1 + h^2) = 1$. Hence, once the radius

a and the wave number k of the helix are given, all the other constants h, ω, λ are determined. The helix translates towards positive z axis and rotates clockwise (seen from above), whereas the circulatory fluid motion about the helical filament is counter-clockwise.

7.7 : Setting $\mathbf{v} = \mathbf{U} + \mathbf{u}_*$ in the expression (7.8) for K, we immediately obtain (7.83) by using the definition (7.12) of $\mathbf{P}$.

7.8 : For the variation δR of the ring radius, we have

$$\delta K = \frac{1}{2}\Gamma^2\left[\log\frac{8R}{a} - \frac{7}{4} + 1 + O\left(\epsilon^2\log\frac{1}{\epsilon}\right)\right]\delta R,$$

$$\delta P = 2\pi R\Gamma[1 + O(\epsilon^2)]\delta R,$$

$$U = \frac{\partial K}{\partial P} = \frac{\Gamma}{4\pi R}\left[\log\frac{8R}{a} - \frac{1}{4} + O\left(\epsilon^2, \epsilon^2\log\frac{1}{\epsilon}\right)\right].$$

7.9 : (i) Steady solution must satisfy

$$ar\partial_r\omega + 2a\omega + \nu r^{-1}\partial_r(r\partial_r\omega) = r^{-1}\partial_r r(ar\omega + \nu\partial_r\omega) = 0.$$

This leads to $ar\omega + \nu\partial_r\omega = 0$ by setting integration constant to zero. Hence, we obtain $\omega = C\exp[-ar^2/(2\nu)]$ with a constant C $(= \omega_B(0))$.

(ii) Using $A'(t) = aA$, $\partial_r = A\partial_\sigma$ and $\partial_t = A^2\partial_\tau + a\sigma\partial_\sigma$, we obtain

$$\partial_t\omega - ar\partial_r\omega = A^2\partial_\tau\omega, \quad r^{-1}\partial_r(r\partial_r\omega) = A^2\sigma^{-1}\partial_\sigma(\sigma\partial_\sigma\omega).$$

Hence, the equation (7.77) reduces to

$$\partial_\tau\omega - 2a\omega/A^2 = \nu\sigma^{-1}\partial_\sigma(\sigma\partial_\sigma\omega).$$

Setting $\omega = A^2(\tau)W$ and using $\partial_\tau A^2 = 2a$ (since $dA = aAdt$ and $d\tau = A^2dt$), we obtain $\partial_\tau W = \nu\sigma^{-1}\partial_\sigma(\sigma\partial_\sigma W)$. The Laplacian $\partial_\xi^2 + \partial_\eta^2$ reduces to $\sigma^{-1}\partial_\sigma(\sigma\partial_\sigma)$ for axisymmetric problems.

(iii) Let us write a solution $W(x, y, t)$ to the diffusion equation $W_t = \nu(W_{xx} + W_{yy})$ by the product $X(x, t)Y(y, t)$. If X and Y satisfy $X_t = \nu X_{xx}$ and $Y_t = \nu Y_{xx}$ respectively, then $W = X(x, t)Y(y, t)$ satisfies the above diffusion equation. In fact, substituting $W = XY$ into the diffusion equation and collecting all the terms in the left-hand side, we obtain

$$(X_t - \nu X_{xx})X + (Y_t - \nu Y_{yy})Y,$$

which in fact vanishes by the assumption of $X(x,t)$ and $Y(y,t)$. Writing X and Y (see Problem 2.3) by

$$X(x,t) = \frac{1}{2\sqrt{\pi \lambda t}} e^{-(x-\xi)^2/4\nu t},$$

$$Y(y,t) = \frac{1}{2\sqrt{\pi \lambda t}} e^{-(y-\eta)^2/4\nu t},$$

we define $w(x,y,t : \xi,\eta) = \Omega_0(\xi,\eta) X(x,t) Y(y,t)$. Integration of $w(x,y,t : \xi,\eta)$ with respect to ξ and η gives the solution (7.90). As $t \to +0$, $w \to \Omega_0(\xi,\eta) \delta(x - \xi) \delta(y - \eta)$. Thus, $W \to \Omega_0(\sqrt{\xi^2 + \eta^2}) = \Omega_0(r)$.

Problem 8

8.1 : (i) Using (8.13),

$$\partial_x p = -(\partial_z p)(\partial z_p/\partial x) = \rho g_*(\partial z_p/\partial x).$$

(ii) On the surface of constant p_*, we have grad $P(p_*) =$ grad $\chi_*(p_*)$ because the first term of (8.9) is invariant. Therefore, the first equality of (8.59) is obvious. The hydrostatic relations (8.10) and (8.13) give the next equality (8.60). The expression (8.61) is obtained by using the equation of state. As a result, we obtain the last equality of (8.62) from (8.21).

(iii) The vector grad $\bar{T}$ is directed to the south. Therefore, the product $\mathbf{k} \times$ grad $\bar{T}$ is directed to the east, hence the *westerly*.

Problem 9

9.1 : (i) The present problem is equivalent to setting $U = 0$, $\rho = \rho_1$ for $y > \zeta$ and ρ_2 for $y < \zeta$, in the perturbation equations considered in Sec. 9.2. Thus, we have

$$\zeta_t = \partial_y \phi_1|_{y=0}, \quad \zeta_t = \partial_y \phi_2|_{y=0},$$
$$\rho_1(\partial_t \phi_1)_{y=0} + \rho_1 g\zeta = \rho_2(\partial_t \phi_2)_{y=0} + \rho_2 g\zeta,$$

from (9.8)~(9.10), with including the gravitational potential $\rho g y$
at $y = \zeta$.

(ii) The equation for the nontrivial solution corresponding to (9.13)
yields

$$\sigma^2 = \sigma_*^2 \; (> 0), \quad \sigma_* := \sqrt{\frac{|\rho_1 - \rho_2|}{\rho_1 + \rho_2} \, gk} \; (> 0).$$

($\rho_1 > \rho_2$). Hence, $\sigma = \pm \sigma_*$. Therefore, the solution includes a
growing mode $\sigma = +\sigma_*$, and the basic state is unstable. (This is
called the *Rayleigh–Taylor* instability.)

(iii) Even if a light fluid is placed above a heavy fluid ($\rho_1 < \rho_2$), we
have the same perturbation equations, but we obtain $\sigma^2 = -\sigma_*^2$,
and we have $\sigma = \pm i \sigma_*$. Therefore, the separation surface oscillates
with the angular frequency σ_*.

[This is called the *interfacial wave*, observed in the ocean (or at
the estuary) at the interface of salinity discontinuity. Often this is
also seen in a heated room at temperature discontinuity visualized
by the tabacco smoke.]

9.2 : (i) Multiplying both sides of Eq. (9.26) by the complex conjugate ϕ^*
of ϕ, and integrating from $-b$ to b, and performing integration by
parts for the first term, we obtain

$$\int_{-b}^{b} (|\phi'|^2 + k^2 |\phi|^2) \mathrm{d}y + \int_{-b}^{b} \frac{U''}{U - c} |\phi|^2 \, \mathrm{d}y = 0,$$

by using the boundary condition (9.24). Setting $c = c_r + i c_i$ and
taking the imaginary part of the above expression, we obtain

$$c_i \int_{-b}^{b} \frac{U''(y)}{|U - c|^2} |\phi|^2 \, \mathrm{d}y = 0,$$

since the first term is real. For the instability, we should have
$c_i > 0$, which means $c_i \neq 0$ anyway. Therefore, the integral
must vanish. For its vanishing, $U''(y)$ must change its sign. Thus,
the profile $U(y)$ should have inflexion-points (where $U''(y) = 0$)
within the flow field for the instability.

(ii) Suppose that we are given the eigenfunction ϕ_e and eigenvalue
c_e to the eigenvalue problem (9.23) and (9.24). Because all the
coefficients of Eq. (9.23) are real, the complex conjugates ϕ_e^* and

c_e^* constitute the eigenfunction and eigenvalue of the same problem, as well. This can be confirmed by taking complex conjugate of (9.23) and (9.24). Therefore, once we have a stable solution with $\Im c_e < 0$, we have another solution of unstable solution with $\Im c_e > 0$ corresponding to the complex conjugate. Thus, in the problem of inviscid parallel flows, stable solution means only *neutrally* stable with $\Im c_e = 0$.

9.3 : Using the expression (9.62) of ψ, we obtain $\nabla^2 \psi = -k\psi$ with $k = (1 + a^2)(\pi/d)^2$. Hence the second convection term of (9.64) vanishes since

$$(\partial_y \psi \partial_x - \partial_x \psi \partial_y)\nabla^2 \psi = \frac{\partial(\nabla^2 \psi, \psi)}{\partial(x, y)} = \frac{\partial(-k\psi, \psi)}{\partial(x, y)} = 0.$$

Writing $\sin(\pi/d)y$ shortly as $S(y)$, the remaining terms are

$$\partial_t \nabla^2 \psi = A\lambda k^2 \partial_\tau X S(ax) S(y),$$
$$\nu \nabla^2 \nabla^2 \psi = -A\nu k^2 X S(ax) S(y),$$
$$-g\alpha \partial_x \theta = (\sqrt{2}a/d^4)\lambda \nu R_c Y S(ax) S(y).$$

where $\tau \equiv k\lambda t$. Using $R_c = \pi^4(a^2 + 1)^3/a^2$ (with $a = 1/\sqrt{2}$), we obtain $(\sqrt{2}a/d^4)\lambda \nu R_c = \nu k^2 A$ for the coefficient of the third term, where $R_c = k^3 d^6/(\pi a)^2$ and $A = \sqrt{2}\lambda(1 + a^2)/a = (\sqrt{2}\lambda k/a)(d/\pi)^2$. Thus, Eq. (9.64) reduces to

$$\partial_\tau X = -\sigma X + \sigma Y, \quad \sigma = \nu/\lambda.$$

Therefore, the first equation has been derived.

Regarding the temperature equation (9.65), each term is written as follows:

$$\partial_t \theta = k\lambda(B\partial_\tau Y C(ax) S(y) - (B/\sqrt{2})\partial_\tau Z S(2y)),$$

$$\frac{\partial(\theta, \psi)}{\partial(x, y)} = \left(\frac{\pi}{d}\right)^2 aABXY C(y) S(y)$$

$$-\left(\frac{\pi}{d}\right)^2 \frac{a}{\sqrt{2}} ABXZ C(ax) S(y)C(2y),$$

$$\lambda \nabla^2 \theta = -\lambda \left(\frac{\pi}{d}\right)^2 (1 + a^2) BY C(ax) S(y)$$

$$+ \lambda \left(\frac{2\pi}{d}\right)^2 (B/\sqrt{2}) \, Z S(2y),$$

$$-\beta \, \partial_x \psi = \beta \left(\frac{a\pi}{d}\right) AX C(ax) S(y),$$

where $C(y) = \cos(\pi/d)y$ and $2C(y)S(y) = S(2y)$ in the second line. Substituting these in (9.65) and collecting the coefficients of $S(2y)$, we obtain

$$\partial_\tau Z = XY - bZ, \quad b = 4/(1 + a^2).$$

Next, collecting the coefficients of $C(ax)S(y)$, we obtain

$$\partial_\tau Y = -XZ + rX - Y, \quad r = R_a/R_c,$$

where $2S(y)C(2y) = -S(y) + S(3y)$ was replaced with $-S(y)$ by the mode truncation.

9.4 : (i) When we substitute the given normal modes to the linear perturbation equation (9.3) (with $\mathbf{f}_1 = 0$), the differential operators are replaced by

$$(\partial_t, \partial_x, \partial_y, \partial_z) = (-i\alpha c, i\alpha, i\beta, \mathrm{D})$$

and $(\mathbf{v}_0 \cdot \nabla) = U\partial_x$, $(\mathbf{v}_1 \cdot \nabla)U(z) = wU'(z)$. In particular, the Laplacian is $\nabla^2 = \mathrm{D}^2 - \alpha^2 - \beta^2$. Normalization is made with the channel half-width b and the maximum velocity U_m. Thus, we obtain the first three equations (9.71)$\sim$(9.73). The fourth equation is the continuity equation: $\mathrm{div}\mathbf{v}_1 = \partial_x u_1 + \partial_y v_1 + \partial_z w_1 = 0$. This reduces to $i\alpha u + i\beta v + \mathrm{D}w = 0$.

Next, multiplying $i\alpha$ to (9.71), $i\beta$ to (9.72) and applying D to (9.73), we add the three resulting equations. Owing to the continuity condition (9.74), many terms cancel out and we obtain the equation for $\nabla^2 p$:

$$R_e \nabla^2 p = R_e(-\alpha^2 - \beta^2 + \mathrm{D}^2)p = -2i\alpha R_e U'(z)w.$$

In order to obtain (9.76), we apply ∇^2 to the third equation (9.73) and eliminate $R_e \nabla^2 p$ by using the above equation. Thus, we obtain finally (9.76) for w only. The boundary condition is the usual no-slip condition at the walls.

(ii) Noting that the Reynolds number R_e is included always as a combination αR_e and the y wave number β is included in ∇^2 as a combination $\sqrt{\alpha^2 + \beta^2}$ in (9.76), the eigenvalue problem determines the exponential growth rate as a function of αR_e and $\sqrt{\alpha^2 + \beta^2}$, say $\alpha c_i = G(\alpha R_e, \sqrt{\alpha^2 + \beta^2})$. In particular, a neutral stability curve is given by $c_i = 0$, i.e. $G(\alpha R_e, \sqrt{\alpha^2 + \beta^2}) = 0$. Solving this, we have $\alpha R_e = F(\sqrt{\alpha^2 + \beta^2})$, which should be positive. The critical Reynolds number R_c is by definition the lowest value of such R_e for all possible values of the wavenumbers α and β, thus represented by (9.78).

(iii) Critical Reynolds number R_c^{2D} for two-dimensional problem is represented by (9.78) with $\beta = 0$:

$$R_c^{2D} = \min_\alpha(F(\alpha)/\alpha).$$

The critical Reynolds number considered in the previous (ii) is R_c^{3D}. Using $k = \sqrt{\alpha^2 + \beta^2}$ and defining an angle θ by $\cos\theta = \alpha/k$, we obtain

$$R_c^{3D} = \min_{\alpha,k}\frac{F(k)}{\alpha} = \min_{\alpha,k}\frac{k}{\alpha}\frac{F(k)}{k} = \min_{\alpha,k}\frac{1}{\cos\theta}\frac{F(k)}{k}$$

$$\geqq \min_{\alpha,k}\frac{F(k)}{k} = R_c^{2D}.$$

Thus, the critical Reynolds number of parallel flows is given by R_c^{2D}.

Problem 10

10.1 : (i) Since ∂_j on Fourier representation can be replaced by ik_j, we have $\hat{\omega}_i(\mathbf{k}) = i\,\varepsilon_{ijk}k_j\hat{u}_k(\mathbf{k})$. Therefore, we obtain

$$|\hat{\boldsymbol{\omega}}(\mathbf{k})|^2 = \varepsilon_{ijk}\varepsilon_{i\alpha\beta}k_jk_\alpha\hat{u}_k\hat{u}_\beta = k^2|\hat{\mathbf{u}}(\mathbf{k})|^2 - (\mathbf{k}\cdot\hat{\mathbf{u}}(\mathbf{k}))^2$$
$$= k^2\,|\hat{\mathbf{u}}(\mathbf{k})|^2,$$

since $\mathbf{k}\cdot\hat{\mathbf{u}}(\mathbf{k}) = 0$ by (10.9).

(ii) By replacing $\mathbf{u}(\mathbf{x})$ with $\boldsymbol{\omega}(\mathbf{x})$ in (10.11), we have

$$\langle |\boldsymbol{\omega}(\mathbf{x})|^2 \rangle_C = \int |\hat{\boldsymbol{\omega}}(\mathbf{k})|^2 \, d^3\mathbf{k} = \int k^2 |\hat{\mathbf{u}}(\mathbf{k})|^2 \, d^3\mathbf{k}$$

$$= 2 \int_0^\infty k^2 \, E(k) dk,$$

since $\frac{1}{2}|\hat{\mathbf{u}}(\mathbf{k})|^2 = \Phi(|\mathbf{k}|)$ by (10.12), and $\Phi(|\mathbf{k}|) \, 4\pi|\mathbf{k}|^2 = E(|\mathbf{k}|)$ by (10.13) for isotropic turbulence.

10.2 : (i) Choosing two arbitrary constant vectors (a_i) and (b_j), we take scalar product (i.e. contraction) with $B_{ij}(\mathbf{s})$. The resulting $B_{ij} \, a_i b_j$ is a scalar and possibly depends on six scalar quantities (only): $(s_i s_i)$, $(a_i s_i)$, $(b_i s_i)$, $(a_i a_i)$, $(b_i b_i)$ and $(a_i b_i)$ by transformation invariance of scalars. In addition, $B_{ij} \, a_i b_j$ should be bilinear with respect to a_i and b_j, such as $(a_i b_i)$ or $(a_i s_i)(b_j s_j)$. Furthermore, $B_{ij} \, a_i b_j$ may depend on scalar functions of the form $F_1(s)$ or $G(s)$. The $B_{ij} \, a_i b_j$ satisfying these can be represented as

$$B_{ij}(a_i b_j) = G(s)(a_i b_i) + F_1(s)(a_i s_i)(b_j s_j)$$

$$= (G(s)\delta_{ij} + F_1(s)s_i s_j)a_i b_j.$$

This implies (10.40) with $F(s) = F_1(s)s^2$.

(ii) Using $\partial u_j(\mathbf{x}')/\partial x'_j = 0$, we obtain

$$0 = \left\langle u_i(\mathbf{x}) \frac{\partial u_j(\mathbf{x}')}{\partial x'_j} \right\rangle = \frac{\partial}{\partial x'_j} \langle u_i(\mathbf{x}) \, u_j(\mathbf{x}') \rangle = \frac{\partial}{\partial s_j} B_{ij}(\mathbf{s}).$$

Using (10.40) and $e_i = s_i/s$, we have

$$\frac{\partial}{\partial s_j} B_{ij}(\mathbf{s}) = G'(s)\frac{\partial s}{\partial s_j} + \frac{F(s)}{s^2}\frac{\partial(s_i s_j)}{\partial s_j}$$

$$+ \frac{d}{ds}\left(\frac{F(s)}{s^2}\right)\frac{\partial s}{\partial s_j} s_i s_j$$

$$= G'(s)\frac{s_i}{s} + \frac{F(s)}{s^2}(s_i + 3s_i) + \frac{d}{ds}\left(\frac{F(s)}{s^2}\right)\frac{s_j}{s} s_i s_j,$$

where $\partial s/\partial s_j = s_j/s$ (since $s^2 = s_j s_j$) and $\partial s_i/\partial s_j = \delta_{ij}$. The above must vanish. Therefore, we obtain (10.41).

(iii) Since longitudinal component u_l can be expressed by $u_i e_i$,

$$\bar{u}^2 f(s) = \langle u_i(\mathbf{x}) e_i \, u_j(\mathbf{x}') e_j \rangle = B_{ij}(\mathbf{s}) e_i e_j.$$

Using (10.40) and $e_i e_i = 1$, we have $B_{ij} e_i e_j = G(s) + F(s) = \bar{u}^2 f(s)$.

(iv) In view of (10.45),

$$\left\langle \frac{u^2}{2} \right\rangle = \frac{1}{2} \langle u_i(\mathbf{x}) \, u_i(\mathbf{x}) \rangle = \frac{1}{2} B_{ii}(0) = \frac{1}{2} \int F_{ii}(\mathbf{k}) \, d^3\mathbf{k}$$

$$= \int F_{ii}(|\mathbf{k}|) \, 2\pi k^2 \, dk.$$

Writing $B_{ii}(\mathbf{s}) = B(|\mathbf{s}|) = B(s)$ and using (10.44),

$$E(k) = 2\pi k^2 F_{ii}(\mathbf{k}) = \frac{2\pi k^2}{(2\pi)^3} \int B(|\mathbf{s}|) \, e^{-i\mathbf{k}\cdot\mathbf{s}} \, d^3\mathbf{s}$$

$$= \frac{k^2}{(2\pi)^2} 2\pi \int_0^\infty B(s) s^2 ds \int_0^\pi e^{-iks\cos\theta} \sin\theta \, d\theta$$

$$= \frac{1}{\pi} \int_0^\infty B(s) \, ks \sin ks \, ds, \quad \text{since} \int_{-1}^1 e^{-iks\,\xi} \, d\xi$$

$$= \frac{2}{ks} \sin ks.$$

where $\xi = \cos\theta$. Using (10.45),

$$B(s) = \int F_{ii}(k) e^{iks\,\xi} \, 2\pi k^2 dk d\xi = \int E(k) \, dk \int_{-1}^1 e^{iks\,\xi} \, d\xi$$

$$= 2 \int_0^\infty E(k) \frac{\sin ks}{ks} \, dk.$$

From (10.40) and using $G(s) + F(s) = \bar{u}^2 f(s)$, $B(s) = B_{ii}(s) = 3G(s) + F(s) = 3\bar{u}^2 f(s) - 2F(s)$. Applying (10.41), we obtain (10.49).

10.3 : (i) Using (10.49), we obtain $d(s^3 \bar{u}^2 f(s))/ds = s^2 B(s)$. Therefore,

$$s^3 \bar{u}^2 f(s) = 2 \int_0^\infty E(k) \, dk \int_0^s \frac{s \sin ks}{k} ds$$

$$= 2 \int_0^\infty E(k) \left(\frac{\sin ks}{k^3} - \frac{s \cos ks}{k^2} \right) dk.$$

(ii) Denoting the integrand of (10.51) as $H(k)$ and using $G'(k) = -E(k)/k^3$ from (10.52),

$$H(k) = -G'(k)\left(\frac{\sin ks}{s^3} - \frac{k\cos ks}{s^2}\right)$$

$$= -\frac{d}{dk}\left[G(k)\left(\frac{\sin ks}{s^3} - \frac{k\cos ks}{s^2}\right)\right] + G(k)\frac{k\sin ks}{s}$$

Integrating $H(k)$ from $k = 0$ to ∞, the first term vanishes since $G(\infty) = 0$, and the second gives (10.52).

(iii) The left-hand side of (10.53) can be written as ($e^{iks} = \cos ks + i\sin ks$),

$$\frac{1}{2\pi}\int_{-\infty}^{\infty}\bar{u}^2 f(s)\,e^{ik_1 s}ds = \frac{1}{\pi}\int_0^{\infty}\bar{u}^2 f(s)\cos k_1 s\,ds.$$

Substituting (10.52) for $\bar{u}^2 f(s)$ on the right, we obtain from (10.50)

$$E_1(k_1) = \frac{2}{\pi}\int_0^{\infty}\int_0^{\infty}G(k)k\,\frac{\sin ks\,\cos k_1 s}{s}\,ds\,dk$$

$$= \int_{k_1}^{\infty}k\,G(k)\,dk,$$

by (10.54). Substituting the definition of $G(k)$ of (10.52), we have

$$E_1(k_1) = \int_{k_1}^{\infty}k\int_k^{\infty}\frac{E(k')}{(k')^3}\,dk'\,dk$$

$$= \left[\frac{k^2}{2}\int_k^{\infty}\frac{E(k')}{(k')^3}\,dk'\right]_{k_1}^{\infty} + \int_{k_1}^{\infty}\frac{E(k)}{2k}\,dk$$

$$= -\frac{k_1^2}{2}\int_{k_1}^{\infty}\frac{E(k)}{k^3}\,dk + \int_{k_1}^{\infty}\frac{E(k)}{2k}\,dk$$

The last expression is the right-hand side of (10.53) itself.

(iv) Differentiating $E_1(k_1)$ with k_1, we have

$$\frac{dE_1(k_1)}{dk_1} = -k_1\int_{k_1}^{\infty}\frac{E(k)}{k^3}\,dk.$$

Dividing both sides with k_1 and differentiating it with k_1 again, and setting $k = k_1$, we obtain

$$\frac{d}{dk}\left(\frac{1}{k}\frac{dE_1(k)}{dk}\right) = \frac{E(k)}{k^3}.$$

Problem 11

11.1 : We write the power series solution beginning from r^p (p: a positive integer) as $A(r) = \sum_{k=p}^{\infty} c_k r^k$. Substituting this in (11.32), the lowest-power term is $(p^2 - n^2)r^{p-2}$. From (11.32), we must have $p = n$. Thus, we set $p = n$. Then, we have $A'(r) = \sum_{k=n}^{\infty} kc_k r^{k-1}$, $A''(r) = \sum k(k-1)c_k r^{k-2}$. Substituting these into (11.32) and collecting the terms including r^{k-2}, we obtain

$$\cdots + ((k^2 - n^2)c_k + c_{k-2})r^{k-2} + \cdots = 0.$$

Hence we find the relation between neighboring coefficients as

$$c_k = (-1)\frac{1}{(k-n)(k+n)}c_{k-2}. \tag{A}$$

If we set the first coefficient as $c_n = 1/(2^n n!)$, we have

$$c_{n+2} = (-1)\frac{1}{2(n+1)2}c_n = (-1)\frac{1}{(n+1)!}\frac{1}{2^{n+2}}.$$

These are consistent with the first two terms of (11.33). In general, the coefficients satisfying $c_{n+2k} = -[(k(n+k)2^2]^{-1}c_{n+2(k-1)}$ derived from (A) are determined as

$$c_{n+2k} = (-1)^k\frac{1}{k!(n+k)!}\frac{1}{2^{n+2k}},$$

and $c_{n+1} = c_{n+3} = \cdots = 0$.

11.2 : (i) Writing $\mathcal{A} = c\zeta + d\zeta^2 + e\zeta^3 + \cdots$, we substitute it into (11.31). Then we obtain $d = 0$ and c is a constant undetermined (arbitrary).

(ii) Writing $\mathcal{A} = a + b\zeta^{-1} + c\zeta^{-2} + \cdots$, we substitute it into (11.31). Then we obtain $a = 1$, $b = 0$, $c = -1/2, \ldots$, etc.

Problem 12

12.1 : Using the velocity $\mathbf{v} = \nabla\phi + s\,\nabla\psi$ of (12.127) and the vorticity $\boldsymbol{\omega} = \nabla \times \mathbf{v} = \nabla s \times \nabla\psi$, we have

$$\partial_t \mathbf{v} + \boldsymbol{\omega} \times \mathbf{v} = \nabla(\partial_t\phi + s\,\partial_t\psi) + (D_t s)\,\nabla\psi - (D_t\psi)\,\nabla s.$$

where $(\nabla s \times \nabla\psi) \times \mathbf{v} = (\mathbf{v} \cdot \nabla s)\nabla\psi - (\mathbf{v} \cdot \nabla\psi)\nabla s$ is used. The last two terms vanish due to (12.129). Thus, Eq. (12.126) is obtained by using (12.128).

The vortex lines are the intersections of the families of surfaces $s = $ const. and $\psi = $ const. These surfaces are moving with the fluid by (12.129).

If the scalar product $\boldsymbol{\omega} \cdot \mathbf{v}$ is integrated over a volume V including a number of closed vortex filaments, the helicity $H[V]$ vanishes:

$$H[V] \equiv \int_V \boldsymbol{\omega} \cdot \mathbf{v}\, \mathrm{d}^3\mathbf{x} = \int_V (\nabla\lambda \times \nabla\psi) \cdot \nabla\phi\, \mathrm{d}^3\mathbf{x}$$
$$= \int_V \nabla \cdot [\phi\boldsymbol{\omega}]\, \mathrm{d}^3\mathbf{x} = 0,$$

if the normal component of vorticity $\boldsymbol{\omega}$ vanishes on bounding surfaces at infinity. Thus, the helicity H vanishes in this case.

References

[Ach90] Acheson, DJ, 1990, *Elementary Fluid Dynamics* (Oxford Univ. P).

[AFES04] *J. Earth Simulator* (see the footnote in Sec. 8.5 of this book), Vol. 1, 8–34 (2004).

[Bat53] Batchelor, GK, 1953, *Theory of Homogeneous Turbulence* (Cambridge Univ. Press).

[Bat67] Batchelor, GK, 1967, *An Introduction to Fluid Dynamics* (Cambridge Univ. Press).

[Br70] Bretherton, FP, 1970, A note on Hamilton's principle for perfect fluids, *J. Fluid Mech.* **44**, 19–31.

[CKP66] Carrier, GF, Krook, M and Pearson, CE, 1966, *Functions of a complex variable* (McGraw-Hill).

[Cha61] Chandrasekhar, S, 1961, *Hydrodynamic and Hydromagnetic Stability* (Oxford Univ. Press).

[Coc69] Cocke, WJ, 1969, Turbulent hydrodynamic line stretching: Consequences of isotropy, *Phys. Fluids* **12**, 2488–2492.

[Dar05] Darrigol, O, 2005, *Worlds of Flow: A History of Hydrodynamics from the Bernoullis to Prandtl* (Oxford Univ. Press).

[DaR06] Da Rios, LS, 1906, *Rend. Circ. Mat. Palermo* **22**, 117–135.

[Don91] Donnelly, RJ, 1991, *Quantized Vortices in Helium II* (Cambridge Univ. Press).

[DR81] Drazin, PG and Reid, WH, 1981, *Hydrodynamic Stability* (Cambridge Univ. Press).

[Fr97] Frankel, T, 1997, *The Geometry of Physics* (Cambridge Univ. Press).

[Has72] Hasimoto, H, 1972, A soliton on a vortex filament, *J. Fluid Mech.* **51**, 477–485.

[HDC99] He, GW, Doolen, GD and Chen, SY, 1999, Calculations of longitudinal and transverse structure functions using a vortex model of isotropic turbulence, *Phys. Fluids* **11**, 3743–3748.

[HK97] Hatakeyama, N and Kambe, T, 1997, Statistical laws of random strained vortices in turbulence, *Phys. Rev. Lett.* **79**, 1257–1260.

[Hol04] Holton, JR, 2004, *An Introduction to Dynamic Meteorology*, 4th edition (Elsevier).

[Hon88] Honji, H, 1988, Vortex motion in stratified wake flows, *Fluid Dyn. Res.* **3**, 425–430.

[Hou77] Houghton, JT, 1977, *The Physics of Atmospheres* (Cambridge Univ. Press).

[Ka84] Kambe, T, 1984, Axisymmetric vortex solution of Navier–Stokes equation, *J. Phys. Soc. Jpn.* **53**, 13–15.

[Ka03a] Kambe, T, 2003a, Gauge principle for flows of a perfect fluid, *Fluid Dyn. Res.* **32**, 193–199.

[Ka03b] Kambe, T, 2003b, Gauge principle and variational formulation for flows of an ideal fluid, *Acta Mech. Sin.* **19**, 437–452.

[Ka04] Kambe, T, 2004, *Geometrical Theory of Dynamical Systems and Fluid Flows* (World Scientific, Singapore).

[Ka06] Kambe, T, 2006, Gauge principle and variational formulation for ideal fluids with reference to translation symmetry, *Fluid Dyn. Res.*

[KD98] Kambe, T and Drazin, PG, 1998, *Fluid Dynamics: Stability and Turbulence* (Univ. of Tokyo Press, ISBN4-13-060601-8) [in Japanese].

[KdV1895] Korteweg, DJ and de Vries, G, 1895, On the change of form of long waves advancing in a rectangular canal, and on a new type of long stationary waves, *Phil. Mag.* Ser. 5 **39**, 422–443.

[KI95] Kambe, T and Ishii, K, 1995, *Fluid Dynamics* (ShoKaBo, Tokyo) [in Japanese].

[KP66] Kawatra, MP and Pathria, RK, 1966, Quantized vortices in an imperfect Bose gas and the breakdown of superfluidity in liquid helium II, *Phys. Rev.* **151**, 591–599.

[KT71] Kambe, T and Takao, T, 1971, Motion of distorted vortex rings, *J. Phys. Soc. Jpn.* **31**, 132–137.

[Lamb32] Lamb, H, 1932, *Hydrodynamics* (Cambridge Univ. Press).

[LL75] Landau, LD and Lifshitz, EM, 1975, *The Classical Theory of Fields*, 4th edition (Pergamon Press).

[LL76] Landau, LD and Lifshitz, EM, 1976, *Mechanics*, 3rd edition (Pergamon Press).

[LL80] Landau, LD and Lifshitz, EM, 1980, *Statistical Physics*, 3rd edition Part 1 (Pergamon Press).

[LL87] Landau, LD and Lifshitz, EM, 1987, *Fluid Mechanics*, 2nd edition (Pergamon Press).

[Lor63] Lorenz, EN, 1963, Deterministic nonperiodic flow, *J. Atmos. Sci.* **20**, 130–141.

[Mak91] Makita, H, 1991, Realization of a large-scale turbulence field in a small wind tunnel, *Fluid Dyn. Res.* **8**, 53–64.

[MCBD01] Madison, KW, Chevy, F, Bretin, V and Dalibard, J, 2001, *Phys. Rev. Lett.* **87**, 190402.

[MY71] Monin, AS and Yaglom, AM, 1971, *Statistical Fluid Mechanics* (MIT Press, USA).

[OFES04] Masumoto, S, Kagimoto, K, Ishida, S, Miyama, M, Mitsudera, T and Sakuma, Y, 2004, A fifty-year eddy-resolving simulation of the world ocean — Preliminary outcomes of OFES (OGCM for the Earth Simulator), *J. Earth Simulator* (see the footnote of Sec. 8.5 of this book), Vol. 1, 35–56.

[Ors70] Orszag, SA, 1970, Comments on "turbulent hydrodynamic line stretching: Consequences of isotropy," *Phys. Fluids* **13**, 2203–2204.

[PS02] Pethick, CJ and Smith, H, 2002, *Bose Einstein Condensation in Dilute Gases* (Cambridge Univ. Press).

[Sa92] Saffman, PG, 1992, *Vortex Dynamics* (Cambridge Univ. Press).

[SCG80] Salwen, H, Cotton, FW and Grosch, CE, 1980, Linear stability of Poiseuille flow in a circular pipe, *J. Fluid Mech.* **98**, 273–284.

[TKA03] Tsubota, M, Kasamatsu, K and Araki, T, 2003, Dynamics of quantized vortices in super-fluid helium and rotating Bose–Einstein condensates, *Recent Res. Devel. Phys.* (Transworld Res. Network, India), **4**, 631–655.

[TMI97] Tanahashi, M, Miyauchi, T and Ikeda, J, 1997, Scaling law of fine scale structure in homogeneous isotropic turbulence, *11th Symp. Turbulent Shear Flows*, Vol. 1, 4–17 and 4–22.

[Tri77] Tritton, DJ, 1977, *Physical Fluid Dynamics* (Van Nostrand Reinhold, UK).

[Tsu03] Tsubota, M, 2003, Dynamics of a rotating Bose condensate, *Kotai Butsuri* (*Solid-State Physics*) (Tokyo) **38**(5), 325–331 [in Japanese].

[Wnb95] Weinberg, S, 1995, *The Quantum Theory of Fields*, Vol. I (1995) and II (1996) (Cambridge Univ. Press).

[YIUIK02] Yokokawa, M, Itakura, K, Uno, A, Ishihara, T and Kaneda, Y, 2002, 16.4TFlops direct numerical simulation of turbulence by a Fourier spectral method on the Earth Simulator, *Proc. IEEE/ACM SC2002 Conf.*, Baltimore (http://www.sc-2002.org/paperpdfs/pap.pap273.pdf);
Kaneda, Y, Ishihara, T, Yokokawa, M, Itakura, K and Uno, A, 2003, Energy dissipation rate and energy spectrum in high resolution direct numerical simulations of turbulence in a periodic box, *Phys. Fluids* **15**, L21–L24.

Index